Mathematical Techniques in Multisensor Data Fusion

Second Edition

For a listing of recent titles in the *Artech House Information Warfare Library,*
turn to the back of this book.

Mathematical Techniques in Multisensor Data Fusion

Second Edition

David L. Hall
Sonya A. H. McMullen

Artech House
Boston • London
www.artechhouse.com

Library of Congress Cataloging-in-Publication Data

A catalog record for this book is available from the Library of Congress.

British Library Cataloguing in Publication Data

A catalog record for this book is available from the British Library.

ISBN: 978-1-58053-335-5

Cover design by Gary Ragaglia

© 2004 ARTECH HOUSE, INC.
685 Canton Street
Norwood, MA 02062

A Library of Congress Catalog Card number is available for this book.

10 9 8 7 6 5 4

This book is dedicated to the memory of Mary Jane M. Hall. For 31 years, she was the loving wife of David Hall. She was the mother of Sonya Anne Hall (McMullen) and Cristin Marie Hall. Her guidance and love for our family will never be forgotten.

Contents

Preface

The original edition of the book, *Mathematical Techniques in Multisensor Data Fusion* was aimed at providing students and working professionals with an overview and introduction to data fusion methods. Data fusion involves combining information from multiple sources or sensors to achieve inferences not possible using a single sensor or source. Applications of data fusion range from military applications such as battle management and target tracking to automated threat assessment. Nonmilitary applications include environmental monitoring, medical diagnosis, monitoring of complex machines, and robotics. The original text was intended to provide a "gentle" introduction to data fusion algorithms across the broad spectrum from data association and correlation to target tracking and identification, automated reasoning, and human-computer interaction. In addition, advice was provided on system implementation including systems engineering methods and utilization of commercial off-the-shelf software tools.

This second edition provides a major update to the original text. New chapters are provided on cognitive-assisted reasoning, human-computer interaction, emerging new applications, and new models for data fusion. All chapters have been upgraded to include new references, sources available on the Internet, new graphics, and new techniques such as intelligent agents, hybrid reasoning, and fuzzy logic processing. The book recognizes new advances in smart sensors, distributed computing, and increased processing capability of personal computers and new interaction devices such as 3-dimensional full-immersion displays and Haptic devices. In addition, references are provided to new research in target tracking, automated reasoning, and cognitive modeling.

Chapter 1

Introduction to Multisensor Data Fusion

This chapter provides an introduction to the concepts and terminology of multisensor data fusion. The chapter also provides a guide to the remainder of this book.

1.1 INTRODUCTION

In the past 15 years, a discipline called multisensor data fusion or distributed sensing has been developed to solve a diverse set of problems having common characteristics. Multisensor data fusion seeks to combine data from multiple sensors to perform inferences that may not be possible from a single sensor or source alone. Applications span military problems such as automatic target recognition, analysis of battlefield situations, or threat assessments. Other applications include remote sensing problems involving the determination of the composition of ground vegetation or the location of mineral resources and industrial applications including the control of complex machinery (e.g., nuclear power plants) or automated manufacturing. Data from different sources and types of sensors are combined using techniques drawn from several disciplines: signal processing, statistical estimation, pattern recognition, artificial intelligence, cognitive psychology, and information theory. Input data from sensors may include parametric positional data, such as angular data (e.g., azimuth, elevation, and image coordinates), range or range-rate information, and data related to object identity (e.g., either actual declarations of identity from a sensor or parametric data that can be related to identity, such as radar cross section or spectral data). Input information may include information from human observers, other data fusion systems, or computational models.

Data fusion is analogous to the ongoing cognitive process used by humans to integrate data continually from their senses to make inferences about the external world. Humans receive and process sensory data–sights, sounds, smells, tastes, and touch–which are then assessed to draw conclusions about the environment and what it means. Recognition of an acquaintance whom one has not seen for a long period of time, for example, may involve assessment of factors such as general facial shape, identification of distinctive visual features (e.g., prominent nose or hair color), identification of voice tonal patterns, or even distinctive ways of walking or gesturing. Comedians mimic distinctive features of famous people to evoke recognition and to caricature others. Recognition and assessment of a situation by a human is greatly affected by training, attention, mood, physical condition, or other factors. A physician may assess a fellow human with much different perceptions than would, say, a friend or casual acquaintance. Nevertheless, humans utilize a natural fusion of sensory data for recognition of external events. Many of the techniques developed for data fusion attempt to emulate the ability of humans (or animals) to perform fusion.

This chapter provides an introduction to data fusion problems and defines the basic terminology. The Joint Directors of Laboratories (JDL) data fusion group model of data fusion [1] is introduced, along with a summary of applications, input data, outputs, types of sensors, and basic implementation issues. The first edition of this book declared that, "Data fusion is not a discipline in the same sense as more well-defined studies such as signal processing or numerical methods. Well-defined techniques, terminology, and a professional community do not yet exist." Since the first edition of this book, significant progress has been made to develop models [2], structured engineering procedures [3], and guidelines for requirements analysis [4], architecture selection [5] and algorithm selection [6, 7]. While data fusion is still not a mature technology, significant progress has been made. That progress is reflected in the updates to this second edition.

Chapter 2 presents the JDL data fusion process, which defines several levels of fusion processing including the following:

- Level 1 processing (object assessment): Fusion of multisensor data to determine the position, velocity, attributes, characteristics, and identity of an entity (such as an emitter or target);
- Level 2 processing (situation assessment): Automated reasoning to refine our estimate of a situation (including determining the relationships among observed entities, relationships between entities and the environment, and general interpretation of the meaning of the observed entities);
- Level 3 processing (impact assessment): Projection of the current situation into the future to define alternative hypotheses regarding possible threats or future conditions;
- Level 4 processing (process refinement): A meta process that monitors the ongoing data fusion process to improve the processing results (namely improved accuracy of estimated kinematics/identity of entities and improved assessment of the current situation and hypothesized threats);
- Level 5 processing (cognitive refinement): Interaction between the data fusion system and a human decision maker to improve the interpretation of results and the decision-making process.

This book is organized using the JDL model structure. Chapter 2 describes the JDL model and identifies algorithms and techniques for fusion. Chapters 3–6 cover different types of processing involved in level 1 fusion. Particular methods involve statistical estimation and pattern recognition. Chapter 7 is devoted to methods for performing level 2 and level 3 fusion. These techniques are drawn from artificial intelligence. Chapter 8 introduces the concept of level 4 fusion involving the control of sensor and information resources and the dynamic refinement of the fusion process. Chapter 9 addresses level 5 fusion. This concept was recently introduced by [8]. Chapter 10 provides a discussion of the implementation of data fusion systems; Chapter 11 presents some emerging applications; and Chapter 12 discusses automated information management.

1.2 FUSION APPLICATIONS

Data fusion applications span a wide domain including the military and nonmilitary applications summarized in Table 1.1. Military applications include ocean surveillance, air-to-air and surface-to-air defense, battlefield intelligence, surveillance and target acquisition, and strategic warning and defense. Nonmilitary applications include law enforcement, remote sensing, and automated monitoring of equipment, medical diagnosis, and robotics. A summary of these applications follows. Additional information about military applications may be found in [9-12]. Chapter 8 provides a survey of implemented data fusion systems and commercial tools related to data fusion.

Table 1.1

Examples of Multisensor Data Fusion Applications

Application	Inferences Sought	Primary Observable Data	Surveillance Volume	Sensor Platforms
Ocean surveillance	Detection, tracking and identification of targets and events	Acoustics Electromagnetic radiation Evidence of nuclear radiation	Hundreds of nautical miles Air, surface, and subsurface	Ships Aircraft Submarines Ground-based Ocean-based
Air-to-air and surface-to-air defense	Detection, tracking and identification of aircraft	Electromagnetic radiation	Hundreds of miles (strategic) Miles (tactical)	Ground-based Aircraft
Battlefield intelligence and surveillance	Detection and identification of potential ground targets	Electromagnetic radiation Acoustics	Tens to hundreds of miles about a battlefield	Ground-based Aircraft
Strategic warning and defense	Detection of indications of	Electromagnetic radiation	Global surveillance	Satellites Aircraft

Table 1.1   (continued)

Examples of Multisensor Data Fusion Applications

Application	Inferences Sought	Primary Observable Data	Surveillance Volume	Sensor Platforms
	impending strategic actions Detection and tracking of ballistic missiles and warheads	Nuclear particles		
Law enforcement	Transportation and location of drug shipments	Electromagnetic radiation Acoustics	Country and state borders	Tethered balloons Aircraft Ground-based
Remote sensing	Identification, location of mineral deposits; crop and forest conditions	Electromagnetic radiation Human reports and observations	Hundreds of miles	Aircraft Satellites Ground-based
Automated monitoring of equipment	Status and health of equipment Identification of impending fault conditions	Electromagnetic radiation Acoustic emissions Vibrations Temperature and pressure Human observations	Volume of the monitored machine or factory	Organic sensors associated with the machine or factory
Medical diagnosis	Diagnosis of disease, tumors, and physical condition	Electromagnetic radiation Nuclear magnetic resonance Chemical reactions Biological data Human observations	Volume of the human body or observed component	Sensors placed in, on, and around the body
Robotics	Identification and location of obstacles	Electromagnetic radiation Acoustics	Near-location about the robot	Robot platform

Ocean surveillance seeks to detect, track, and identify targets, events, and activities. Examples of enemy activities of interest are the launch of a torpedo from a submarine, surface broach of an underwater-launched missile, and communication of a submarine with other vehicles. The ocean surveillance area includes subsurface, surface, and airborne targets in a volume whose dimension may cover a significant portion of the Earth. Numerous ships, submarines, and aircraft may be involved in the surveillance process while also being targets of interest. Observable data spans the entire physical range: from acoustic data obtained via sono-buoys, towed acoustic arrays, and underwater arrays, through the electromagnetic spectrum, nuclear particles, and nonnuclear particles.

Air-to-air defense and surface-to-air defense seeks to detect, track, and identify aircraft at ranges that permit evasion or deployment of weapons [13]. Systems to identify a single aircraft are sometimes termed identification-friend-foe-neutral (IFFN) systems. Primary observable phenomena include electromagnetic radiation (infrared, visible, and radio frequencies) utilizing passive and active sensors such as radar, electronic support measure (ESM) receivers, infrared (IR) cameras, laser radar, and electro-optical sensors (TV). The surveillance volume ranges from hundreds of cubic miles for strategic applications such as the air defense of a country's coastline, to a few-mile radius for defense of a single tactical aircraft. A closely related nonmilitary application is the identification of incoming aircraft at commercial airports.

Battlefield intelligence is aimed at the detection and identification of potential ground targets *(movers, shooters,* or *emitters)* to infer enemy capabilities, tactics, and strategies. Electromagnetic radiation is a primary observable, including IR emission from enemy engines, RF emissions of communications radios, beacons, radars, visible photographs of an area of interest, and ESM receivers. Additional information may be gained from acoustic and chemical sensors. Battlefield intelligence seeks to develop an enemy order of battle (a database containing the accurate identification, location, and characteristics of enemy platforms, emitters, and military units) and to determine the meaning and intent of the order of battle, (i.e., the enemy situation and assessment of threat). The area of interest for battlefield intelligence ranges from tens to hundreds of square miles. The advent of modern weapons extends the tactical area of surveillance, corresponding to the increased range of new weapons.

The mission of strategic warning and defense [14] is twofold: (1) to detect indications of impending strategic actions and, in a worst case, (2) to detect and track ballistic missiles and warheads. The large-scale Strategic Defense Initiative has highlighted the difficulties of such a mission. A global sensor system is required, utilizing satellites and aircraft. Primary observable quantities are the detection of electromagnetic radiation (e.g., actual visual confirmation of launch facilities, IR emission from rocket plumes, heating of reentering warheads), and nuclear radiation. The detection of strategic indications and warnings involves observations of an enemy's military activities such as communication, force dispersion, alert status, and even nonmilitary political activities. Clearly, data fusion for this problem requires very complex observations, combinations of data, and inferences via scenarios and models.

Nonmilitary applications of fusion attack a diverse set of problems. Examples include law enforcement applications such as drug interdiction, remote sensing, medical diagnosis, automated control and monitoring of complex equipment, and robotics.

Law enforcement applications are similar to military intelligence and surveillance. Drug interdiction, for example, may involve patrol of a border to identify and locate drug shipments. Sensory data includes sensors similar to those used in military applications. A unique biological sensor, a specially trained dog, may be used to *sniff-out* the presence of drugs. As drug criminals become more sophisticated, the range of sensors required and processing required for data fusion become more extensive. Law enforcement applications have another commonality with military applications, namely, the use of countermeasures by the criminal to reduce the information content of sensors. The huge profits in drug traffic lead to the purchase of sophisticated technology by criminals to avoid detection.

Remote sensing applications [15] include the surveillance of the Earth to identify and monitor crops, weather patterns, mineral resources, environmental conditions, and threats (e.g., oil spills, and radiation leaks). Here again the entire spectrum of data may be monitored. Special sensors such as synthetic aperture radars allow active surveillance, while passive sensors may monitor visual and IR spectra. Special examples of remote sensing are NASA's use of Landsat satellites to monitor the Earth's surface and space probes to investigate the planets and solar system and the Hubble space telescope. Each of these examples involves a suite of sensors used in coordination to locate, identify, and interpret physical phenomena and events.

Fusion systems are also being developed to monitor and control complex equipment and manufacturing processes [16]. Certain systems, such as nuclear power plants and modern aircraft, require monitoring beyond the ability of a human operator. Semi-automated monitoring is required to ensure that the system continues to operate properly. Data from multiple sensors is monitored to assess the health of a system. A number of data fusion systems have also been developed for automatic fault diagnosis of complex equipment. Inferences range from the simple monitoring of equipment's function (e.g., are the output data— temperature, pressure, speed—within acceptable ranges?) to complex inferences (e.g., are there indications of an impending meltdown?) involving many possible observations and indicators.

Other examples of data fusion are the techniques employed for medical diagnosis [17]. At a basic level, to diagnose common illnesses, a physician may use touch (feeling a patient's skin, or checking the motion of a joint), sight (observing a patient's complexion, or seeking evidence of ear obstructions.) and sound (listening to breathing), as well as a patient's self-reported symptoms. More complex problems may involve obtaining data from multiple sensors (e.g., x-ray images, nuclear magnetic resonance, chemical and biological tests, and ultrasound images) and other data to determine the condition of a patient. Typically one or more physicians uses this data to diagnose a patient's condition. Experiments have been performed to develop expert system computer programs to emulate the physician's diagnosis process. Well-known examples include the program MYCIN, developed at Stanford to diagnose blood diseases, and INTERNIST, an expert system developed to assist a physician's diagnosis of the main diseases of internal medicine [18].

A final example of fusion applications is the integration of multiple sensor data for robotic applications [19]. Industrial robots [20] use pattern recognition and inference techniques to recognize three-dimensional objects, determine their orientation, and guide robotic appendages to manipulate the objects. The problem of robot multisensor fusion is reviewed by Luo and Ming-Hsiumn [21]. Other experiments at Carnegie Mellon University are leading to the development of a mobile robot that can avoid obstacles and follow general directions.

These examples certainly do not exhaust the range of problems for which data fusion techniques are applicable. They do provide, however, an indication of the breadth of potential types of data fusion problems. How do data fusion problems differ from one another in ways other than their specific application? Data fusion applications may be characterized by evaluating the following questions and issues:

1. What specific inferences should be made by a data fusion system? Is the location, identification, or characterization of an event important? Can these inferences be directly related to input data (e.g., input data exceed an a priori threshold), or are more complex patterns or inferences sought? Are humans (or at least specially trained humans) capable of making the required inferences?

2. What data is potentially observable? What are the physical phenomena available for observation—e.g., electromagnetic, nuclear, chemical, biological, magnetic, acoustic?

3. What is the sensing environment? Do interference, noise, or active countermeasures degrade sensor observations? Does the observer have any control over the sensing environment?

4. What sensors are available or utilized? What are the types of sensors, numbers of sensors, quality of data, and susceptibility of the sensors to an adverse sensing environment? Are the sensors active, passive, controlled, or guided by the data fusion system?

5. What are the observed data types and their associated rates?

6. Finally, what is the time line for the inference process? (e.g., what is the time scale between the physical occurrence of an event, the observation of related data, and the time at which an inference is required—milliseconds, seconds, minutes, hours, or days?)

These types of issues characterize a data fusion application. In subsequent discussions, we will see that these affect the selection and utilization of data fusion techniques, as well as the implementation of a fusion system.

1.3 SENSORS AND SENSOR DATA

The inputs to a sensor data fusion system include three basic components: (1) data observed by sensors, (2) data and commands input by human operators or users, and (3) a priori data from a pre-established database. This section provides a brief introduction to sensors, sensor data, and the sensing environment.

Waltz [11] provides an overview of the energy spectrum and the associated sensor systems that are available for each energy range. A general discussion of sensors is provided by Hovanessian [22]. In addition, Swanson [23] provides an excellent discussion of the signal processing that can be performed to translate raw sensor data into parameters suitable for input to a data fusion system. Table 1.2, adapted from Waltz, illustrates the range of detectable characteristics for which sensors have been developed, from acoustical phenomena to nuclear and nonnuclear particle emission. Table 1.2 lists detectable characteristics, the corresponding spectral range, and associated sensor systems.

Table 1.2

Spectrum of Sensors Available for Data Fusion [11]

Detectable Characteristics	Spectral Range	Sensor Systems
Acoustic Frequency	1 Hz–10 KHz	Acoustic detectors Active and passive sonar Seismometers
Electromagnetics	1 Hz–1 MHz (LF)	Magnetometers; passive electronic support measure (ESM) receivers
	10 MHz–100 MHz (HF/VHF/UHF)	High-frequency (HF); very high frequency (VHF); ultra-high frequency radars
	1–10 GHz	Surveillance and fire control radars
	10–50 GHZ (SHF/EHF)	Super high frequency (SHF); extreme high frequency (EHF)
	30–300 GHZ (MMW)	Millimeter wave (MMW) radar and radiometers
Infrared (IR) radiation	300nm–1 micron	Infrared cameras Scanning in search track Focal plane arrays

Table 1.2   (continued)

Spectrum of Sensors Available for Data Fusion [11]

Detectable Characteristics	Spectral Range	Sensor Systems
Visible light	780–380 nanometers	Multispectral arrays Laser radars Electro-optic sensors (TV)
Ultraviolet (UV) light	200–380 nanometers	UV spectrometers
Nuclear Particles	3×10^2 - 3×10^{-4} Angstroms	X-ray detectors Gamma ray detectors
Nonnuclear particles		Mass spectrometers

Throughout the energy spectrum, sensors can utilize either an active or passive approach. An active sensor emits energy with the intent of inducing a detectable phenomenon in a target (or event) to be observed. Examples of active sensors include radars, lasers, x-ray machines, and devices that induce nuclear magnetic resonance or electromagnetic induction. Passive sensors observe natural emissions from a target or event. Examples of passive sensors include visual or IR cameras, passive acoustic sensors, and x-ray or gamma-ray detectors. The choice between utilizing an active or passive sensor strategy depends on many factors, including stealth *(see but not be seen)* for tactical observers, and issues concerning destructive observation (i.e., whether an active sensor may adversely affect the object being observed).

Figure 1.1 illustrates a generic sensor. The sensor uses a collection element to detect incoming energy and observe a combination of intentional energy from a target (namely, electromagnetic radiation, acoustic energy, and nuclear particles) as well as environmental energy such as noise (thermal, galactic, and terrestrial), multi-path signals, and interference. Some of the nontarget energy may be generated by the sensor/collector itself. For example, the thermal noise from the detector elements of an infrared camera partially contributes to an observation. Similarly, radio frequency (RF) antenna elements contribute thermal noise due to the ambient temperature of the antenna element. An additional source of incoming energy may include electronic jamming signals, which are intentionally generated by an enemy to deceive the observer by decreasing the signal-to-noise ratio (SNR) of the observed environment.

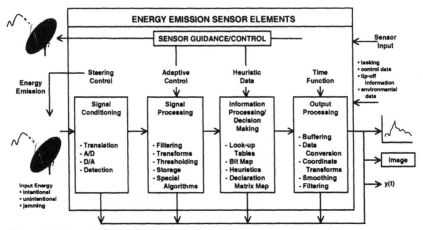

Figure 1.1 Concept of a generic sensor.

The internal functional components of a generic sensor are illustrated in Figure 1.1. The sensor components are partitioned into two groups: (1) the elements required for an active sensor to emit energy, which are indicated by a single box labeled energy emission sensor elements, and (2) the elements required to detect energy and create an output observation. These latter functions are shown in the lower portion of the box and include a signal conditioning process, signal processing, information processing and decision-making, and output processing. The signal conditioning process does not change the information content of a signal (e.g., does not alter the bandwidth), but it does perform translations that facilitate subsequent processing. Examples of signal conditioning functions include frequency shifts, delaying the signal in time, and translations from one energy form into another via heterodyning, and analog-to-digital (A/D) or digital-to-analog (D/A) changes. Often at this point, the signal is referenced to a fundamental frequency or time standard. Associations of time of arrival, frequency of arrival, or Doppler processing may also be performed. A general discussion of signal processing for sensors is provided by Swanson [23] and by Kirianaki and Yurish [24].

A second function of internal sensor processing is signal processing. The goal here is to select or isolate the energy of interest (i.e., typically that associated with a target) from all the energy received by the sensor. Signal processing functions may include transformations from the time to the frequency domain, application of thresholding to detect amplitudes above background noise, filtering, interference canceling or limiting, and band-limiting. At this point, limiting the bandwidths and transformations to detect and isolate the signal of interest from the received energy corrupts the original signal received by the sensor. This data may be output directly from the sensor, or additional information processing may be performed.

When a signal of interest has been isolated, information processing functions can be applied to assemble the signal data into the next higher level of integration. Classification algorithms seek to recognize patterns in the signal to determine issues such as the following: is the received signal a bit stream, return from a target, association of bit data into a bit stream, recognition and measurement of data features (amplitudes, rise times, etc.), assembly of an image from a bit map, and target tracking. Classification processing utilizes techniques from pattern recognition (described in more detail in Chapter 6) to transform the raw observational data into a declaration of target identity. For example, radar cross-section (RCS) data can be analyzed automatically to determine a target size and shape, with the subsequent declaration that an observed target is, for example, an aircraft (of a specified class), a bird, or clouds. Similarly, imagery data may be processed to determine the size, shape, and identity of an object. Methods for classification include simple bit mapping, mapping of observed parameters to identifications via table lookups, cluster analysis methods [25], and adaptive neural networks [26].

If a sensor performs a classification function, the identity declarations utilize one of two basic decision strategies. *Hard-decision* sensors declare target identity only when the parametric classifier data exceeds a predetermined threshold. A hard-decision sensor performs signal processing on incident signal data and utilizes a decision rule to declare target identity. By contrast, *soft-decision* sensors report partial evidence for identity as soon as the signal from a target is detected. Hence, a soft-decision sensor reports object identity and an associated confidence, probability, or evidential-interval. Soft-decision sensors accumulate and integrate evidence, reporting partial evidence and associated uncertainty (via probabilities, fuzzy membership functions, confidence factors, or evidential intervals).

The output processing within a sensor includes data buffering, coordinate transformations, data unit conversion, smoothing, and filtering. In particular, positional data may be processed to form an estimate of an object's position and velocity. Such processing might include ambiguity surface processing or parameter estimation. The resulting output *from* a sensor then has one of several forms:

1. A continuous or sampled waveform consisting of amplitude, frequency, or phase versus time.
2. A two-dimensional image consisting of image coordinates with associated amplitude or spectral data. A vector consisting of parametric positional data (e.g., range, angular information, or range-rate), target state vector (actual estimated target location), parametric data related to target identity or characteristics (e.g., frequency, size, shape, or spectra), or a declaration of target identity (namely, aircraft-type piper-cub).
3. Other useful output data from a sensor includes information about the sensor's state (current pointing angles, equipment, health, status) and characterization of the observing environment, such as background noise thresholds.

Data input *to* a sensor involves information required to control the sensor: commands, pointing or *look-angle* data, tip-off information to assist the classification function, or other data. Clearly the extent to which a sensor performs the functions illustrated in Figure 1.1 depends on its purpose, design, and deployment constraints. Modern microprocessors, application-specific integrated circuits (ASICs), and very high-speed integrated circuit (VHSIC) technology allows sensors to become *smarter*, with increasing functionality built into the sensor rather than having functions allocated to a postprocessing or fusion system [24, 27]. We will see later how these decisions concerning processing within a sensor affect the overall design of a fusion system architecture and the choice of fusion algorithms.

Waltz [11] characterizes sensors by the general attributes shown in Table 1.3. Any sensor may be characterized by specifying detection performance, spatial/temporal resolution, spatial coverage, detection/tracking modes, target revisit rates, measurement accuracy, measurement dimensionality, hard versus soft data reporting, and detection/track reporting.

Table 1.3

Description of Sensor Characteristics

Sensor Characteristic	Description
Detection performance	Detection characteristics (false alarm rate, detection probabilities and ranges) for a calibrated target characteristic in a given noise background
Spatial/temporal resolution	Ability to distinguish between two or more targets in space or time
Spatial coverage	Spatial volume covered by the sensor, for scanning sensors this may be described by the instantaneous field view, the scan pattern volume, and the total field-of-regard achievable by moving the scan pattern
Detection/tracking modes	Search and tracking modes performed: • Staring or scanning • Single or multiple target tracking • Single or multimode (track-while-scan/stare)
Target revisit rates	Rate at which a given target is revisited by the sensor to perform a sample measurement (staring sensors are continuous)
Measurement accuracy	Accuracy of the sensor measurements in terms of measurement statistics (e.g., standard deviation of measurement in specified observation conditions and bias estimates)
Measurement dimensionality	Number of measurement variables (e.g., range, range-rate, and spectral features) between target categories
Hard versus soft data reporting	Sensor outputs are provided either as hard-decision (threshold) reports or as preprocessed reports with quantitative measures of evidence for possible decision hypotheses
Detection/track reporting	Sensors report target detections or maintain a time-sequence representation (track) of the target's behavior

While these characteristics provide a basic specification of a sensor, a much more difficult problem is the evaluation of sensor performance (i.e., ability to locate, characterize, and identify a target or event accurately in the presence of environmental noise and interference). Steinberg [28] provides an example of how different types of sensors perform against different threats for the case of a tactical aircraft against threats such as surface-to-air missiles (SAMs), and antiaircraft artillery. Further, the performance of sensors varies with an engagement time line. Steinberg illustrates the use of on-board sensors (radar warning receiver, electro-optical sensor, forward-looking infrared, and others) to detect, identify, and locate threats against the aircraft as a function of the engagement time line. Detailed analysis of a sensor's performance requires that for various scenarios, simulations must vary the targets to be observed, the observation time line, environmental conditions, and other factors.

Table 1.4 is a summary of the basic measurements for 11 types of sensors, including radar, electronic intelligence (ELINT) receivers, electronic support measures (ESMs), synthetic aperture radars (SARs), IFFN sensors, radar homing and warning devices, infrared and laser warning sensors, communications sensors (e.g., receivers), and electro-optical (EO) and ground-based acoustic/seismic sensors. For each sensor, Table 1.4 provides a summary of the basic measurement and derived observational data. The column labeled *state determination* gives an indication of the extent to which these sensors are typically used to determine the position, velocity, and identification of objects. Comments are noted on the utilization of these sensors. For each sensor type, references are provided for further information.

Table 1.4

Measurements and State Determination from Various Sensors

Sensor Type	Basic Measurements	Derived Measurements	State Determination	Notes	References
Radar	Radar cross section Frequency Time	Range Azimuth Elevation Range-rate Target size and shape Signature	Position Velocity Identity See notes 1 and 2	Active sensor 1. Limited target identification from RCS signature 2. Target size and shape may be obtained in limited cases	[29-31]
Electronic (ELINT) intelligence receivers	Amplitude Frequency Time	Received SNR Polarization Pulse shape Pulse repetition	Position Velocity Identity See notes 3 and 4	3. General and specific identity 4. Analysis yields radar	[32, 33]

Table 1.4 (continued)

Measurements and State Determination from Various Sensors

Sensor Type	Basic Measurements	Derived Measurements	State Determination	Notes	References
		interval (PRI) Radio frequency (RF)		design characteristics	
Electronic support measures (ESMs)	Similar to ELINT	Similar to ELINT	Position Velocity Identity Note 3	Same as note 3 for ELINT	[34, 35]
Synthetic aperture radar	Coherent radar cross section	Target/platform shape Target size Aim point (direction)	Identity	Enhance radar resolution via coherent processing of moving antenna	[36, 37]
Radar homing and warning	Radio-frequency intensity detection	Direction Intensity	Identity	Warning of illumination by a missile guidance radar	[32]
Infrared warning	Infrared intensity detection	Direction Intensity	Identity	Warning of heat sources approaching	[38]
Laser warning	Optical intensity detection	Intensity	Identity	Warning of being illuminated by a laser source	
Electro-optical (EO)	Picture elements (Pixels) Pixel intensity Pixel color	Location Size Shape	Position Identity Note 3	Good for size and shape determination, but requires significant processing time Can use to obtain environmental data	[39]
Infrared (IR)	Same as EO	Location Size Shape Temperature Spectral characteristics	Position Identity (continued)	Good for identity based on spectral characteristics; relatively poor position and velocity determination	
Communica-	External signal	External meas-	Position		

Table 1.4 (continued)

Measurements and State Determination from Various Sensors

Sensor Type	Basic Measurements	Derived Measurements	State Determination	Notes	References
Communications	External signal data Internal signal data	External measurements: radio frequency, modulation, coding, location, operations schedule, emitter function Internal measurements: textual information, language, key words, speaker identification	Position Identity		
Ground sensors	Acoustic sensors Seismic sensors Environment sensors	Location Identity Environmental conditions (temperature, pressure, atmospheric conditions)	Position Identity	Acoustic detection performance is strongly influenced by atmospheric conditions	[40]

Any analysis of sensor performance results in the obvious conclusion that there is simply no one *perfect* sensor, (i.e., no single sensor or type of sensor accurately detects, locates, and identifies targets in all circumstances)–namely, for all ranges and sensing environments. Some sensors are more accurate at locating and tracking objects, while others are best suited for providing identity information. Moreover, the performance versus time line (or equivalently, engagement distance) varies widely. This result is not surprising; however, it points to the need for, and utility of, combining data from multiple sensors via data fusion techniques. In Section 1.7, we will illustrate the qualitative improvements made available by combining multisensor data.

The environment in which a sensor operates to collect data includes a natural physical environment (e.g., the Earth's atmosphere or ocean) and, for military applications, potentially hostile attempts to reduce sensor effectiveness via countermeasures. The natural physical environment affects the propagation of wave phenomena via refraction, diffraction, and attenuation. In the atmosphere, for example, the presence of water vapor rapidly attenuates electromagnetic radiation by factors ranging from 1 to 10 dB/km, depending on the rain rate. Fog provides even greater attenuation, exceeding 100 dB/km in the IR and visible wavelengths. For radio frequencies, propagation through the atmosphere is a strong function of wavelength and is particularly affected by the charged particles in the Earth's ionosphere. Figure 1.2 illustrates representative effects of the atmosphere on wave propagation. Observation models that seek to predict the emission (or reflection) of energy from a target, propagation of waves or particles through an intervening media, and detection by a sensor can be exceedingly complex [41]. Such models may entail the use of large-scale computer programs such as NASA's global reference atmosphere model [42].

For military applications, active or passive countermeasures may be used by an enemy to degrade the information content of the friendly sensors [43]. Two basic strategies may be used: (1) the addition of random signals to decrease the observed SNR and hence the information content (Jamming) [44] or (2) the addition of special signals (deception) to increase the sensor error rate. Electronic countermeasures (ECMs) may be classified into three categories: (1) the use of active radiators such as jammers or the use of deceptive techniques such as false target generators or track breakers, (2) the use of medium modifiers such as chaff or absorbing aerosols, and (3) the modification of the reflectivity of a target via vehicle design, radar-absorbing material, echo enhancers, or comer reflectors. The possibility of deception and jamming greatly complicates the data fusion problems in military applications [45]. In subsequent chapters, we will discuss how these factors may be addressed algorithmically.

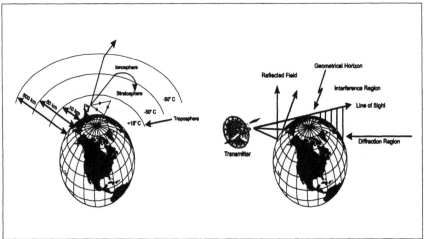

Figure 1.2 Representative atmospheric effects.

1.4 THE INFERENCE HIERARCHY: OUTPUT DATA

While input to a data fusion system consists of sensor data, commands, and a priori data, the output from a fusion system represents the combined or fused data.

This output data represents specific and accurate estimates of the location and identity of a target, entity, activity, or situation. Often, a hierarchy of inferences may be sought as shown in Figure 1.3. At the most basic level, a sensor suite and fusion system seek to detect energy and to determine the existence of an entity or situation. Here the term *entity* will be used to indicate a physical phenomenon, platform or vehicle, enemy unit, target, presence and identity of an equipment fault, source of illness, or other situation the fusion system was intended to observe. Clearly the entity referenced depends on the specific application as summarized in Section 1.2.

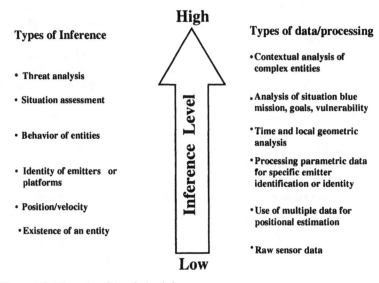

Figure 1.3 Hierarchy of data fusion inferences.

Given the suspected existence of an entity, the next level of inference to be developed is the position and velocity of the entity. Ideally, a complete positional state vector is sought having six dimensions: three components specifying position (x, y, and z), and three components specifying velocity. Other parameters may include object orientation, and body angular rates. The concept of state will be defined in mathematical terms in Chapter 4. Briefly, the state vector is comprised of those parameters that, if known accurately, would allow the prediction of the future of a system. For a dynamic system, knowledge of the system's state and equations of motion at the same epoch time, t_0, allows the prediction of the state of the system at an arbitrary future time, t. Hence, at a minimum, the state of a moving entity can be represented by a six-dimensional vector.

Closely related to position is the identity of the entity. Identity might consist of a general classification of which the observed aircraft is a member; a class of fighter aircraft; a specific classification within a general class (e.g., the observed aircraft is an F-16); or a specific identity (e.g., the observed aircraft is an F-16 having tail number 7632SC). Inference of an entity's identity involves transformations from parametric data to identity declarations utilizing pattern recognition methods.

Higher levels of inference involve the behavior of entities, intent, situation assessment, and (for military systems) assessment of threat. The highest levels of inference mimic the reasoning performed by human experts. Progression up the inference hierarchy pictured in Figure 1.3 involves utilization of techniques ranging from signal processing algorithms and statistical estimation methods for combining parametric data, to heuristic methods such as templating, or expert systems for situation assessment and threat analysis. The specific inferences sought are dependent upon the particular fusion application as described in Section 1.2 and Table 1.1. For example threat analysis for machinery monitoring may involve assessment of the evolution of a mechanical fault or failure condition. The data utilized to ascend the hierarchy of inferences ranges from raw parametric sensor data to nonparametric data such as collateral intelligence information, assessments, and judgments provided by human experts and tip-off information.

It is important to realize that ultimately, the output from a data fusion system is aimed at supporting a human decision process. In designing fusion systems, designers sometimes act as if the aim is to create a design that "satisfies the sensors" (i.e., it efficiently processes all of the sensor input data) rather than one that supports a process aimed at making accurate high-level inferences for human users of the fusion system. For example, the aim of a medical diagnosis system is to obtain a correct specific diagnosis of patient illness, not to process x-ray images rapidly and automatically. Hence, the utility of a fusion system must be measured by the extent to which the system supports the intended decision process.

1.5   A DATA FUSION MODEL

Several views can be developed to represent the data fusion process: A functional model can illustrate the primary functions, relevant databases, and interconnectivity to perform data fusion; an architectural model can specify hardware and software components, associated data flows, and external interfaces; and a mathematical model can describe algorithms and logical processes. This section describes a functional model for data fusion. Chapter 2 will introduce a taxonomy of algorithms and techniques, and Chapter 8 will present an architectural model. Chapter 10 thus draws together several different representations of data fusion systems. The functional model introduced here will also be illustrated with specific algorithms in Chapter 10.

Figure 1.4 illustrates a functional model for the data fusion process. This functional model is based upon a presentation by Hall and Llinas [46], and it incorporates the level 1, 2, and 3 terminology adopted by the data fusion subpanel.

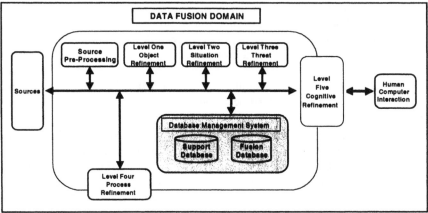

Figure 1.4 The Joint Directors of Laboratories (JDL)data fusion process model.

To the left, the model shows input data received from multiple sensors and types of sensors. Collateral information such as environmental data, a priori data, and human guidance/inferences may also be input to the illustrated fusion system. Major fusion functions include preliminary filtering, collections management, level one processing, level two processing, level three processing, a database management system, and creation and maintenance of situation and support databases. External interfaces provide for a man-machine interface and continuous or off-line evaluation of system performance. These functions are summarized below.

Preliminary filtering of input data provides an automatic means of controlling the flow of data into a fusion system. Typically, any fusion system can be computationally overwhelmed by data from multiple sensors. The sensor data rates frequently exceed the computational ability of a fusion system. Thus preliminary filtering may be performed to sort data according to observation time, reported location, data or sensor type, and identity or signature information. Filtering may also utilize environmental or sensor quality information (e.g., signal-to-noise ratio, reported quality indicators, and covariance error estimates). This information provides a means to associate data into general categories and prioritize data for subsequent processing.

Five levels of processing are shown in the center of Figure 1.4. Level one processing fuses data to establish the position, velocity, and identity of low-level entities or activities. The term entity refers here to a target, platform, emitter, or geographically (locationally) distinct object observed by multiple sensors. Level one processing combines positional and identity data from multiple sensors to establish a database of identified entities, target tracks, and uncorrelated raw data. Level one processing may be partitioned into four functions: (1) data alignment, (2) association, (3) tracking, and (4) identification.

Data alignment functions transform data received from multiple sensors into a common spatial and temporal reference frame. Specific alignment functions include coordinate transformation (e.g., from topocentric noninertial coordinates to geocentric inertial coordinates), time transformations (e.g., mapping from reported observation times to actual physical events), and unit conversions. Alignment may also require significant computational efforts. For example, transforming from image coordinates for an infrared sensor to an absolute direction in space may involve integral deconvolution transformations.

Data association tackles the problem of sorting or correlating observations from multiple sensors into groups, with each group of representing data related to a single distinct entity. In a dense tracking environment in which many targets of unknown identity are in close physical proximity, this problem requires multiple strategies. Association thus compares observation pairs (i.e., observation N vs. observation N-1, N-2, et al.) and determines which observations "belong" together as being from the same entity, object, event, or activity. Association also seeks to determine which, if any, new observations belong to an observation or group of observations already contained in the fusion database. Clearly, the association problem scales as N^2, where N is the current number of observations or tracks in the database.

The tracking problem refers to a process in which multiple observations of positional data are combined to determine an estimate of the target/entities position and velocity. A common tracking problem involves direction finding, in which multiple angular observations (e.g., lines of bearing) are combined to estimate the location of a nonmoving target. Another common tracking problem involves the use of observations from radar to establish the track of an aircraft. Tracking algorithms begin with an a priori estimate of an object's location and velocity (namely either an initial estimate based on a minimum data set, or an a priori track established in the database) and update the estimated position to the time of a new observation. The new observation is utilized to refine the position estimate to account for the new data. Tracking algorithms are closely coupled to association strategies, particularly in a dense tracking environment.

Identification completes the category of level one processing functions. Identification methods seek to combine identity information, analogous to the estimation of position utilizing positional data. Identity fusion combines data related to identity (i.e., either actual declarations of identity from sensors, or parametric data that can be related to identity). Identity fusion techniques include cluster methods, adaptive neural networks, templating methods, and Dempster-Shaffer and Bayesian inference methods.

Level two processing, or situation refinement, seeks a higher level of inference above level one processing. Thus, for example, while level one processing results in an order of battle (e.g., location and identity of low-level entities such as emitters or platforms), level two processing aims to assess the meaning or patterns in the order of battle data. To obtain an assessment of the situation, the data is assessed with respect to the environment, relationships among entities, and patterns in time and space.

Level three processing (for military or intelligence fusion systems) performs threat refinement. The purpose of threat refinement is to determine the meaning of the fused data from an adversarial view. Threat refinement functions include determination of lethality of friendly and enemy forces, assessment of friendly ("blue force") and enemy ("red force") compositions, estimate of danger, evaluation of indications and warnings of impending events, targeting calculations, and weapon assessment calculations. Level three processing is an inferential process to assess the enemy's threat.

In order to manage the ongoing data fusion process, a level four meta-process is often used. This is a process that monitors the data collection and the fusion processing to optimize the collection and processing of data to achieve accurate and useful inferences. The level four process is deliberately shown as being partly inside and partly outside the data fusion process. The reason is that actions could be taken by the level four process (e.g., the use of active sensors) that could impede an operational mission. Hence, level four processing must be cognizant of mission needs and constraints. Within level four processing, the collections management function manages or controls the data collection assets (sensors) available to a fusion system. Specific functions involve determining the availability of sensors, tasking sensors to perform collections, prioritization of tasking, and monitoring sensor health/status in a feedback loop. Sensor tasking in particular requires predicting the future location of dynamic objects to be observed, computing the pointing angles for a sensor (namely from the predicted position of an object and location of the sensor, it computes parameters such as the directions and ranges required for a sensor to point to a target), computing the controls for sensor pointing, scheduling the sensor's observations, and optimizing the use of sensor resources (e.g., power requirements and movement from one target to another). For a large network of sensors, the collection management function can be extremely challenging. Ideally the collection management function should work in concert with other data fusion functions to optimize the rate and accuracy of inferences output from the fusion system.

Finally, a fifth level of processing was proposed by [8] to recognize the importance of addressing humans in the loop decision processes. Level five processing involves functions such as human-computer interaction, cognitive aids to reduce human biases, adaptation of information needs and system control to individual user preferences and constraints, automated interpretation of data using multiple human sensory mechanisms, and tools such as search engines. The concept of level five processing has been adopted by the data fusion community and highlighted at the 2002 conference on multisensor data fusion held in San Diego, California.

Last, functions in a fusion system include database management, man-machine interface, and evaluation functions. These functions may actually comprise a major portion of a fusion system. Chapter 10 provides more details on these ancillary functions.

1.6 BENEFITS OF DATA FUSION

Fusion of data from multiple sensors results in both qualitative and quantitative benefits. Certainly the primary aim of data fusion is to estimate the location and identity of entities or to make inferences, which may not be feasible from a single sensor alone. Table 1.5 (adapted from [11]) summarizes the qualitative benefits of multisensor fusion. Benefits include improved operational performance, extended spatial coverage, extended temporal coverage, increased confidence (e.g., higher probability of correct inference), reduced ambiguity of inferences, improved detection, enhanced spatial resolution, improved system reliability, and increased dimensionality. Table 1.5 summarizes these benefits, describing the impact of the benefit and the operational advantages that result. While the benefits listed in Table 1.5 have a military or intelligence bent, most of the benefits (e.g., increased confidence, reduced ambiguity, and improved system reliability) are applicable to most of the data fusion applications summarized in Section 1.2.

Table 1.5

Benefits of Multisensor Data Fusion (adapted from [11])

Category of Benefits	General Benefit	Operational Advantages
Robust operational performance	One sensor can contribute information while others are unavailable, denied (jammed), or lack coverage of a target or event	Allows continued operation despite jamming Graceful degradation Increased probability of detection
Extended spatial coverage	One sensor can look where another cannot	Increased survivability Probability of detection increased
Extended temporal coverage	One sensor can detect/measure a target or event when others cannot	ECM, stealth, RF management, and IR suppression Increased detection probability
Increased confidence	One or more sensors can confirm the same target or event	Rules of engagement require positive target identification Assurance that the correct counter-measure is employed given a threat condition
Reduced ambiguity	Joint information from multiple sensors reduces the set of hypotheses about the target or event	Establish target identification for employment of extended range weapons

Table 1.5   (continued)

Benefits of Multisensor Data Fusion (adapted from [11])

Category of Benefits	General Benefit	Operational Advantages
		Reduced pilot workload
		Target prioritization
Improved detection	Effective integration of multiple measurements of the target/event increases the assurance of detection	Increased reaction time
		Increased detection range for weapons deployment
		Increased survivability
Enhanced spatial resolution	Multiple sensors can geometrically form a synthetic aperture capable of greater resolution than a single sensor can form	Improved positional data on target supports both defensive reaction and selection of attack profiles
Improved system reliability	Multiple sensor suites have an inherent redundancy	Redundant systems
		Graceful system degradation
Increased dimensionality	A system employing different sensors to observe different physical phenomena are less vulnerable to disruption by enemy action or by natural phenomena	Allows continued operation
		Improves probability of survival

Quantitatively, the aim of data fusion is to improve the accuracy of inferences such as the estimation of location or identity declaration. Figure 1.5 illustrates the concept of utilizing multiple sensors to improve the accuracy of location for forward-looking infrared (FLIR) and radar data (adapted from [11]). The surveillance accuracy of a radar is illustrated in the upper left. Typically, a pulsed radar has the ability to accurately determine slant range (namely radial range from the radar antenna to the object being observed), but it has a lower accuracy in determining the angular position (i.e., the direction) of a target. Hence, the uncertainty associated with a radar observation is a volume whose dimensions are relatively large perpendicular to the observed line of site (LOS), but narrow along the LOS. Conversely, a FLIR accurately determines the angular location of an observed object (i.e., its location within a two-dimensional image), but has a large uncertainty in the range at which the target is observed. Hence, the uncertainty of a FLIR observation may be represented as a volume whose dimensions are relatively small perpendicular to the LOS, but elongated along the LOS. This is illustrated in the right side of Figure 1.5.

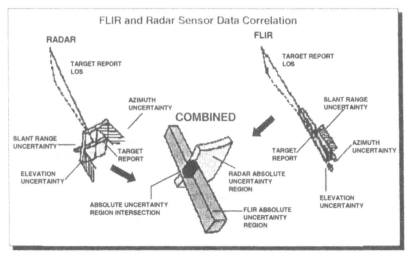

Adapted from W.G. Pemberton, M.S. Dotterweich, and L.B. Hawkins, "An Overview of Fusion Techniques",
Proc. of the 1987 Tri-Service Data Fusion Symposium, vol. 1, 9-11 June 1987, pp. 115-123.

Figure 1.5 Forward-looking infrared (FLIR) and radar sensor data correlation.

When the FLIR and radar data are combined, the uncertainty of the estimated position of the target is smaller than the uncertainty of either observation alone. In particular, as illustrated in the lower portion of Figure 1.5, the combined uncertainty is relatively small in both angle and direction of the LOS as well as being small in slant range. Thus the fused estimate of location takes advantage of the relative strengths of each sensor, resulting in an improved estimate of location and reduced positional uncertainty.

The particular quantitative improvement in estimation that results from using multiple sensors depends of course on the performance of the specific sensors involved (namely data collection rates and observational accuracy), environmental effects, and the specific algorithms used in the data fusion estimation process. Accurate determination of the improvements gained via data fusion requires numerical simulations, Monte Carlo simulations, or covariance error analysis. The latter technique will be introduced in Chapter 4.

An example of a Monte Carlo simulation for two sensors is illustrated in Figure 1.6, from [11]. Waltz performed a Monte Carlo simulation to show the benefits of combining data (via a Bayesian inference algorithm) for a tactical air-to-air identification friend-foe-neutral (IFFN) problem. Two sensors, labeled sensor 1 and sensor 2 in Figure 1.6, onboard a tactical aircraft have different performance characteristics in determining the presence of an enemy aircraft as a function of distance between the tactical aircraft and the enemy craft. Figure 1.6 shows the probability of correct identification as a function of engagement time or (equivalently) distance. Sensor 1 has an identification performance that is nearly independent of distance but that exhibits at best a detection probability of 0.5. Sensor 2 shows the ability to accurately identify an enemy aircraft at close range, but its inference accuracy degrades rapidly with increasing distance.

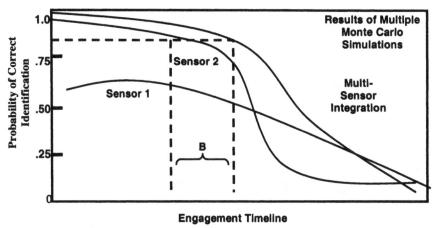

Engagement Timeline

Increased range at which identity reaches decision threshold for a specified probability of correct identification (ID).

Figure 1.6   Example of improved target identification using decision-level fusion [11].

In Figure 1.6, the curve labeled multiple sensor integration (MSI) represents the inference accuracy based on combining the results from sensors 1 and 2. Notice that the inference accuracy is always better than either sensor 1 or sensor 2 alone. Hence at any given engagement distance (or equivalently time-of-engagement), the MSI inference is more accurate than that obtained from either sensor alone. In addition, if the tactical rules of engagement require that a particular certainty threshold (e.g., probability of enemy > 0.9) must be exceeded to allow deployment of weapons, then the MSI inference provides an added engagement distance or time (i.e., the distance at which the MSI inference exceeds an a priori threshold) up to 25% greater than the distance at which either sensor alone exceeds the threshold. This added engagement distance is shown in Figure 1.6 by the distance labeled "B." The benefit of data fusion for this example is clear—added engagement distance equates to increased tactical options and to an improved probability of survival for the friendly aircraft and pilot.

Another attempt to illustrate the quantitative benefit of data fusion is provided by [47]. Figure 1.7 illustrates the marginal gain in correct classification with an increasing number of sensors. Nahin and Pokoski [47] assume that N identical sensors are utilized to observe a phenomenon with an aim towards correctly identifying or classifying the phenomenon. Assuming that the sensors are statistically independent, and that the a priori probabilities are equal (namely the principle of indifference described in Chapter 5), Figure 1.7 plots the increased probability of correct inference when the number of sensors is increased from 5 to 7 (the bottom curve in Figure 1.7), from 3 to 5 sensors (middle curve) and from 1 to 3 sensors (top curve). The sensor data is combined by a majority vote decision rule.

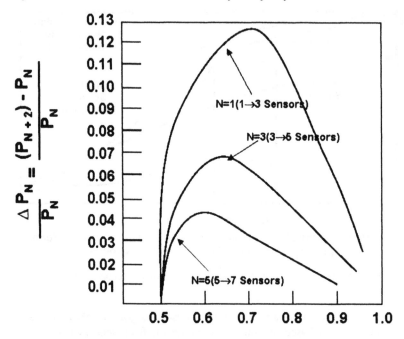

P_N Single Sensor Probability of Correct Classification

Figure 1.7 Marginal gain in correct classification with additional sensors [47].

Nahin and Pokoski's results are illustrative only, but do demonstrate the quantitative benefits of multisensor fusion. The results also provide some conceptual rules of thumb, specifically the following:

1. Combining data from multiple inaccurate sensors (having individual probability of correct inference less than 0.5) does not provide a significant overall advantage. (See the left side of Figure 1.7.)
2. Combining data from multiple highly accurate sensors (having individual probability of correct inference greater than 0.95) does not provide a significant increase in inference accuracy. (See the right side of Figure 1.7.)
3. When the number of sensors (N) becomes large (e.g., greater than 8 or 10), adding additional identical sensors does not provide a significant improvement in inference accuracy. Note, however, that adding a new sensor type may have a very significant impact in inference capability, because of an added dimensionality of observational data.

4. The greatest marginal improvement in sensor fusion occurs for a moderate number of sensors (i.e., 1 to 7), each having a reasonable probability of correct identification.

While none of the results presented here is a compelling argument for the benefits of fusing data from multiple sensors, it does do illustrate that there are both qualitative and quantitative benefits to be derived from data fusion. The covariance error analysis techniques described in Chapter 4 provide a means for determining the specific quantitative benefits for particular applications, sensor suites, and observational conditions.

1.7 ARCHITECTURAL CONCEPTS AND ISSUES

Transformation of the functional model described in Section 1.5 into a physical data fusion system involving computer hardware and software requires significant effort and resolution of many issues. In particular, implementation requires development of ancillary support functions such as database management, man-machine interfaces, communications software, utility functions, and many others. For fusion systems implemented via software, the portion of an implemented system, which is specific to data fusion algorithms, is commonly less than 20% of the total software. Figure 1.8 provides a profile of the relative amounts of software required for various functions in typical data fusion systems comprising 50 thousand to 70 thousand executable lines of software higher order language [48].

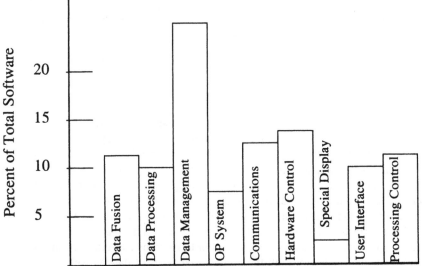

Figure 1.8 Profile of software required for data fusion systems.

An implementation view of a fusion system is illustrated in Figure 1.9. Figure 1.9 illustrates that fusion processing requires support from functions such as communications, message processing, database management, man-machine interface, and sensor management. Each of these functions is addressed below. Communications functions receive data from distributed sensors and send control and look angle (pointing) data to the sensors. Communications hardware and software handle the mechanics of data transport. The contents of transported data (e.g., from sensors and from human input) are processed via message processing software in order to detect special events, alert a system user of the arrival of particular data, and route data to appropriate processes or system operators. Input message processing strips out message headings, text, and end-of-message data, extracts data for insertion into the database, and performs archival functions. Output data processing constructs messages for outgoing communications, and appends appropriate message headers, addresses, and end-of-message control information. For distributed systems, message processing verifies and controls the mapping between message addresses and addressees.

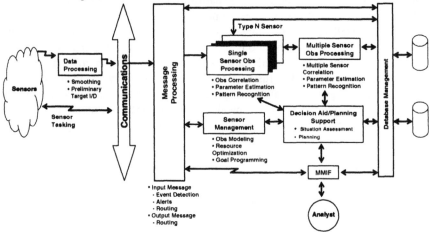

Figure 1.9 Implementation view of a data fusion system.

Another major function required in implemented systems is database management. Numerous data files must be maintained in fusion systems, including the input sensor data, the data associated with sensor and dynamic models, state vectors and track files, data associated with the sensing environment (e.g., terrain models, and atmospheric data), and knowledge such as emitter characteristics, enemy tactics, and doctrine (for military applications). Required data management functions include data storage, retrievals, archiving, validation of data integrity, and data transformations to and from internal storage formats to external display and output formats. Database management tends to be especially complex for fusion systems because of both the number and types of data as well as the rate at which data enters the system. The database software must trade off efficient storage of data and rapid insertion of sensor data into the database against providing general retrieval capabilities such as rapid Boolean queries. If an expert system is used in a fusion system, the database software must also manage a knowledge base comprising rules, frames, or other knowledge representation formats.

The man-machine interface (MMIF) for a fusion system is another key area of functionality. Data displays often include maps, overlays of data such as the locations of emitters or platforms, environmental data, graphical representations, and numerical and alphanumeric data. Modern interfaces utilize a combination of window-type displays, graphics overlays, command language inputs, menus for selections of options, and fill-in-the-blank templates (e.g., for creating reports and messages). Man-machine interfaces must provide flexibility of displays while assisting a user by displaying only the data that is required at a particular time. Much research needs to be done in this area to tailor man-machine interfaces for operational systems in which an operator is bombarded with data and is required to make rapid decisions in a generally stressful setting.

The final major ancillary function shown in Figure 1.9 is sensor management. This entails a number of numerically demanding tasks, such as observation prediction for each sensor type and entity in the database (e.g., track file). Based on sensor models and differential equations of motion, sensor management schedules sensors to ensure observational coverage of activities of interest. This function also monitors the health and status of all sensors in the system. After the required ancillary functions have been identified and appropriate models or an appropriate approach has been established, a number of implementation issues remain. Some are common to the development of other large-scale hardware and software systems, such as complex system architectures involving multiprocessing with distributed data and processing, external interfaces to real-time sensors or communications devices, stringent throughput and response time requirements, platform constraints (e.g., mobile platforms on which the system must operate), and extensive reliability and maintainability requirements. Some issues specific to data fusion systems include algorithm selection, choice of when (in the processing flow) to fuse data, and the role of the human in the loop. These topics will be summarized here and addressed in detail in Chapter 10.

The problem of algorithm selection depends strongly upon the nature of the inferences sought by the data fusion system, the specific application, and the sensor data available. For example, systems employing imagery data require algorithms that are fundamentally different than, say, those whose data inputs are declarations of identity. The taxonomy developed in Chapter 2 provides an overview of applicable algorithms. However, given a specific application, sensor suite, and inference hierarchy to be achieved, there are generally a number of competing algorithms that are applicable to process and fuse the data. Some trade-offs can be made based on the mathematical nature of the algorithms and the availability or lack of availability of a priori information. Other trade-off considerations include throughput constraints, computer resource requirements (e.g., memory and computational complexity) of algorithms, and operational constraints. We argue in

Chapter 10 that a structured process can be applied to algorithm selection based on four separate perspectives: a systems analysis view, an operational view, a mathematical or numerical view, and the end-user's view.

After algorithms have been selected for data fusion, a fundamental issue involves where in the processing flow to perform fusion. Stoltz [49] describes three basic architectural approaches to data fusion for a tracker-correlator as illustrated in Figure 1.10. Stoltz's centralized architecture transmits raw (unprocessed) data from several sensors to a central fusion process that performs the functions of data alignment and association, followed by correlation, tracking (dynamic estimation), and target classification. This is illustrated in Figure 1.10(a). At the opposite extreme, Stoltz's autonomous architecture shown in Figure 1.10(b), allows each sensor to perform a maximum amount of preprocessing to generate state vectors and declarations of identity, which in turn are transmitted to a fusion process that fuses these data. In this case, data alignment, association, correlation, filtering, and classification are performed on state vectors rather than on raw data. Stoltz's [49] third architecture is a hybrid combination of the autonomous and centralized architectures [shown in Figure 1.10(c)]. In the third architecture added multiplex (MUX) and select and merge functions are required. Trade-offs among these architectures involve factors such as the availability of "smart" sensors that perform data preprocessing (single-source target tracking and identity declaration), the availability of communications links that can support the large data rates necessary for sending raw data to a central processor, and the computational abilities of the central processor. A detailed discussion of these trade-offs is provided in Chapter 10.

Finally the issue of human in the loop is fundamental to the implementation of a data fusion system. To what extent is the fusion system meant to be autonomous in its inference processes? For a tracking system, for example, a totally autonomous system would assign observations to target tracks, perform estimates of object locations (and velocities), and introduce new tracks or discontinue (or purge) old tracks without human intervention. Decisions concerning such factors as data association, validity of target tracks, and rejection of data would be made completely automatically. Conversely, a tracking system that was human-intensive would allow (require) a human to intercede and evaluate the decision processes. For higher level inference systems, such as a medical diagnosis system, the human-in-the-loop question may entail determining whether a system is meant to be advisory (e.g., act as an advisor or reference system for a physician) or whether the system is truly meant to act as a diagnostician. The operational concept for a fusion system significantly affects the systems design and implementation.

The remainder of this book focuses on the algorithms required for data fusion. Chapter 10 describes some implementation concepts in more detail.

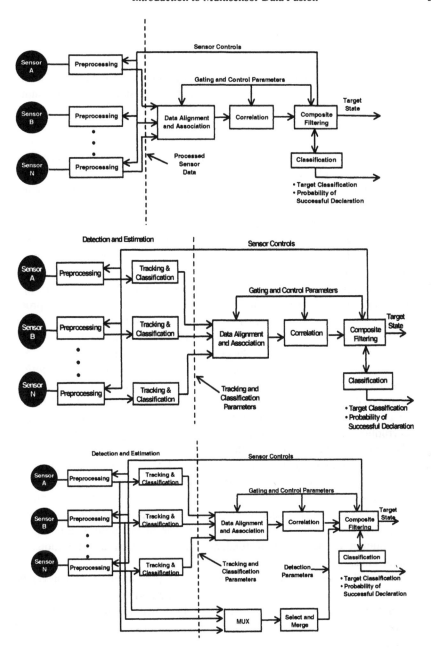

Figure 1.10 Data fusion architectures: (a) centralized data fusion architecture, (b) autonomous data fusion architecture, and (c) hybrid data fusion architecture.

1.8   LIMITATIONS OF DATA FUSION

It is only fair in this introduction to data fusion to identify some limitations of data fusion systems. Hall and Steinberg [50] have discussed these limitations and identified a number of issues or "dirty secrets" in data fusion systems. These are summarized below. Additional information on how to mitigate these issues is presented throughout this book.

- *There is no substitute for a good sensor:* No amount of data fusion can substitute for a single accurate sensor that measures the phenomena that you want to observe. Thus, while a combination of sensors may provide robust system operation, these sensors will not necessarily observe the feature of interest. For example, in a mechanical system, conditions such as operating speed, vibration levels, and lubrication pressure may provide valuable information about a system's operation, these do not provide direct evidence of the system operating temperature.

- *Downstream processing cannot make up for errors (or failures) in upstream processing:* Data fusion processing cannot correct for errors in processing (or lack of pre-processing) of individual sensor processing. For example, failure to identify and extract the correct features from a signal cannot be corrected by a sophisticated pattern recognition scheme.

- *Sensor fusion can result in poor performance if incorrect information about sensor performance is used:* A common failure in data fusion is to characterize the sensor performance in an ad hoc or convenient (but incorrect) way. Failure to accurately model sensor performance will result in a corruption of the fused results, because the sensor weights (or effects of the data from individual sensors) will be incorrect.

- *There is no such thing as a magic or golden data fusion algorithm:* Despite claims to the contrary, there is no perfect algorithm that is optimal under all conditions. Often, real applications do not meet the underlying assumptions required by data fusion algorithms (e.g., available prior probabilities or statistically independent sources). A major theme of this book is a continual admonition to be aware of fundamental assumptions or limitations of algorithms.

- *There will never be enough training data:* In general, there will never be sufficient training data for pattern recognition algorithms applications such as automatic target recognition. Hence, hybrid methods must be used (e.g., model-based methods, syntactical representations, or combinations of methods).

- *It is difficult to quantify the value of a data fusion system:* A challenge in data fusion systems is to quantify the utility of the system at a mission level. While measures of performance can be obtained for sensors or processing algorithms, measures of mission effectiveness are difficult to define. How does the application of data fusion algorithms assist in how well a system (e.g., machine, factory, or weapon system) will perform in an operational environment?

- *Fusion is not a static process:* The data fusion process is not static, but rather an iterative dynamic process that seeks to continually refine the estimates about an observed situation or threat environment. The level four process seeks to optimize the ongoing fusion process.

These factors should be kept in mind in studying and applying methods in data fusion.

REFERENCES

[1] Kessler, O., *Functional Description of the Data Fusion Process*, Warminster, PA: Office of Naval Technology, Naval Air Development Center, 1992.

[2] Steinberg, A., C. Bowman, and E. E. White, Jr., "Revisions to the JDL Data Fusion Model," *3rd NATO/IRIS Conference*, Quebec City, Canada, 1998.

[3] Bowman, C., and A. Steinberg, *A Systems Engineering Approach for Implementing Data Fusion Systems*, in *Handbook of Multisensor Data Fusion*, ed. by J. Llinas, New York, NY: CRC Press, 2001.

[4] Hall, D. L., and E. Waltz, *Requirement Analysis for Data Fusion Systems*, in *Handbook of Multisensor Data Fusion*, Boca Raton, FL: CRC Press, 2001.

[5] Bowman, C., "The Data Fusion Tree Paradigm and Its Dual," *Proceedings of the 7th National Symposium on Sensor Fusion*, 1997.

[6] Hall, D., and J. Llinas, "Algorithm Selection For Data Fusion Systems," *1987 Tri-Service Data Fusion Symposium*, Johns Hopkins Applied Research Laboratory, Laurel, MD, June, 1987, pp. 100-110.

[7] Llinas, J., and D. Hall, "A Survey Of Techniques For CIS Data Fusion," *2nd Annual Conference on Command, Control, Communications and Management Information Systems*, Bournemouth, United Kingdom, 1987.

[8] Hall, M. J., S. A. Hall, and T. Tate, "Removing The HCI Bottleneck: How The Human Computer Interface (HCI) Affects The Performance Of Data Fusion Systems," *2000 MSS National Symposium on Sensor and Data Fusion*, San Diego, CA, June, 2000, pp. 89-104.

[9] Hall, D., and J. Llinas, eds. *Handbook of Multisensor Data Fusion*, New York, NY: CRC Press, 2001.

[10] Blackman, S. S., and T. J. Broida, "Multiple Sensor Data Association and Fusion in Aerospace Applications," *Journal of Robotic Systems*, vol. 7, 1990, pp. 445-485.

[11] Waltz, E. L., *Data Fusion for C^3I*, in *Command, Control, Communications Intelligence (C^3I) Handbook*, Palo Alto, CA: EW Communications, Inc., 1986, pp. 217-226.

[12] Waltz, E., and J. Llinas, *Multi-sensor Data Fusion*, Norwood, MA: Artech House, Inc., 1990.

[13] Macfadzean, R. H. M., *Surface-based Air Defense System Analysis*, R. H. M. Macfadzean, 2000, 404.

[14] Naveh, B.-Z., and A. Lorber, eds. *Theatre Ballistic Missile Defense*, Progress in Astronautics and Aeronautics, Vol. 192, Reston, VA: American Institute of Astronautics and Aeronautics (AIAA), 2001.

[15] Lillesand, T. M., and R. Kiefer, *Remote Sensing and Image Interpretation*, 4th ed., New York, NY: John Wiley & Sons, 1999.

[16] Davies, A., ed. *Handbook of Condtion Monitoring: Techniques and Methodology*, London: Chapman and Hall, 1998.

[17] Spichiger-Keller, U. E., *Chemical Sensors and Biosensors for Medical and Biological Applications*, New York, NY: John Wiley & Sons, 1998.

[18] Jackson, P., *Introduction to Expert Systems*, Reading, MA: Addison-Wesley, 1986.

[19] Everett, H. R., *Sensors for Mobile Robots: Theory and Practice*, Natick, MA: A. K. Peters, Ltd., 1995.

[20] Kumara, S. T., R. L. Kashyap, and A. L. Soyster, *AI in Manufacturing: Theory and Practice*, Atlanta, GA: Institute of Industrial Engineers, 1989.

[21] Luo, R. C., and L. Ming-Hsiumn, "Robot Multi-Sensor Fusion And Integration: Optimum Estimation Of Fused Sensor Data," *1988 IEEE International Conference of Robots and Automation*, Philadelphia, PA, April, 1988, pp. 24-29.

[22] Hovanessian, S., *Introduction to Sensor Systems*, Norwood, MA: Artech House, Inc., 1988.

[23] Swanson, D. C., *Signal Processing for Intelligent Sensing Systems*, New York, NY: Marcel Dekker, Inc., 2000.

[24] Kirianaki, N. V., S. Y. Yurish, and N. Shpah, eds. *Data Acquisition and Signal Processing for Smart Sensors*, New York, NY: John Wiley & Sons, 2002.

[25] Fukunaga, K., *Introduction to Statistical Pattern Recognition*, New York, NY: Academic Press, 1990.

[26] Haykin, S., *Neural Networks: A Comprehensive Foundation*, 2nd ed., New York, NY: Prentice Hall, 1998.

[27] Frank, R., *Understanding Smart Sensors*, Norwood, MA: Artech House, Inc., 2002.

[28] Steinberg, A. N., "Threat Management System For Combat Aircraft," *1987 Tri-Service Data Fusion Symposium*, Johns Hopkins University, Baltimore, MD, June, 1987, pp. 532-554.

[29] Skolnick, M. L., *Radar Handbook*, New York, NY: McGraw-Hill, 1970.

[30] Stein, A., "Bistatic Radar Applications In Passive Systems," *Journal of Electronic Defense*, vol. 13, 1990, pp. 55-61.

[31] Barton, D. L., *Modern Radar System Analysis*, Norwood, MA: Artech House, Inc., 1988.

[32] Long, M., ed. *Airborne Early Warning System Concepts*, Norwood, MA: Artech House, Inc., 1992.

[33] Wiley, R. G., *Electronic Intelligence: The Analysis of Radar Signals*, Norwood, MA: Artech House, Inc., 1982.

[34] Hall, D., P. Lapsa, and C. Voas, "Improving The Performance Of ESM Using A Hierarchical Data Fusion Approach," *Proceedings of the Combat Identification Systems Conference*, Naval Postgraduate School, Monterey, California, 1994.

[35] *EW Design Engineers Handbook 1989/1990*, vol. Supplement to January 1990 Journal of Electronic Defense, Norwood, MA: Horizon House Microwave, 1990.

[36] Hovanessian, S., *Introduction to Synthetic Array and Imaging Radar*, Norwood, MA: Artech House, Inc., 1980.

[37] Fitch, J. P., *Synthetic Aperature Radar*, New York, NY: Springer-Verlag, 1988.

[38] Soref, R., *Threat Warning And Target Identification With Ge/Sige Multispectral Quantum Well Infrared Sensors*: Air Force Research Laboratory, 2001.

[39] Blouke, M. M., and D. Pophal, "Optical Sensors and Electronic Photography," *SPIE*, Los Angeles, CA, January 16-18, 1989.

[40] Hall, D., and D. C. Swanson, "Real-Time Data Fusion Processing For Internetted Acoustic Sensors For Tactical Applications," *1994 IEEE International Conference on Multisensor Data Fusion and Integration for Intelligent Systems*, Las Vegas, NV, October 2-5, 1994, pp. 443-446.

[41] Brussaard, B., and P. Watson, *Atmospheric Modeling and Millimetre Wave Propogation*, London: Chapman and Hall, 1995.

[42] *Global Reference Atmospheric Model - 1988*, Washington, DC: National Aeronautics and Space Administration, 1991, p. 35.

[43] Fitts, R. E., ed. *The Strategy of Electromagnetic Conflict*, Los Altos, CA: Peninsula Publishers, 1980.

[44] Lothes, R. N., M. Szymanski, and R. G. Wiley, *Radar Vulnerability to Jamming*, Norwood, MA: Artech House, Inc., 1990.

[45] Blake, B., ed, *International Electronic Countermeasures Handbook*, Norwood, MA: Artech House, Inc., 1997.

[46] Hall, D. L., and J. Llinas, "Multi-sensor Data Fusion," *Three Day Seminar in London*, London, England, 1985.

[47] Nahin, P. J., and J. L. Pokoski, "NCTR Plus Sensor Fusion Equals IFFN," *IEEE Transactions on Aerospace Electronic Systems*, vol. AES-16, 1980, pp. 320-327.

[48] Hall, D., J. J. Gibbons, and M. Knell, "Quantifying Productivity: A Software Metrics Data Base to Support Realistic Cost Estimation," *Journal of Parametrics*, vol. VI, 1986, pp. 61-75.

[49] Stoltz, J. R., and D. C. Cole, "Design and Testing of Data Fusion Systems for the U.S. Customs Service Drug Interdiction Program," *SPIE*, 1991.

[50] Hall, D., and A. Steinberg, "Dirty Secrets in Multisensor Data Fusion," *National Symposium on Sensor Data Fusion*, San Antonio, TX, June, 2000.

Chapter 2

Introduction to the Joint Directors of Laboratories (JDL) Data Fusion Process Model and Taxonomy of Algorithms

This chapter provides an introduction to the JDL data fusion process model, developed to assist in characterizing and understanding data fusion systems. A taxonomy of algorithms is presented for all levels of data fusion. A brief introduction is provided to alternative data fusion process models.

2.1 INTRODUCTION TO THE JDL DATA FUSION PROCESSING MODEL

One of the products of the JDL data fusion sub-panel was the creation of a process model to describe the data fusion process [1]. A current version of that model is shown in Figure 2.1 (identical to Figure 1.4). Figure 2.1 incorporates updates of the model recommended by Steinberg [2] and by Hall et al. [3]. The model is intended to define the basic components of the data fusion process and to assist in communications about data fusion applications. The top level of the model is summarized below and in Table 2.1. The remaining sections of this chapter address each function in more detail and provide taxonomy of applicable algorithms. On the left-hand side of the model, multiple sources of data and information provide input to the fusion process. These sources could include sensor data (either colocated with a particular platform such as a ship or an aircraft, or geographically distributed sensors), information from databases, input from other fusion processors or smart sensors, or human input and directions. On the right-hand side of Figure 2.1, the data fusion process human-computer interaction (HCI) is shown. The HCI provides the mechanism for displaying the results of the fusion process (e.g., situation maps, threat alerts, and alarms) and for human input and control of the fusion process. Within the fusion process, six basic functions are shown. These are summarized as follows. (For critics of the JDL data fusion process model, it should be noted that the JDL subpanel that developed the original process model did not intend for the model to be used for system implementation. Indeed, the original discussions concerning the model considered issues related to showing a functional decomposition, communications and interfaces between functions, and the challenge of trying to explicitly show recursion and blurring of functionality among components. Despite a number of limitations, the model has proven to be useful for discussions among researchers and system developers. The model has been extended to nondefense applications [4-7].)

- *Level 0 processing (source preprocessing)*: Data from sensors and databases is often preprocessed before fusion with other data. Examples of source pre-processing include image processing, signal processing or conditioning, alignment of the data in time or space, filtering of data, and other operations to prepare the data for subsequent fusion. Because this processing is specific to individual sensors and data types, this topic is not addressed in detail in this book. Information on image processing is provided by many sources including [8]. Similarly, signal processing techniques are described by sources such as [9].

- *Level 1 processing (object refinement)*: Level one processing is aimed at combining sensor data to obtain the most reliable and accurate estimate of an entity's position, velocity, attributes, and characteristics. The term entity involves a spatially or geographically localized object such as a target (a tank or small military unit), a fault condition in a mechanical system, or a localized tumor in a human. Level one processing focuses on the physical entities and the attempt to refine knowledge about the entity's existence, location, movement, characteristics, attributes, and identity. Level one processing involves classic target tracking and pattern recognition (for identity determination).

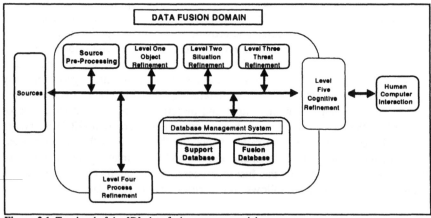

Figure 2.1 Top-level of the JDL data fusion process model.

Table 2.1

Summary of JDL Data Fusion Model Components

JDL Model Component	Description	Example Functions
Level 0 processing (source preprocessing/subobject refinement)	Preprocessing of data from sensors and databases to correct biases, standardize inputs, and extract key information; preprocessing is dependent upon the individual characteristics of the observing platform, sensor and data sources	Signal processing and conditioning Image processing Platform corrections Bias corrections Unit conversions Feature extraction
Level 1 processing (object refinement)	Combining data to obtain estimates of an entity's location, motion, attributes, characteristics, and identity	Data alignment Association and correlation Position, kinematic, and attribute estimation Identity estimation
Level 2 processing (situation refinement)	Development of a description or interpretation of an evolving situation based on assessment of relationships among entities and their relationship to the environment	Object aggregation Event/activity detection and aggregation Context-based interpretation Multiperspective reasoning
Level 3 processing (impact assessment)	Projection of the current situation into the future to assess inferences about alternative futures or hypotheses concerning the current situation; assessment of threats, risks, and impacts	Estimate and aggregate force capabilities Predict enemy intent Identify threat opportunities Estimate implications Multiperspective assessment
Level 4 processing (process refinement/resource management)	Monitoring the ongoing data fusion process to optimize utilization of sensors or information sources and algorithms to achieve the most useful set of information	Look-angle computations Computation of measures of performance Optimization of resource utilization Sensor and resource tasking
Level 5 processing (cognitive refinement)	Monitoring the ongoing interaction between the data fusion system and a human decision-maker; optimization of displays, interaction commands, focus of attention to	Focus of attention management Search engines

Table 2.1 (continued)

Summary of JDL Data Fusion Model Components

JDL Model Component	Description	Example Functions
	improve the human/computer effectiveness	Cognitive aides Query decomposition
Database management	Management of the data and information related to the data fusion system; types of data managed include sensor data, supporting databases, knowledge bases, and interim processing results	Query decomposition Boolean query management Data ingestion Data record management
Human-computer interaction	Human-computer interface processing such as provision for graphical displays, and processing of input commands, alerts, and system queries.	Geographical information system (GIS) functions Natural language interfaces Haptic interface management

- *Level two processing (situation refinement):* Level two processing uses the results of level one processing and attempts to develop a description of current relationships among entities and their relationships to the environment to determine the meaning or interpretation of the situation. Thus while level one processing is focused on the existence of entities such as enemy sensors, weapon systems, and airfields, level two processing is focused on interpreting the meaning of these entities (i.e., to determine the military order of battle and the existence of high-level military organizations). Thus, the observation of an emitter, transport system, power supply, and communications equipment might be interpreted as a mobile military weapon system. For a data fusion system focused on observing a mechanical system, the observation of excessive heat, vibration, and noise might signal the existence of a condition such as a broken gear tooth or misaligned shaft. Level two techniques are drawn from automated reasoning and artificial intelligence.
- *Level three processing (threat refinement/impact assessment):* Level three processing concerns the projection of the current situation into the future to determine the potential impact or threats associated with a current evolving situation. For military systems, level three processing seeks to draw inferences about potential enemy threats, probable courses of actions, friendly and enemy vulnerabilities, and opportunities for operations. For a nonmilitary application such as monitoring a mechanical system, level three

processing is aimed at developing alternative predictions about the future conditions and health of a machine, based on anticipated machine utilization and load conditions. Level three processing would seek to estimate the time of remaining useful life for the machine in question. Level three processing utilizes methods from automated reasoning, artificial intelligence, predictive modeling and statistical estimation.

- *Level four processing (process refinement/resource management)*: Level four processing is a metaprocess (namely a process that addresses a process) that monitors the overall data fusion process to assess and improve the real-time performance of the ongoing data fusion. The level four process involves functions such as sensor modeling, look-angle prediction, platform dynamic modeling, computation of measures of performance, and optimization of resource utilization. In the JDL model, the level four subprocess is shown as partially inside, and partially outside, the fusion process. This is a deliberate representation to indicate awareness that the optimization of the ongoing fusion process must account for both the needs of the fusion system (e.g., for improved accuracy and reduced ambiguities), as well as for operational needs (e.g., the need for stealth). Often these needs are conflicting and must be resolved by the level four process or a human decision-maker. For example, a fusion system that monitors an operating machine might detect the onset of a failure condition. The optimal approach for recovery might be to shut down the machine and perform preventive maintenance. However, this might not be possible because the machine is required to operate to complete a mission.

- *Level five processing (cognitive refinement):* The concept of level five processing was introduced by [3]. The authors argued that many data fusion systems involve a human-in-the-loop decision maker. The effectiveness of the sensors/processing/human system is ultimately affected by how the human perceives the information being produced by the fusion system and the intervention and control effected by the human user. Because of these factors, they introduced the concept of a level five process that acts similarly to the level four process. That is, the process transforms the ongoing results of the fusion system into displays and meaningful information for the user. This may involve traditional human-computer interface (HCI) functions such as geographical displays, displays of data and overlays, processing of input commands, and utilization of nonvisual interfaces such as the use of sound and haptic interaction. In addition, Hall et al. argued for the development of advanced functions such as deliberate synesthesia, time compression and expansion functions, cognitive aids to support negative reasoning and focus/defocus of attention, and new methods for representing uncertainty. More recent research by McNeese et al. [10] incorporates concepts of affective computing, in which a computer uses special sensors to monitor the physiological condition of the user and adapts the interface to address the user's emotional state and mental condition.

- *Database management*: A key element of the data fusion process is the management of extensive amounts of data [11]. This data includes the sensor reports, supporting databases (e.g., terrain data, doctrine, friendly order of battle, and political and geophysical information) [12]. Database management functions may require a significant amount of software for their performance. Typically commercial off-the-shelf (COTS) database systems are unable to address the complete data fusion problem, as a result of the extent and diversity of the data to be managed. More details on database issues for data fusion systems are provided in Chapter 10.

- *Human-computer interaction*: Another key element of the data fusion process (but considered as an external support function) is human-computer interface (HCI) functions. Military data fusion systems may require the display of situation maps (featuring terrain information, political information, roads and man-made features) with overlays associated with the dynamic data fusion results (e.g., icons associated with targets, ellipses to show location uncertainty, information about level two processing and level three results). In addition, the HCI supports command of the data fusion system, information queries, control of the displays, and related functions. Modern geographical information systems such as ARCINFO [13] contain many of the functions required to support data fusion systems.

2.2   LEVEL 1 FUSION ALGORITHMS

Perhaps the most developed area of data fusion processing is level one processing. In the early 1800s Gauss and Legendre originally (and independently) addressed the fundamental problem of level one fusion in developing the method of least squares [14]. In the United States, the problem was also addressed by Robert Adrain, who published the approach at roughly the same time as Gauss's publication [15]. Gauss and Legendre addressed the problem of combining multiple observations related to an asteroid, and attempting to estimate the elliptical orbit of the asteroid. Adrain developed the method of least squares to address the problem of improving the accuracy of land surveys using multiple measurements.

Figure 2.2 shows a partitioning of level one processing algorithms into four main functions: (1) data alignment, (2) data/object correlation, (3) position, kinematic, attribute estimation, and (4) object identity estimation. Each of these areas is described in the following sections. The general concept of level one fusion is shown in Figure 2.3 for the case of tracking aircraft with multiple sensors.

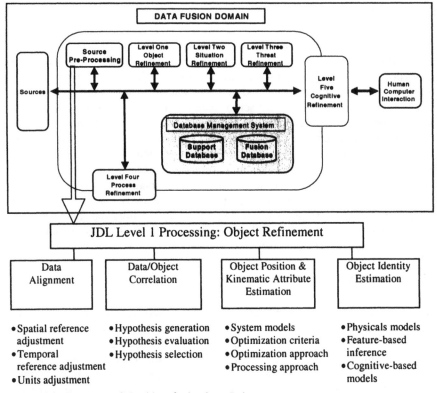

Figure 2.2   Taxonomy of algorithms for level one fusion.

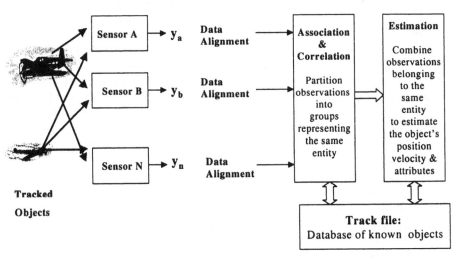

Figure 2.3   Concept of level one fusion.

2.2.1   Data Alignment

Prior to the fusion of any data, the sensor data from individual sensors must be processed to transform the information into a form that is suitable for subsequent processing. Types of algorithms for data alignment include spatial reference adjustments, temporal adjustments, and unit adjustments. These transformations or adjustments prepare the individual sensor data for subsequent processing. For example, information received from two radar sensors at different locations could be transformed into a common spatial reference system to translate the sensor data from sensor coordinates (e.g., range, azimuth, and elevation) to a common coordinate frame. Similarly the time when an observation is received at a fusion system must be distinguished from the time when the observation was actually made. Data alignment can be a very complex process, particularly if sensors of widely different types (e.g., image and nonimage sensors) are involved. Some general formulae for data alignment and observation modeling for satellite tracking sensors are provided by [16] and [17].

2.2.2   Data/Object Correlation

In dense tracking environments with numerous, closely spaced targets, it can be challenging to determine which observations belong to (or are associated with) which targets. In general the number of targets are unknown a priori. Association and correlation links observations from multiple sensors (or even a single sensor) to individual platforms or entities. That is, a sensor may observe multiple targets during a given time period. How can we group the observations together to ensure that the observations "belong" to a single physical entity such as an aircraft, ship, or emitter? A detailed description of the complexities involved in multisensor tracking in a dense target environment is provided by [18]. Parametric association seeks to associate observations to other observations, or observations to existing tracks. This is accomplished by defining a measure of association that quantifies the closeness between observation (or observation-to-track) pairs. Commonly used association measures include correlation coefficients, distance measures, association coefficients, or probabilistic similarity measures. Reference [19] provides a discussion of the use of association measures. The term association is commonly used to mean the process of quantifying the "closeness" of observations or tracks, while the term correlation refers to the decision process of grouping observations or tracks.

Parametric association seeks to determine the relationship between observation pairs, namely to determine if y_{Ai} in Figure 2.3 (i.e., the ith observation from sensor A) is related to y_{Bj} (the jth observation from sensor B). In order to perform this comparison, two issues must be addressed. First, if sensors A and B observe different types of data, then transformations must be made to determine the physical proximity of y_{Ai} and y_{Bi}. Second, if the observation y_{Ai} is made at a different time than observation y_{Bi} (i.e., if $t_i \neq t_j$), then the observations must be synchronized via equations of motion that allow the propagation of an observation at time t_i to a subsequent time t_j. Such observation propagation is based on an assumed dynamic model of the observed entity. These transformations and synchronization are dependent upon the specific sensors utilized and the dynamics of the objects being observed.

An alternative goal of parametric association is to link sensor observations (y_{Ai}) to existing tracks, x_k, of known objects in the track file. The subscript, k, refers to the kth track or state vector. In this case, the association process requires that for each known track $x_k(t)$, in the database, the track must be propagated or updated via dynamic models, to the time, t_i of the candidate observation, y_{Ai}. Subsequently, an observation model is used to predict an observation that would be observed by sensor A, for object k, *if* in fact that object were in the sensor's field of view at time t_i.

$$y_{A\,predicted}\left(t_i\right) = g\left[x_k\left(t_i\right)\right] + n \tag{0.1}$$

The function g represents an observation model that typically includes various coordinate transformations, and n represents (unknown) observational noise. For this case, an association measure compares the actual observation, $y_{Aactual}$, with the predicted observation $y_{Apredicted}$.

After an association measure is computed to determine the closeness between two observations, special association strategies or logic must be applied to determine whether or not to actually declare two observations as belonging together. Gating techniques establish boundaries or limits to provide an initial determination of whether two observations could be physically related. For example, kinematic models are used along with an a priori knowledge of the maximum speeds possible for a platform, to set limits on which observation pairs may be physically related. The use of gating techniques reduces the combinations of observation pairs that must be considered as potentially belonging to the same entity. Subsequently, assignment strategies are used for final correlation of observations to observations, or observations to tracks. Blackman [18] provides an extensive description of these strategies and their implications to performance and processing requirements. The simplest technique (nearest neighbor) simply assigns an observation to its nearest neighboring observation or track, with the distance measured by the association measure. Other methods may entail simultaneously investigating multiple hypotheses. Additional details on these techniques are provided in Chapter 3.

2.2.3   Object Position, Kinematic, and Attribute Estimation

Position, kinematic, and attribute fusion utilizes a combination of physical models (such as differential equations of motion and observation models) and statistical assumptions about observation noise processes to map observational data to a state vector. The term *state vector* is used to mean an independent set of variables such as position and velocity, which, if known, would allow accurate prediction of the future behavior (e.g., dynamical motion) of an entity. Positional fusion algorithms such as a Kalman filter provide an estimate of the state vector that best fits (in a defined mathematical sense) the observed data. These algorithms provide incremental refinements or corrections to the state vector based on the differences between the actual (observed) data and the predicted observations (based on the current estimate of the state vector and associated sensor observation equations). The multiple sensor data is processed by utilizing a separate physical observational model for each sensor. These observation models predict the observations for specific sensors given estimated states.

The concept of position and kinematic fusion is illustrated in Figure 2.3. Multiple sensors observe (potentially multiple) dynamic objects. The sensors provide parametric data such as angular data (e.g., azimuth, elevation, or picture coordinates), range, or range-rate. The data may include parametric vectors, imagery, or actual state vectors (an estimate of the three-dimensional position and velocity) of the observed platforms. Observed objects may be mobile, such as rapidly moving aircraft or ballistic objects, or stationary objects such as entities fixed with respect to the Earth's surface. Data from the sensors is preprocessed within a fusion system to perform data alignment. Data alignment transforms incoming sensor data into a common frame of reference via coordinate transformations and unit conversions. Subsequent to data alignment, two functions are performed: parametric association and state-vector estimation. Each of these functions is described below.

Once observations have been sorted into groups that are associated with separate dynamic entities (platforms), then estimation techniques may be applied to actually combine or fuse the data. The concept is illustrated in Figure 2.3. The estimation techniques determine the value of a state vector (e.g., position and velocity) that best fits the data in a defined mathematical sense. For example, the method of least squares seeks to find the value of an (unknown) state vector, **x,** that minimizes the summed-squared difference between the actual observations and the predicted observations. Specifically, an objective function is defined that is a measure of the *degree-of-fit* to the data. For the least-squares method, we define a scalar quantity L(x)

$$L(x) = \sum \left[y_i - y_{i\,\text{predicted}}(x) \right]^2 \tag{2.2}$$

L(x) is the sum of the squares of the residuals, the difference between the *actual* sensor observations, y_i, and the *predicted* observations, $y_{i\,\text{predicted}}$. The predicted observations are a function of the unknown state vector, **x,** as represented by (2.2). Any of a number of objective functions may be defined including a least-squares function, weighted least squares, likelihood functions, and

Bayesian formulations. Details of these objective functions are provided in Chapter 4.

Solution of the estimation problem involves systematically varying the state vector, **x,** to meet the conditions specified by the objective function. Two basic strategies may be used: either process all of the data simultaneously in a batch mode (namely after all data from all sensors have been collected), or process the data sequentially, updating the estimate of the state vector with each new arriving observation. Well-known techniques for sequential estimation include the Kalman filter and the extended Kalman filter. A description of statistical estimation techniques is provided in Chapter 4. Blackman [18] and Gelb [20] provide an excellent introduction to these methods. Fundamental design choices in estimation involve the selection of an objective function (the mathematical definition of *best fit),* the strategy for solving the objective function to find the state vector, and the method of modeling the noise (observation noise and process noise).

Selection of these alternatives is dependent upon the specific fusion application and computational resources available. The result of the combination of association and estimation processing is fused data that accurately determines the position, velocity, and attributes of dynamic entities. These alternatives are shown in the taxonomy provided by Figure 2.4.

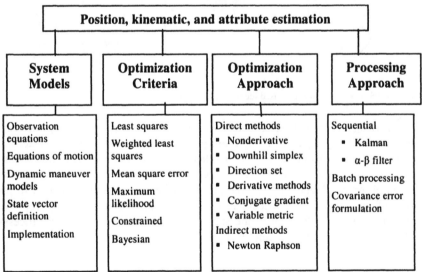

Figure 2.4 Overview of positional and kinematic fusion algorithms.

2.2.4 Object Identity Estimation

Identity fusion poses the problem of combining identity declaration data from multiple sensors to obtain a joint estimate of identity. Ideally, the combined declaration is both more specific and more accurate than any of the declarations from an individual sensor.

The concept of identity fusion is illustrated in Figures 2.5 and 2.6. Figure 2.5 illustrates how data collected by a sensor can be transformed into a declaration of target identity using pattern recognition techniques. Signal data (such as acoustic information) is processed to extract key features. In turn, these features are used in a pattern recognition algorithm or classifier to transform the data into a declaration of identity. This data may be preprocessed by one of a variety of transformations to extract a feature vector, y_i, that represents the observed data. Examples of feature extraction transformations include sampling functions, transformations from the time to frequency domain, and image processing algorithms for recognition of patterns (e.g., edge detectors, segmenting algorithms, and pattern recognizers).

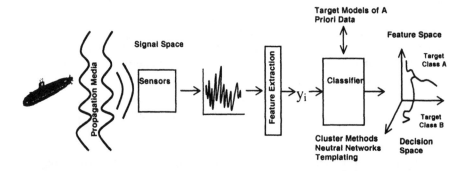

Figure 2.5 Concept of feature-based identity declaration.

Having extracted a set of features or feature vector, y_i, an identity declaration may be made by a pattern recognition process. Techniques such as cluster algorithms, adaptive neural networks, or other statistical pattern recognition methods may be used to transform from a feature vector, y, to a declaration of the identity of an observed object. Just like the problem of association process used in positional fusion, these identity declarations must be partitioned into groups representing observations belonging to the same observed entity. Association processing affects this sorting/partitioning function.

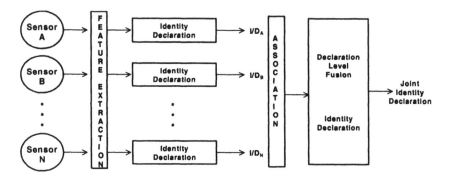

Figure 2.6 Concept of identity fusion using decision-level.

The identity fusion concept shown in Figure 2.6 indicates fusion of identity declarations. This is analogous to fusing state vectors for location fusion. Alternatively, the identity fusion could occur at the raw data level (prior to feature extraction), at the feature vector level (prior to identity declaration), or at the decision level (after each sensor has made an independent declaration of identity). The choice of when to perform the fusion is dependent upon the types of sensor data available and the types of preprocessing performed by the sensors.

Another simplification in Figure 2.6 is the presumption that only identity fusion is being performed (i.e., we have ignored positional fusion). In practice, positional fusion and identity fusion can occur in a simultaneous or interleaved fusion.

Conceptually, identity fusion could be performed in the same manner as positional fusion; however, several difficulties typically prohibit such a straightforward approach.

First, physical models for identity declaration generally do not exist or are at best very difficult to develop. For example, it is possible to model the radar cross-section (RCS) return versus aspect angle for different types of aircraft. Similarly, in principle, an infrared (IR), spectral model (intensity versus wavelength) could be developed for various types of man-made objects (e.g., tank engines or automobiles). Thus, one could envision an identity estimation process, analogous to positional fusion, combining IR and radar data. The process would make use of the observational models (namely the RCS model and IR spectra model) and match observed data versus predicted data to determine an entity's identity. In practice, however, this is not feasible. Each model requires significant computational resources, and these models must be developed for each type of object observed in realistic conditions. This is not feasible in complex tactical environments with hundreds or thousands of potential objects to identify.

A second difficulty with identity fusion is that identity is generally hierarchical. At the lowest level of inference, objects to be identified may include individual emitters (radars, radios, beacons), or physical entities (such as an engine on a jet airplane, a tank, or a portion of a ship). At higher levels of inference, the identity of battlefield entities (units, divisions) may be sought. Thus, identity may imply complex interrelationships among multiple components. Further, the identity inference may seek intangibles such as the intention of an enemy unit. Finally, identity may seek membership in a general class (e.g., observation of a missile-guiding radar versus a weather radar).

These complexities make the identity declaration process more challenging than that of positional fusion. As a result, a number of techniques have been developed to address fusion of identity data. This section provides an overview of algorithms that have been utilized for positional fusion.

Precise and unique categories of algorithms utilized for identity fusion do not exist. Techniques successfully used for identity fusion range from well-known statistical-based algorithms such as classical inference and Bayesian methods, to ad hoc methods such as templating, voting, and evolving techniques such as adaptive neural networks. Comparison of the complete range of techniques smacks a bit of an *apples and oranges* comparison, and any partitioning of algorithms into categories may be arguably arbitrary. Nevertheless, in Table 2.2 we provide a conceptual taxonomy of identity fusion algorithms. More details on these techniques are provided in subsequent chapters of this book. Three major categories are shown: (1) physical models, (2) feature-based inference techniques, and (3) cognitive-based models.

Physical models attempt to accurately model sensor-observable data (e.g., RCS, and IR spectra) and estimate identity by matching predicted (modeled) observations with actual data. Techniques in this category include simulation and estimation methods such as Kalman filtering. While we noted previously that modeling of identity may be difficult, it is conceptually possible to perform identity estimation utilizing classical estimation techniques. The process for such estimation is completely analogous to positional data fusion.

Table 2.2

Taxonomy of Identity Fusion Algorithms

Types of Models	Categories of Techniques	Method/Techniques
Physical Models	Simulations	Physics-based models of target signatures including signal propagation and sensor performance
	Estimation	Kalman filtering Maximum likelihood estimation Least squares Bayesian methods

Table 2.2   (continued)

Taxonomy of Identity Fusion Algorithms

Types of Models	Categories of Techniques	Method/Techniques
Feature-based inference techniques	Parametric methods	Statistical-based algorithms
		Classical inference
		Bayesian methods
		Dempster-Shafer's method
		Cluster algorithms
		Hierarchical agglomerative (K-means s_link, Ward's method)
		Hierarchical divisive methods
		Iterative partitioning (e.g., K-means and hill climbing)
		Density search methods
		Factor analytic methods
		Clumping algorithms
		Graph theoretic methods
	Nonparametric methods	Parametric templates
		Adaptive neural networks
		Binary input nets
		Continuous value input
		Voting methods
		Entropic techniques
		Figure of merit
		Measure of correlation
		Thresholding logic
Cognitive-based models	Logical templates	Knowledge representation
		Uncertainty representation
		Inference process
	Knowledge-based systems	Knowledge representation
		Scripts
		Rules
		Frames

Table 2.2   (continued)

Taxonomy of Identity Fusion Algorithms

Types of Models	Categories of Techniques	Method/Techniques
		Semantic nets
		Analogical techniques
		Inference methods
		GPS
		Action/object
		Production systems
		Blackboard architectures
		Agent-based inference
		Search techniques
		Uncertainty representation
		Dempster-Shafer
		Probability
		Confidence factors
		Fuzzy membership
	Fuzzy set theory	Fuzzification functions
		Defuzzification functions
		Definition of fuzzy connectives

Feature-based inference techniques seek to make identity declarations based on attribute data, without utilizing physical models. Attributes of an entity are observable characteristics such as shape, temperature, size, or emissions that can be related to entity identity. A direct mapping is made between attribute data and a declaration of identity. We divide these techniques into two broad categories: (1) parametric (or statistic-based) techniques that require a priori assumptions about the statistical properties (e.g. distributions) of the identity data, and (2) nonparametric techniques that do not require a priori statistical information. Statistic-based techniques include classical inference, Bayesian inference, and Dempster-Shafer's method, as well as cluster methods. Nonparametric techniques include templating, adaptive neural nets, voting methods, and entropy methods.

Finally, *cognitive-based models* are a third major category of identity fusion algorithms. These methods seek to mimic the inference processes of human analysts in recognizing identity. Techniques in this category include logical templates, knowledge-based (expert) systems, and fuzzy set theory. These methods in one way or another are based on a perception of how humans process information to arrive at conclusions regarding identity of entities.

Table 2.3 presents an overview of the techniques identified in Table 2.2. It lists each algorithm and summarizes the kernel process, the character of the required input, and the character of the output. Further details on each of these algorithms are given by the references listed in the right-hand column of Table 2.3, and in subsequent chapters of this book.

Table 2.3

Overview of Techniques for Identity Fusion

Method	Kernel Process	Input Characteristics	Characteristics of Output	References
Classical inference	P (observations $\mid H_0$)	Empirical probability population distribution for the statistic	P (Error dec. on H_0)	[21]
Bayesian inference	A posteriori $P(H_0 \mid \text{evidence})$ Updates belief on H_0 given new data or evidence	Empirical or subjective probability; exhaustive definition of possible causes A priori P (causes)	Updated likelihood of the occurrence of an event	[22, 23]
Dempster-Shafer's method	P (H_i multiple evidence) and P (any H_i true); General uncertainty and Dempster's combination rule(s)	Empirical or subjective probabilities; exhaustive $P(H_i \mid \text{evidence})$ including disjunction	Updated likelihood of the occurrence of an event	[24]
Fuzzy set theory	Set algebra where set elements have a membership function property	Subjective membership functions for all set elements	Profile of goal set elements and membership function	[25]
Cluster analysis	Sorting of observations into natural groups based on similarity measures and cluster heuristic	Training labeled set; selection of similarity metric	Cluster elements and similarity measures	[26-28]
Estimation theory	Best state estimate for the given observations (e.g., least squares)	Quantitative observations; dynamic state model; observation model (given state vector)	Estimated state vector	[20]
Entropy	Computes measure of information content	Empirical subjective probability	Optimal probability	[29]

Table 2.3 (continued)

Overview of Techniques for Identity Fusion

Method	Kernel Process	Input Characteristics	Characteristics of Output	References
Figure of merit	Computes degree of similarity between two reports or entities	Two reports or attribute vectors	Numerical value of similarity	[30]
Expert systems theory	Computer program to mimic human inference process; requires knowledge representation, search engines, pattern recognizer	Observation data records; knowledge base (e.g., rules)	Declaration of an inference or conclusion with associated uncertainty	[31-33]
Templates	Pattern matching technique for complex associations; utilization of parametric pattern matching and reasoning heuristics	Observed parametric data	Declaration that the observations support (matches) a template describing a target, entity, event, or activity	[34, 35]
Adaptive neural networks	Nonlinear transformation from observations to recognition or classification space; use of layered nodes and activation functions	Observed parametric data (feature vector)	Match of input data to prespecified categories (i.e., declaration of identity or pattern match)	[36, 37]

2.3   LEVEL 2 FUSION ALGORITHMS

Level 2 fusion involves processing the results of level one processing to understand the context of the level one products. In a sense, level 2 processing involves understanding the relationships among entities, the relationship of those entities to the environment, and aggregation of entities in time and space for a higher level or more abstract reasoning. Figure 2.7 shows a conceptual view of the key functions including object aggregation, event/activity aggregation, contextual interpretation, and multiperspective assessment. Each of these is summarized as follows.

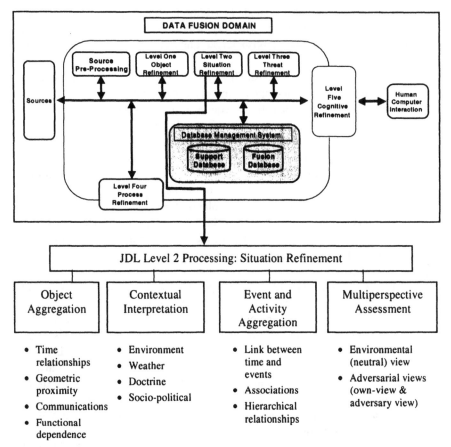

Figure 2.7 Functions and algorithms for level 2 processing.

- *Object aggregation*: One of the keys to understanding a situation (whether
  it be a battlefield situation or a medical condition) is to aggregate in space
  and time the entities identified and located by level 1 processing. The level
  1 process tends to detect, locate, characterize, and identify individual entities. For military applications these entities might include entities such as
  tanks, aircraft, emitters, or sensor systems. For a medical application, this
  might include the location of abrasions, tumors, or painful areas. Object
  aggregation seeks to identify groups of entities that belong together and
  make up a larger entity or unit. For military applications, a surface-to-air
  missile (SAM) military unit might consist of a missile launcher, surveillance radar, an acquisition/missile launch guidance radar, a communications system, and facilities for the military operators. Object aggregation
  analyzes objects that are in geographical proximity to determine if these

objects are functionally related. In addition, object aggregation analyzes patterns of communications (namely which entities are communicating with other entities) to establish larger units and even identify chains of command.

- *Event/activity aggregation*: Another type of reasoning performed in situation refinement is an attempt to analyze events and activities in time. In many situations, certain actions require a special sequence of events or subactivities (e.g., logistic support, data collection via sensors, and communications of plans and intentions). Some of these events may be very mundane. In the early 1990s, for example, the *Washington Post* reported that information about Dominos pizza deliveries to the Pentagon allowed them to accurately predict when a major military operation or crisis was being planned or analyzed (because planners would work late and order pizza). Their key indicator was an increase in the nighttime delivery of pizza to the Pentagon. Other types of indicators for impending events may include actions that must occur sequentially, in parallel, or other relationships.

- *Contextual interpretation:* A third type of reasoning to assist situation refinement involves the analysis of the context in which level 1 entities are being viewed. How does the environment (e.g., weather and terrain) affect the placement and deployment of military units? For the analysis of the health of a complex machine, what are the operating conditions? Is the interpretation of the system health conditions different if it is known that the machine is in a start-up versus a steady-state operation? Are there external conditions that affect how we would understand the current situation?

- *Multiperspective assessment*: The situation assessment is sometimes assisted by considering potential adversarial relations. For military applications there is generally an adversarial relationship between the "friendly (blue) forces" and the "enemy (red) forces." Thus, it is helpful to analyze a situation from three perspectives:

1. The blue view: How does the situation appear from the point of view of the blue forces? What are the blue force objectives, what is their focus, and how does the environment (terrain and weather) affect the ability to achieve blue objectives?
2. The red view: How does the situation appear from the point of view of the red forces? What are the red force objectives, what is their focus, and how does the environment (terrain and weather) affect the ability to achieve red objectives?
3. The white view: This is the "neutral" view of the situation. What factors such as environment, trafficability, observation environment, and communications situation affect both red and blue forces?

While this is a distinctly military point of view, such analysis concepts can be used in nonmilitary situations. Even for applications such as the health monitoring of a machine, one can consider a "white view" to analyze what can be observed, and how the operating environment affects our understanding of the situation. Similarly, the anticipated mission or use of the machine significantly affects our interpretation of an impending failure condition. An aircraft with low tire pressure is assessed differently if it is sitting in a hanger being stored compared to an aircraft preparing to land with a load of passengers.

What are the algorithms that can be used to support these functions? In general, techniques are drawn from the fields of pattern recognition (e.g., cluster analysis, neural networks, decision trees, and parametric templates) and automated reasoning (e.g., expert systems, blackboard systems, logical templates, case-based reasoning, and intelligent agents). These methods are described in subsequent chapters.

2.4   LEVEL 3 FUSION ALGORITHMS

Level 3 fusion, threat refinement, or consequence prediction involves projecting the current situation into the future to understand the consequences including threats, opportunities, or risks. For military applications, threat prediction involves attempting to determine what the opponent(s) will do, where they'll be, what their intentions are, and what are the consequences relative to the blue or friendly forces. For applications such as health monitoring for machinery, level 3 fusion involves attempting to predict the time to failure, evolution of failure conditions, consequences for operational capability, and uncertainties. Figure 2.8 provides an overview of the types of functions that are required for level 3 data fusion. The key functions involved in level 3 processing include: (1) estimation/aggregation of force capabilities, (2) prediction of enemy intent, (3) identification of threat opportunities, (4) estimation of implications, and (5) multi-perspective assessment. Each is summarized as follows.

- *Estimate/aggregate force capabilities*: For military applications, a basic element of threat refinement or prediction is to estimate the location and capabilities of enemy (and friendly) forces. This includes not merely identification and counting of weapon systems and military units, but also assessment of the ensemble of forces, sensors, weapon systems, and logistical components. This requires an understanding and representation of the relationship among military entities, their capability to work together, communications and supply requirements, and related knowledge. For nonmilitary applications such as monitoring a complex machine, one is interested in aggregating information about mechanical components, subsystems, and

systems. How do these components work together? What are their
interdependencies?

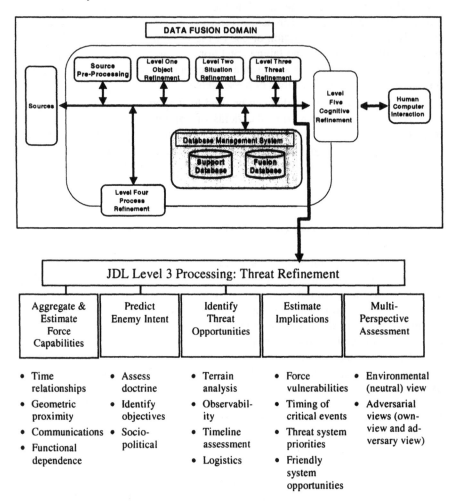

Figure 2.8 Functions and algorithms for level 3 processing.

- *Predict enemy intent*: Perhaps the most difficult area for military applications is predicting enemy intent. To be performed accurately this would literally require "reading the minds" of the opponents. Clearly this is not possible. However, by understanding sequences of actions, dependencies, logistical requirements, and typical courses of action, one can gain knowledge about the probable intent of an opponent.

- *Identify threat opportunities*: As part of a military mission planning process, one can identify threat opportunities. For example in planning an aircraft sortie to engage a target, the flight path may encounter surface-to-air (SAM) systems and anti-aircraft artillery (AAA) systems. This analysis seeks to identify the potential opportunities for an opponent to cause harm to friendly forces or systems. Interestingly, there are analogies to this analysis in many applications. For condition-based monitoring of machinery there is a formal type of analysis called failure methods, effects and causal analysis (FMECA) [38]. In this approach system designers and maintenance planners seek to identify potential types of failures, assess the anticipated course of the failure evolution, and determine the effects of such failure trajectories. This is analogous to identifying threat opportunities in tactical military situations.
- *Estimate implications*: Given a hypothesized future action (e.g., by friendly forces and their opponents for a military situation or failure trajectory for a complex machine), what are the implications of such actions? What are the anticipated effects such as losses, damage, and reaction by an opponent? Analogs to military situations are available for applications such as diagnosis of medical conditions and condition-based maintenance of mechanical systems.
- *Multiperspective assessment*: Just like the analysis performed for level 2 fusion, multiple perspectives are useful to fully assess predicted actions and consequences. These include the white or neutral (environmental) view, the blue view related to our own objectives and mission requirements, and the red view symbolizing the viewpoint of an adversary.

Again, analogous to level 2 processing, what are the algorithms that can be used to support these functions? In general, techniques are drawn from the fields of pattern recognition (e.g., cluster analysis, neural networks, decision trees, and parametric templates) and automated reasoning (e.g., expert systems, blackboard systems, logical templates, case-based reasoning, and intelligent agents). These methods are described in subsequent chapters.

2.5   LEVEL 4 FUSION ALGORITHMS

The level 4 process in the JDL data fusion process model is a metaprocess. That is, it is a process that monitors the ongoing data fusion process and seeks to optimize that process to achieve improved fusion products (e.g., improved estimates of the position, velocity, attributes and identity of entities, and improved situation refinement). This is achieved by controlling the sensors and the algorithms in the fusion process. Historically, level 4 processing was focused on the control of single sensors having limited parametric agility. Under these circumstances, the level 4 process entailed the computation of so-called look angles, used to determine where to point a sensor in order to track objects, and a straightforward simple control process.

The level 4 process has become increasingly complex. First, data fusion systems frequently involve multiplatform, multisensor systems. Second, intelligent sensors exhibit wide variability in parameter space control (e.g., different selectable waveforms) and processing parameters. Third, evolving concepts of human-in-the-loop computer assisted cognition suggest that the data fusion system should adapt to the dynamic preferences and needs of the human decision-maker.

Figure 2.9 provides an overview of basic functions performed in level 4 processing. These categories of functions are described as follows.

- *Mission management:* In the top level of the JDL data fusion process model, the level 4 process is usually shown as being partially inside, and partially outside of, the data fusion process. This is deliberate. This representation explicitly recognizes the fact that level 4 processing cannot be performed solely to satisfy the needs of the data fusion process. Instead, the level 4 process must also account for the constraint and needs of the overall mission that is being supported by the fusion system. For example, in order to obtain an improved estimate of a target's position, the level 4 process may recommend that active radar be utilized. However, if the mission involves stealth, then this recommendation would be counter to the needs of the mission. Similarly, monitoring a complex machine involves understanding the context in which the machine is currently operating.

- *Target/entity prediction:* In one sense, a data fusion system is unable to perceive targets or entities that it does not expect to observe. In order to compute "look angles" (namely to tell the sensor system or information source where to look for data) models are required about the entities to be observed. For mobile entities this includes models of target motion. In addition, models are needed to predict the emissions or observable characteristics or attributes of the entities to be observed.

- *Source requirements (sensor and platform modeling)*: Given specific entities to be observed, models are needed to characterize the sensors or information sources. For sensors mounted on moving platforms, it is necessary to model the flight paths of the hosting platform and model the sensor's performance (namely to predict what the sensor would observe given a hypothesized target or entity). Such models include signal propagation models, sensor/target interaction, and sensor noise. Algorithms for such modeling span the range from detailed physical models of entities, to signal propagation models (e.g., to describe the path of electromagnetic radiation from the sensor to the target and its return), statistical noise models, models of internal sensor processing (to characterize signal conditioning and feature extraction processing), and even hierarchical syntactical models of targets.

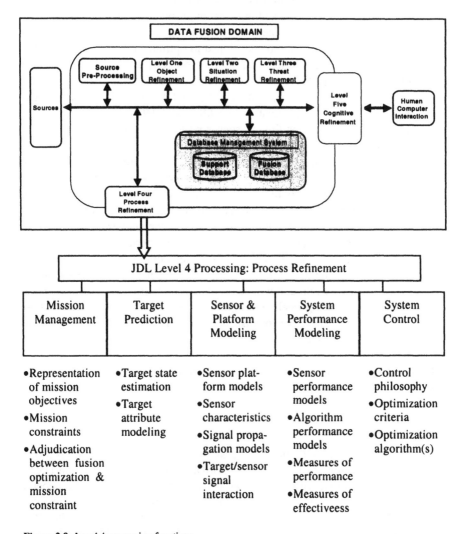

Mission Management	Target Prediction	Sensor & Platform Modeling	System Performance Modeling	System Control
•Representation of mission objectives •Mission constraints •Adjudication between fusion optimization & mission constraint	•Target state estimation •Target attribute modeling	•Sensor platform models •Sensor characteristics •Signal propagation models •Target/sensor signal interaction	•Sensor performance models •Algorithm performance models •Measures of performance •Measures of effectiveess	•Control philosophy •Optimization criteria •Optimization algorithm(s)

Figure 2.9 Level 4 processing functions.

- *System performance modeling:* In addition to the specific models for sensor performance, the overall data fusion system needs to be modeled. How does the system perform including sensors, communications components between the sensors and the processing components, computer computational performance, algorithm performance, and (most challenging) the human in the loop processing? We seek to develop measures of performance (MOP) and measures of effectiveness (MOE) to use as a characterization of how the data fusion system is performing (given the observed targets, utilization of sensors and information resources, specified fusion algorithms, and the human in the loop). Given this characterization of the system, control strategies can be developed to optimize the performance. This is analogous to characterizing an information system via quality of service (QoS) parameters. There have been numerous MOP and MOE developed for target tracking systems. Waltz and Llinas [39] provide an excellent discussion of the issues associated with MOP and MOE, and they present a hierarchical concept for linking sensor performance ultimately to system effectiveness.

- *System control:* Finally, a basic component of the level 4 processing is the actual control function. This entails classic feedback and control in which a controller attempts to optimize an objective function (e.g., maximize the accuracy of the fusion products and minimize utilization of system resources). Fusion control requires knowledge of the entities being observed, the sensor/source performance, the system performance, and mission/system constraints. Techniques for fusion control are drawn from classic methods in multiobjective control and optimization. More recent research has focused on utilizing emerging concepts from e-commerce and e-business (e.g., using intelligent agents as resource bidding agents). These are particularly attractive for distributed systems involving heterogeneous information resources.

More details on level 4 processing are provided in Chapter 8.

2.6   LEVEL 5 FUSION TECHNIQUES

The concept of a level 5 data fusion process was introduced at the 2000 National Symposium on Sensor and Data Fusion (NSSDF) conference held in San Diego. The concept was introduced by Hall et al. [3] and simultaneously (and independently) by Blasch. Both of these papers argued that, analogous to a level 0 processing (to highlight individual sensor processing) a level 5 process should be introduced to acknowledge functions required to support a human-in-the-loop decision process. Hall et al. cited research they conducted in a training environment to illustrate how individual differences in decision styles and information access modes affect the efficacy with which computer systems are used. They also identified a number of recommended areas of research to enhance computer-aided cognition.

Figure 2.10 provides a hierarchy of functions associated with level 5 processing. These are summarized below. More details on level 5 processing are provided in Chapter 9. This is a very preliminary taxonomy of algorithms. Historically, the human-computer interaction (HCI) for data fusion systems consisted primarily of traditional geographical information displays and menu/dialog interaction. Recent research is greatly expanding these types of interaction to include enhanced modality (e.g., not only visual displays, but also gesture recognition, haptic interfaces, and natural language interaction). In addition, research is beginning to develop cognitive aides to enhance the effectiveness of the human-computer dialog. The basic functions performed in level 5 processing are described as follows.

- *HCI utilities:* A basic element of level 5 processing involves utilities to support the human-computer interaction (HCI). These include the traditional functions related to display and interaction of menus and command languages (e.g., database queries) and geographical information system (GIS) functions. As previously indicated, commercial GIS systems have rapidly evolved to provide quite sophisticated displays and interactions. In addition, progress is being made to establish natural language dialog interfaces, utilization of haptic interactions, and more recently, gesture recognition. HCI utilities involve all of those functions to recognize a human input (via touch, voice, keyboard, or haptic device), parse the input, transform the input for internal use by the computer programs, and prepare output displays and responses. Utilities may also be developed to support on-line help and explanation features (e.g., the universally annoying *paperclip character* used by Microsoft on their products).

- *Dialog and transaction management:* The overall dialog between the human and the computer needs to be managed. Historically this dialog and transaction management was performed in a way that was not adaptive to individual differences in human users. Dialogs with computers were relatively static with the burden of learning how to control the computer levied on the human user. Advances in computer interaction include obtaining explicit and implicit information from the user to adapt the interaction to the individual [40]. Explicit information involves simply developing a user profile by asking the user to "fill in the blanks" and respond to queries about such factors as preferences and level of training. Implicit information involves programs that monitor a user's computer interactions and adapt the interaction based on the development of a user model created by the computer program. Competition on Internet Web sites, search engines, and commercial programs is driving new concepts in this area.

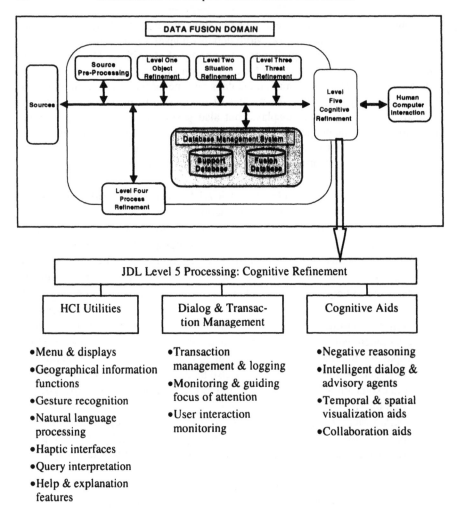

Figure 2.10 Hierarchy of level 5 functions.

- *Cognitive Aids:* The final area associated with level 5 fusion involves development of cognitive aids. This is a new research area in information fusion. Examples include;
    - o Development of aids to mitigate known cognitive biases (e.g., the tendency of a human to focus on positive information related to a hypothesis but to ignore contradictory evidence);
    - o Focus of attention aids to assist a user in balancing the need to focus on a particular area or topic of interest while maintaining general surveillance;

o Intelligent agents to act as advisors (including agents with an attitude or particular point of view);

o Collaboration aids to assist in interaction among different users or analysts;

o Special aids to assist in the understanding of data using different visual techniques or other modalities such as sound or touch.

The functions identified in this section are meant to be illustrative. As this area evolves, numerous changes are anticipated in functions and processes. For example, McNeese [10] has performed research in the area of affective computing. In this approach, the physiological conditions of a human user are monitored by the computer system. The displays and interaction are adapted to react to the specific human conditions (e.g., level of stress and fatigue).

2.7 ANCILLARY SUPPORT FUNCTIONS

In addition to algorithms for positional and identity fusion, a number of functions are required to support level 1 processing. Figure 2.11 summarizes typical functions required, including a basic numerical methods library, data alignment algorithms, data preprocessing, database management, and human-computer interface (HCI) techniques. In an operational data fusion system, these support functions may account for as much as 80% of the total software.

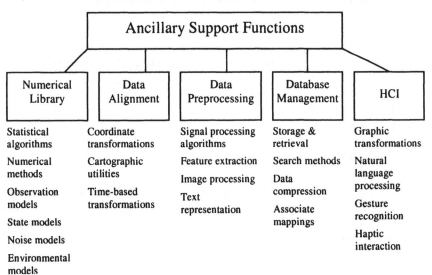

Figure 2.11 Taxonomy of ancillary support functions for data fusion.

Two areas deserve special comment: data alignment and data preprocessing. Data alignment refers to the functions required to transform observed data from sensors (e.g., range, azimuth, and elevation) into a common coordinate system utilized in the data fusion system for estimation, state vector propagation, and storage in the database. The data alignment processing is specific to each sensor and may require a significant amount of processing, particularly for transforming nonlinear image coordinates (pixels) into geocentric noninertial vectors. Such alignment may attempt to address sensor calibration and biases. The choice of fundamental coordinates for data fusion processing generally involves a trade-off between the computational resources required to transform from coordinate space into the central coordinate frame (e.g., transform topocentric noninertial observations into geocentric inertial coordinates) versus the selection of a fundamental coordinate system that simplifies the formulation and solution of the equations of motion for tracking.

Finally, data preprocessing functions are those algorithms applied to raw sensor data. These may include thresholding, averaging techniques, the application of signal processing methods (smoothing, filtering, and frequency domain transformations) or image processing methods for imagery sensors. Such processing may also require very large computational resources for sensors such as synthetic aperture radar. Selection of these techniques is also specific to the sensors utilized in a data fusion system.

2.8   ALTERNATIVE DATA FUSION PROCESS MODELS

Before concluding this chapter, it should be acknowledged that there are models other than the JDL data fusion process model that have been used to describe the data fusion process and functions. These include Dasarathy's functional model [41], Bedworth and O'Brien's omnibus process model [42], the Boyd decision loop, and the TRIP model. These are summarized briefly below. They are presented here to acknowledge that the JDL model is not the only way of viewing the data fusion process.

2.8.1   Dasarathy's Functional Model

Dasarathy [41] defined a useful category of data fusion functions based on the types of data and information that are processed, and the types of information that result from the process. Table 2.4 illustrates the input/output interpretation [43] based on Dasarathy's model.

Table 2.4

Dasarathy's Input/Output Data Fusion Characterization

Input	Output Data		
	Data	*Features*	*Objects*
Data	Signal detection (DAI-DAO)	Feature extraction (DAI-FEO)	Gestalt-based object characterization (DAI-DEO)
Features	Model-based detection/feature extraction (FEI-DAO)	Feature refinement (FEI-FEO)	(Feature-based) object characterization (FEI-DEO)
Objects	Model-based detection/estimation (DEI-DAO)	Model-based feature extraction (DEI-FEO)	Object refinement (DEI-DEO)

Dasarathy describes processes corresponding to the cells in the table using the abbreviations; DAI-DAO, DAI-FEO, FEI-FEO, FEI-DEO, and DEI-DEO. Note that Dasarathy does not address the cells FEI-DAO, DEI-DAO, and DEI-FEO combinations. Steinberg and Bowman [43] point out that Dasarathy's model provides a natural way of identifying types of algorithms based on this input-output view. For example, algorithms that accept input features and provide output features (namely transform an input feature vector into an output feature vector) tend to be pattern classification or associative memory algorithms such as cluster algorithms or neural networks. Steinberg and Bowman also provide an expanded view of Dasarathy's model and link Dasarathy's view to the JDL level 0-4 processes.

2.8.2   Boyd's Decision Loop

Boyd's decision loop [the observe, orient, decide, and act (OODA) loop] has been used to assist in formalizing concepts of tactical command and decision-making. This loop, shown on the inside of the omnibus model in Figure 2.12, is an iterative and cyclic process. The concept has also been used for common training in driver's education and other applications. The concept simply involves focusing on four steps:

- *Observe*: Gather data via human and related senses to determine information about a situation. Be cognizant of the need for systematic and continuous data gathering – including general surveillance and gathering information to confirm or refute an evolving hypothesis.
- *Orient*: Perform a situation assessment related to the gathered data. What does the data mean on an individual and synthesized level? How does the

data fit with a priori hypotheses? What has changed in the past decision cycle? How does the data fit together?

- *Decide*: Having observed data and sought sources of information, and having oriented oneself to achieve an assessment of the situation, the third step involves making a decision. Such decision-making is performed considering the likelihood of alternate hypotheses and the consequences of each alternate hypothesis. Decisions may utilize heuristics such as "make a decision based on the most likely hypothesis having the most serious consequences (or threats)."

- *Act:* Finally, the decision-maker should act based on the decision. This may entail military operations, collection of additional data, and redirection of information resources, additional modeling, or other activities. Modern military focus is on "actionable knowledge," that is, the search for knowledge that is useful to assist making decisions and taking actions.

This model of decision-making and cognition can be applied to human processes as well as to computer-assisted cognition.

2.8.3   Bedworth and O'Brien's Omnibus Process Model

Bedworth and O'Brien reviewed [42] different data fusion models including the JDL model, the waterfall system engineering process model, Boyd's decision loop, and the intelligence cycle and synthesized them to develop an "Omnibus" model. A summary of their comparison of these models is shown in Table 2.5 and their process model is shown in Figure 2.12. Notice how the omnibus model combines aspects of the OODA loop with components of sensor management, signal processing, feature-based fusion, and decision-level fusion.

Table 2.5

Bedworth and O'Brien's Comparison of Fusion Models [42]

Activity	Waterfall Model	JDL Model	Boyd Loop	Intelligence Cycle
Command execution	Not applicable	Not applicable	Act	Disseminate
Decision-making process	Decision making	Level 4	Decide	Evaluate
Threat assessment		Level 3	Orient	
Situation assessment	Situation assessment	Level 2		Collate
Information processing	Pattern processing & feature extraction	Level 1		
Signal processing	Signal processing	Level 0		
Source/sensor acquisition	Sensing		Observe	Collect

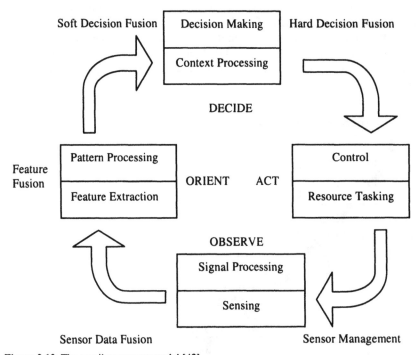

Figure 2.12 The omnibus process model [42].

2.8.4   TRIP Model

A team led by William Fabian under a contract for the Defense Advanced Research Projects Agency (DARPA) developed the most recent Transformation of Requirements for the Information Process (TRIP) model of the data fusion process [44]. The model was developed to understand the transformation of information needs by a tactical commander to tasking of sensor resources. Kessler and Fabian indicate that the model was developed to accomplish four goals:

1. To describe the process for developing collection tasks from information requirements,
2. To understand relationships between collection management and the situation estimation process,
3. To understand where the "human in the loop" is required,
4. To understand the internal and external drivers for the intelligence, surveillance, and reconnaissance (ISR) process.

The model has been developed in some detail and is readily relatable to the JDL data fusion process model. The TRIP model, however, attempts to address both identification of processing functions, and the detailed information interfaces between the functions. This is done via the formal system engineering IDEF hierarchical decomposition process. The TRIP model links human information requirements to data collection. A conceptual view of the TRIP model is shown in Figure 2.13.

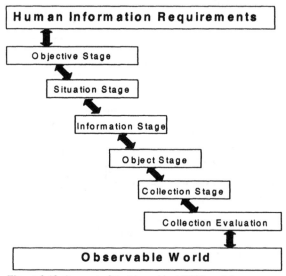

Figure 2.13 Concept of the hierarchical TRIP model.

REFERENCES

[1] Kessler, O., *Functional Description of the Data Fusion Process*, Warminster, PA: Office of Naval Technology, Naval Air Development Center, 1992.

[2] Steinberg, A. N., C. L. Bowman, and E. E. White, Jr., "Revisions to the JDL Data Fusion Model," *3rd NATO/IRIS Conference*, Quebec City, Canada, 1998.

[3] Hall, M. J., S. A. Hall, and T. Tate, "Removing the HCI bottleneck: How the Human Computer Interface (HCI) Affects the Performance of Data Fusion Systems," *2000 MSS National Symposium on Sensor and Data Fusion*, San Diego, CA, June, 2000, pp. 89-104.

[4] Hall, D., "Multisensor Data Fusion on the Internet," *Symposium on Multisensor Data Fusion*, Atlanta, GA, 1998.

[5] Zaleski, J., and D. Hall, "Extending the Joint Directors of Laboratories (JDL) Fusion Processing Hierarchy for Commercial Applications," *International Council on Systems Engineering Annual Conference*, Melbourne, Australia, July 1-6, 2001.

[6] Llinas, J., and D. Hall, "Dual-Use of Data Fusion Technology: Applying the JDL Model to Non-DoD Applications," *Fourth Annual IEEE Dual-Use Technology and Applications Conference*, SUNY Institute of Technology, Rome, NY, May 23-25, 1994.

[7]   Hall, D., "Intelligent Monitoring of Complex Systems," in *Advances in Intelligent Systems for Defence*, ed. by L.C. Jain, N.S. Ichalkaranje, and G. Tonfonis, London, England: World Scientific Publ. Co., 2003.

[8]   Russ, J. C., *The Image Processing Handbook*, Springer-Verlag, 1998.

[9]   Swanson, D. C., *Signal Processing for Intelligent Sensing Systems*, New York, NY: Marcel Dekker, Inc., 2000.

[10] McNeese, M. D., E. Salas, and M. Endsley, eds. *New Trends in Cooperative Activities: System Dynamics in Complex Environments*, Santa Monica, CA: Human Factors and Ergonomics Society, 2001.

[11] Antony, R., *Principles of Data Fusion Automation*, Norwood, MA: Artech House, Inc., 1995.

[12] Antony, R., "Data Management Support to Tactical Data Systems," in *Handbook of Multisensor Data Fusion*, ed. by J. Llinas, Boca Raton, FL: CRC Press, 2001, p. 18-1, 18-25.

[13] ARCINFO, http://www.zizoft.com/html/arcinfo/arcinfo.html: Internet Site by Zizoft company, 2002.

[14] Sorenson, H. W., "Least Squares Estimation: From Gauss to Kalman," *IEEE Spectrum*, 1970, pp. 63-68.

[15] Hayes, B., "Science on the Farther Shore," *American Scientist*, vol. 90, 2002, pp. 499-502.

[16] Cappellari, J. O., C. E. Velez, and A. J. Fuchs, *Mathematical Theory of the Goddard Trajectory Determination System*, Greenbelt, MD: NASA Goddard Space Flight Center, 1976.

[17] Escobol, R. R., *Methods of Orbit Determination*, Melbourne, FL: Krieger Pub. Co., 1976.

[18] Blackman, S. S., *Multiple-Target Tracking with Radar Applications*, Norwood, MA: Artech House, Inc., 1986.

[19] Clifford, H., and W. Stephenson, *An Introduction to Numerical Taxonomy*, New York, NY: Academic Press, 1975.

[20] Gelb, A., *Applied Optimal Estimation*, Cambridge, MA: MIT Press, 1974.

[21] Henkei, R. E., *Tests of Significance*, Beverly Hills, CA: Sage, 1976.

[22] Berger, J., *Statistical Decision Theory: Foundations, Concepts and Methods*, New York, NY: Springer-Verlag, 1980.

[23] Bayes, T., "Essay Towards Solving a Problem in the Doctrine of Chances," *Royal Soc. Philos. Trans.*, vol. 53, 1763, pp. 370-418.

[24] Lawrence, J. D., and T. D. Garvey, "Evidential Reasoning: A Developing Concept," *Proceedings of the International Conference on Cybernetics and Society, IEEE*, 1982.

[25] Yen, J., and R. Langari, *Fuzzy Logic: Intelligence, Control, and Information*, Upper Saddle River, NJ: Prentice Hall, 1999.

[26] Aldenderfer, M. S., and R. K. Blashfield, *Cluster Analysis*, vol. 07-044, London: Sage Publications, 1984.

[27] Anderberg, M., *Cluster Analysis for Applications*, New York, NY: Academic Press, 1973.

[28] Breiman, L., *Classification and Regression Trees*, Boca Raton, FL: CRC Press, 1984.

[29] Pugh, G. E., *An Information Fusion System for Wargaming and Information Warfare Applications*: Decision-Science Applications, Inc., 1981.

[30] Wright, F. L., "The Fusion of Multisensor Data," *Signal*, 1980, pp. 39-43.

[31] Charniak, E., C. Riesback, and D. McDermott, *Artificial Intelligence Programming*, Hillsdale, NJ: Lawrence Erlbaum Associates, 1987.

[32] Giarratano, J. C., *Expert Systems: Principles and Programming*, Stamford, CT: Brooks/Cole Pub. Co., 1998.

[33] Jackson, P., *Introduction to Expert Systems*, 3rd. ed., Reading, MA: Addison-Wesley Pub. Co., 1999.

[34] Hall, D., and R. J. Linn, "Comments on the Use of Templating for Multisensor Data Fusion," *Tri-Service Data Fusion Symposium*, John Hopkins University, Laurel, MD, 1989, pp. 152-162.

[35] Noble, D. F., "Template-Based Data Fusion for Situation Assessment," *1987 Tri-Service Data Fusion Symposium*, John Hopkins University, Laurel, MD, 1987, pp. 152-162.

[36] Prieve, C., and D. Marchette, "An Application of Neural Networks to a Data Fusion Problem," *1987 Tri-Service Data Fusion Conference*, John Hopkins University, Laurel, MD, 1987, pp. 226-236.

[37] Widrow, R., and R. Winter, "Neural Nets for Adaptive Filtering and Adaptive Pattern Recognition," *IEEE Computer*, 1988.

[38] Moubray, J., ed. *Reliability-Centered Maintenance (RCM) Analysis*, 2nd ed., New York, NY: Industrial Press, Inc., 2001.

[39] Waltz, E., and J. Llinas, *Multisensor Data Fusion*, Norwood, MA: Artech House, Inc., 1990.

[40] Mauavoglu, E., et al., "Modeling and Personalization of Users," in *The Power of One - Leverage Value From Personalization Technologies*, ed. by A. Raugaswamy, New York, NY: Amacom Books, American Management Association, 2002.

[41] Dasarathy, B., *Decision Fusion*, New York, NY: IEEE Computer Society Press, 1994.

[42] Bedworth, M., and J. O'Brien, "The Omnibus Model: A New Model of Data Fusion?," *2nd International Conference on Information Fusion*, 1999.

[43] Steinberg, A., and C. Bowman, "Revisions to the JDL Data Fusion Model," in *Handbook of Multisensor Data Fusion*, ed. by J. Llinas, Boca Raton, FL: CRC Press, 2001, p. 2-1 - 2-19.

[44] Kessler, O., and B. Fabian, *Estimation and ISR Process Integration*, Washington, D.C.: Defense Advanced Projects Research Agency, 2001.

Chapter 3

Level 1 Processing: Data Association and Correlation[1]

This chapter introduces the problem of data association and correlation and describes techniques for correctly associating and correlating data and information. This is part of the level 1 process in the JDL model.

3.1 INTRODUCTION

A problem that is fundamental to data fusion involves assignment or correlation. The problem is simple to state, but relatively difficult to solve. The problem involves how to determine which observations from multiple sensors *belong* together, representing observations of the same physical object. This is illustrated conceptually in Figure 3.1, in which an aircraft is tracking two objects. Data must be sorted into separate files, each file containing the data corresponding to a single observed target. A related problem involves how to determine which observations should be assigned as *belonging to* existing tracked objects. The ability to correctly solve this problem affects *all* of the subsequent processing in a data fusion system.

---

[1] Materials from this chapter were originally developed by a team effort that included Dr. James Llinas, Dr. David Hall, and Dr. Chris Bowman for U.S.A.F. project correlation.

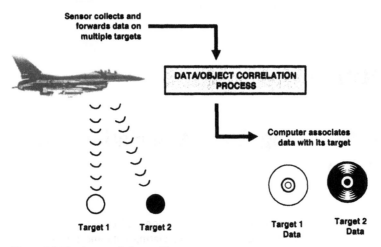

Figure 3.1 The data correlation problem.

The correlation problem is illustrated in Figure 3.2, which shows three dynamic *targets* denoted by a circle, square, and a star. These targets are shown as being observed by several sensors, including S_1, S_2, and S_3. The resulting observations, denoted y_1, y_2, etc., must be sorted into groups and *assigned as belonging to* tracks that are being developed in an evolving situation database (shown on the right-hand side of Figure 3.2). In this example we have made the problem relatively easy, since we clearly show the identity of each target using a geometric icon (circles, squares, and stars).

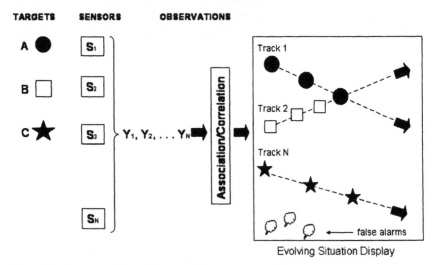

Figure 3.2 A conceptual view of the association/correlation problem.

Ordinarily, however, this identity is not known in advance and a number of complications arise, such as the following:

- It is unknown a priori how many targets actually exist (namely how many tanks, aircraft, or military units there are in the battlefield);
- The targets may be closely spaced and dynamically maneuvering (indeed, the targets may maneuver deliberately to avoid detection, such as the case in which one aircraft "hides" behind another aircraft to avoid radar detection);
- There may be poor observing conditions (e.g., low signal-to-noise ratios, multipath effects, and interference or complex signal propagation conditions) that cause false alarms;
- An intelligent opponent may deliberately create confusion via electronic countermeasures (ECM) or use of stealth techniques;
- Sensors may have relatively low resolution, causing an inability to distinguish between closely spaced targets.

Solving the correlation problem requires that we sort observations into groups which *explain* the data. Uhlman [1] provides the example illustrated in Figure 3., which shows a multitime frame, multiobject (i.e., multitarget) data correlation problem. On the left-hand side of the figure, a single target is shown. At sequential times, t_1, t_2, t_3, additional observations are obtained. On the left-hand side of Figure 3.3 (looking down from the frame at the top containing a single dot, to the frame at the bottom containing five dots and an interconnecting arrow), these are shown as belonging to a single moving target. After five time intervals, for example, we may interpret the five observations as all belonging to a single target, moving in a straight line. This interpretation of the data is illustrated by a line that "connects the dots," as shown in the lower left-hand side of Figure 3.3.

By contrast, a more complex situation is shown on the right-hand side of Figure 3.3. At each snapshot in time (i.e., at times t_1, t_2, t_3,), multiple observations are available. For example, at time t_1 (shown on the upper right-hand side of Figure 3.3) five observations are available. At time t_2, an additional five observations are available (for a total of 10 observations at time t_2). How can we interpret these observations? Are these new targets; are these (or some combination of these) observations possible false alarms (e.g., representing ECM effects, or multipath signal propagation effects)? The lower right-hand side of Figure 3.3 shows one possible interpretation of these observations as *belonging to* five tracked objects traveling in straight lines. It must be emphasized, however, that even though this interpretation of the data is very plausible, it is by no means the only possible interpretation. For example, the targets could be maneuvering, there could be more than five targets, or some of the data could be spurious. Based on this example, it can be seen that the data correlation problem can be challenging, even at the object level.

Section 3.2, develops a process model to solve the data correlation problem, and Sections 3.3-3.5 describe specific solution techniques. Engineering trade-offs and advice on how to select specific algorithms, based on the characteristics of the application problem, are described in Chapter 10. One question that naturally arises is the following: What is the impact or consequences in data fusion processing if we do not correctly solve the association problem (namely, what happens if data are incorrectly associated or miscorrelated)? The answer to this question depends upon a number of factors such as the relative accuracy of the sensors, whether the sensors are unbiased (i.e., have zero mean observational error), the rate of observations, and the dynamics of the observed targets. To answer this question quantitatively would require covariance error analysis [2] or Monte Carlo simulations for a specific application. These models could incorporate the details of the target dynamics, observation rates, sensor performance, signal propagation environment, and effects of countermeasures.

While it is not feasible to provide a general quantitative answer to the miscorrelation question, a qualitative sense of the effect of miscorrelations is illustrated in Figure 3.4. Figure 3.4 shows two tracked objects, denoted by track 1 and track 2, moving in straight lines and apparently converging. If a new position report is available as shown, the correlation process attempts to determine to which tracked object the observation should be assigned. There is ambiguity because the observation could equally well be assigned either to track 1 or to track 2. Suppose we arbitrarily assign the observation as belonging to track 1. In subsequent fusion processing (e.g., via a Kalman filter), the state vector for object 1 would be updated to include the effect of this observation. The result would be that the trajectory of object 1 would be *moved*, so that the object would change direction and "move" in the direction indicated by the dashed line in Figure 3.4. Similarly, if the observation is arbitrarily assigned as belonging to track 2, then this would change the state vector estimate for object 2, and the trajectory for object 2 would be moved to traverse along the dashed line shown in Figure 3.4.

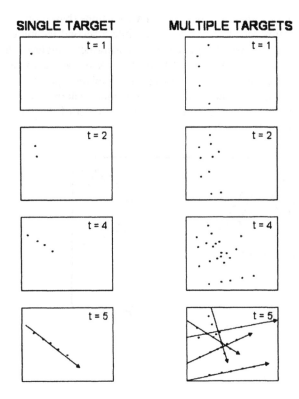

SINGLE TARGET   MULTIPLE TARGETS

Figure 3.3 Example of the correlation problem [1].

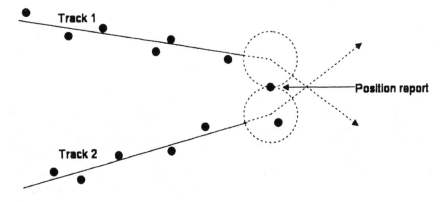

Figure 3.4 Ambiguity in observation track assignment.

Blackman describes a similar situation, and provides a numerical example as shown in Figure 3.5 which compares two correlation techniques. The first technique is an optimal technique that provides *correct* associations (based on the information available at the time of the observations), while a second technique provides *suboptimal* associations (which are sometimes incorrect). Figure 3.5 shows that over time, incorrect associations lead to more incorrect associations, which corrupt the state vector estimates. Thus, solving the correlation problem is key to providing the best estimates of the state vectors (position, velocity, attributes, and identity) to characterize entities and targets. The remainder of this chapter is focused on how to develop these solutions.

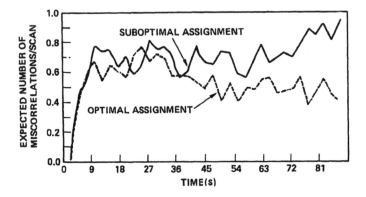

Figure 3.5 Effect of miscorrelations [3].

3.2 PROCESS MODEL FOR CORRELATION

The solution of the data correlation problem occurs within a broader context of a multisensor data fusion system. Fusion of multisensor data may occur in a hierarchical fashion with many data fusion *nodes*. Each node may accept data directly from sensors, from human input, from internal and external databases, and from other data fusion *nodes*. For example, a data fusion system onboard a single aircraft may involve multiple interacting fusion *nodes*, one fusion node associated with a multisensor identification-friend-foe (IFF) system, another node corresponding to an automatic target recognition (ATR) system, and input data from external intelligence sources (each of which involves multiple fusion nodes). Within each of these individual nodes, data fusion processing involves three key functions: (1) data preparation (i.e., transforming the incoming data to a common reference system); (2) data correlation (sorting the observations into groups, with each group representing a separate physical entity or event); and (3) state estimation and prediction (in which the data are actually fused to determine the position, velocity, attributes, and identity of observed entities). The processing components for a single data fusion node are illustrated in Figure 3.6.

DATA FUSION TREE NODE

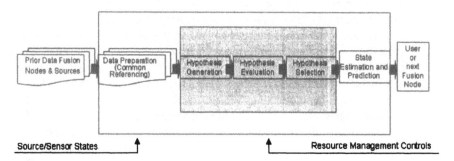

Figure 3.6 Functions within a data fusion processing node.

Data correlation occurs within each data fusion-processing *node*. Each processing node accepts input from sensors, databases, and other nodes, and performs data fusion, to estimate the position, velocity, attributes and identity of targets (i.e., fused information to produce tracks). Within a single processing node, the data correlation function acts as an intermediary between the data preparation function (which transforms input data into a common reference frame), and the state estimation and prediction function. This latter function is the point at which data fusion actually occurs, typically using sequential estimation techniques such as Kalman filters, and identity estimation processes (e.g., Bayesian identity fusion methods). Data correlation accepts input data and sorts the observations in groups (e.g., groups of observations or assignments of observations to existing tracks). Each group of data represents data that *belongs to* a single physical entity (target, platform, emitter). The state estimation function subsequently combines or fuses the data.

To provide a description of correlation processing, we have identified three required functions: (1) hypothesis generation, (2) hypothesis evaluation, and (3) hypothesis selection. Each of these functions is summarized as follows, and described in the following sections:

- *Hypothesis generation* (HG): HG processes the incoming data (e.g., sensor observations, reports, tracks), either sequentially or in batches (groups) to develop a list of one or more hypotheses that could explain or interpret how the data are related. The word *hypothesis* is used here to mean a possible explanation or interpretation of the data. For example, a hypothesis might indicate that *a received observation, O_l, may be correlated with a previously observed tracked object, T_a.* These hypotheses are usually grouped into a matrix form called an association matrix, which establishes links between the input data and possible hypotheses (or explanations of the data).

- *Hypothesis evaluation* (HE): HE processes the association matrix to evaluate the likelihood of the feasible hypotheses output from hypothesis generation. The output of the HE function is a quantified evaluation or ranking of the feasible hypotheses. Thus, the hypothesis evaluation task processes the association matrix, and develops quantitative measures to determine how feasible or likely these explanations are. Many different measures can be used including probabilistic metrics, similarity measures, distance calculations, and likelihood functions. The HE function populates or *fills in* the association matrix with numerical values representing the *feasibility* or *relative likelihood* of the alternative hypotheses explaining the incoming data.

- *Hypothesis selection* (HS): HS uses the results from the HE function to guide a search for the best association hypotheses for all the input data. The association hypotheses (e.g., assignment of observation-to-observation, observation-to-track, and track-to-track) output from the HS function are used in subsequent state estimation techniques to combine the data for improved estimation and prediction of the state (position, velocity, parametrics, attributes, identity) of entities. The HS function sorts through the feasible hypotheses (identified by the HG function), compares the likelihood of these hypotheses (quantified by the HE function), and finds an optimal set of hypotheses that *best fit* or *provide the best explanation of* the incoming data.

In introducing these three functions, it must be noted that for a real fusion system, these functions are interrelated and integrated. We will address each of these functions separately, but the reader should remember that they must be integrated in a real system implementation. Sections 3.3-3.5 describe these functions in detail, including a discussion of alternate solutions and algorithms. It addition, it should be noted that this chapter treats data correlation as a separate process from the state estimation process.

3.3 HYPOTHESIS GENERATION

The hypothesis generation function accepts input data from sensors or other data fusion nodes, and develops feasible hypotheses or *explanations* of the input data. The result is an association matrix that links input data to feasible hypotheses. The first step to perform data correlation is the identification of alternative hypotheses or feasible *explanations* of the data. The concept is illustrated in Figure 3.7. Incoming data are evaluated to determine how to interpret the data. These interpretations depend upon factors such as the sensor data (i.e., the observed quantities, knowledge of the accuracy of the data, and whether or not there is information about target identity), the information in an evolving fusion database (evolving tracks, estimates of the location and identity of targets and enemy units), and a priori information about the environment.

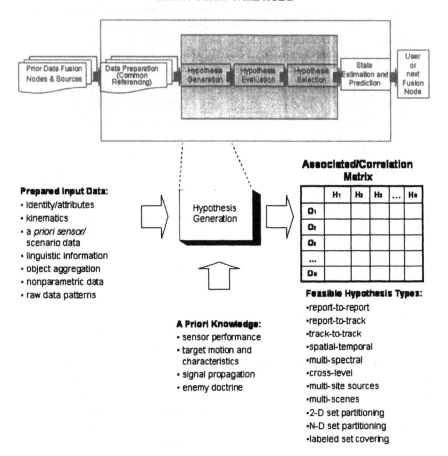

DATA FUSION TREE NODE

Figure 3.7 An overview of the hypothesis generation (HG) function for correlation.

In order to develop methods to perform hypothesis generation, we must first address three basic *philosophical* issues or choices that affect the available solutions. These include: (1) assignment uniqueness, (2) number of scans, and (3) hypothesis testing approach. Each of these is described as follows, followed by a discussion of HG techniques.

- *Assignment uniqueness*: A basic question in performing correlation, and developing feasible hypotheses involves whether incoming data in one list can be assigned to one (and only one) possible data in another list. For example, should we assign each incoming report or observation uniquely to one existing track in the database, or could the observation be assigned to multiple existing tracks? Uhlman [1] provides an example of this question, as shown in Figure 3.8, in which two tracked objects are shown, denoted by track 1 and track 2. These objects appear to be converging as time progresses. At some subsequent time, a new position report is obtained. As shown, it is feasible for the position report to *belong* either to tracked object 1 or to tracked object 2. The uncertainties in the predicted positions of objects 1 and 2, illustrated by the dashed circles, do not allow us to easily dismiss track 1 or track 2 from further consideration. There are two basic approaches. In the first approach, we assume assignment *uniqueness*, and consider that the position report can only *belong* either to track 1 or to track 2, but not to both. The *nonunique* approach allows the report to be considered as *belonging* both to track 1 and to track 2. Bar-Shalom and Tse [4] have extensively developed this latter approach via techniques such as the probabilistic data association model.

- *Number of scans:* The second basic decision for correlation involves the issue of how to ingest or process the observations. Typically, data fusion systems process observations in a sequential way (i.e., accepting observations as they are obtained from the sensors or other fusion nodes) and incorporating these observations into a dynamic estimation process to continually update a situation database with evolving or refined estimates of the state (position, velocity, attributes, and identity) of targets. Even though these observations are typically processed sequentially, there is still an option of whether to actually process the observations one-at-a-time, or whether to wait until several observations are available (e.g., at times t_1, t_2, or t_3). This choice is sometimes referred to as the *single scan* versus *multiple scan* choice. The word *scan* describes how some sensors dynamically *pan* or *scan* an observation area and collect *snapshots* of observations (note that strictly speaking a single *scan* may itself contain multiple observations). The advantage of a multiple scan approach is straightforward; by waiting until multiple observations are available, more information can be used to detect effects such as the initiation of a new track, or crossing targets. By contrast, a single scan approach is computationally more efficient, and easier to implement, but provides more ambiguity in interpretation of the observed data.

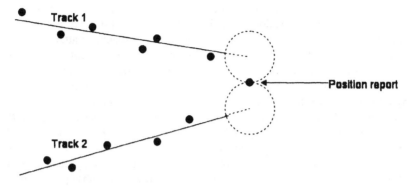

Figure 3.8 An example of assignment ambiguity [1].

- *Multiple versus single hypotheses*: The third basic choice involves the concept of single versus multiple hypothesis processing. Consider the ambiguous situation shown at the top of Figure 3.9. At a snapshot in time, the estimated location of three objects denoted P_1, P_2, and P_3, are shown. The exact location of these objects is not known, and a dashed circle about each point, P, denotes the associated uncertainty in their locations. Suppose that in a given snapshot, we are provided with three observations, O_1, O_2, and O_3. The basic correlation question to be resolved involves determining which observations (O_1, O_2, and O_3) *belong to* tracked objects, P_1, P_2, or P_3. The problem is compounded because there is some ambiguity in possible assignments. For example, while observation O_3 could only be assigned to tracked object P_2 (because it lies within the location uncertainty region about P_2), it is feasible for observation O_2 to be assigned to either object P_1, P_2 or to P_3. Similarly, observation O_1 could be assigned either to tracked object P_1 or to object P_3.

A traditional way to perform association involves a single hypothesis approach. In this approach, we develop a single (evolving) interpretation concerning the assignment of the data to tracks. For example, based on measures of association, we might decide that observation O_3 *belongs to* tracked object P_2, and that observation O_1 *belongs to* tracked object P_1 and that observation O_2 *belongs to* object P_3. This is a single hypothesis that provides an explanation for (or interpretation of) the observational data. The single hypothesis approach seeks to develop a single hypothesis to interpret the available data. In subsequent data fusion processing, these observations are incorporated into an estimation process that would update the previous estimates of the locations of objects P_1, P_2, and P_3, based on the newly assigned observations.

Clearly, however there are alternative hypotheses that could explain the observational data. For example, any one of these observations may be a false alarm (namely a spurious observation generated by noise that does not belong to any existing track), or observation O_2 could equally well be assigned to either tracked object, P_1, P_2, or to P_3. In the multiple hypothesis approach, an attempt is made to explicitly enumerate these alternative hypotheses. This concept is shown in the bottom part of Figure 3.9, which shows a hypothesis tree identifying different interpretations of the data. Each branch of the hypothesis tree represents a separate hypothesis. For example, the dashed line indicates the branch of the tree corresponding to the following hypothesis: Observation *O_1 belongs to object P_1* and *observation O_2 belongs to object P_3* and *observation O_3 belongs to object P_2*. In the multiple hypothesis approach, these alternative hypotheses are systematically developed (until a preset number of observations have been processed, or until a preset number of hypotheses have been identified) and are subsequently evaluated to select the most likely hypothesis. The advantage of the multiple hypothesis approach is that it allows explicit consideration and evaluation of alternative explanations of the data. On the other hand, it can be seen from Figure 3.9, that the number of alternative hypotheses grows rapidly with increasing number of observations (and existing tracks in the database). Hence such an approach may become computationally infeasible for real-time applications. A general discussion of the multiple hypothesis approach is provided by Blackman [3], including approaches to develop hypotheses, evaluate the hypotheses, and a common representation scheme.

These three basic processing choices (i.e., assignment uniqueness, number of scans, and single versus multiple hypothesis tracking) are relatively independent of each other. For example, a multiscan, multiple hypothesis approach could be implemented using either a unique or nonunique assignment scheme. Additional comments regarding these choices are provided in Section 3-6. The remaining portions of this chapter describe how to characterize the hypothesis generation problem, and provide an overview of solution techniques. The discussion of how to select from among these techniques is provided in Chapter 10.

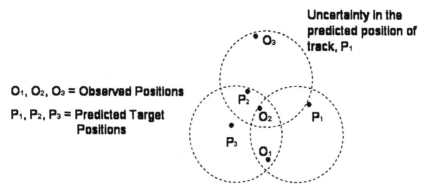

AMBIGUOUS OBSERVATION-to-TRACK ASSOCIATION

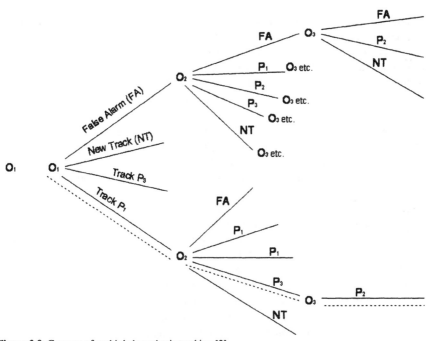

Figure 3.9 Concept of multiple hypothesis tracking [3].

3.3.1   Characterizing the Hypothesis Generation Problem

An understanding of the hypothesis generation problem space (namely the general boundaries and character of the problem), depends upon three basic factors:

1. The characteristics of the input data (which includes our understanding of the nature of the sensors, the signal propagation environment, and the characteristics of the targets or entities being observed);
2. The feasible hypothesis types (i.e., which types or categories of hypotheses that we are willing to consider as possible explanations of the input data);
3. The measures of desired performance for hypothesis generation.

The selection of appropriate HG techniques is based upon these factors. Although there are numerous possible techniques for hypothesis generation, the applicable techniques and methods depend upon the character of the input data, the available a priori information, and the nature of the feasible hypotheses.

3.3.1.1   Input Data Characteristics

The inputs to the HG function are the observations, reports, and a priori information about the sensors, the environment, and entities being observed. In general, the input-to-hypothesis generation involves one or more sets of sensed information (e.g., raw sensor data, pre-processed sensor data such as feature vectors or attributes, or complex records that contain multiple pieces of data about an entity). A variety of information may be available for each observed entity. These are summarized in Table 3.1.

The available data strongly influences the HG problem definition. First, the ability to distinguish between observations (or to link observations together) based on parameters such as location, velocity, attributes, or identity, depends upon whether those data are available (both in the incoming reports and their availability for existing data and tracks). Second, the extent to which a designer is willing (or able) to make assumptions about target motion, sensor accuracy, signal propagation, target constituency, environmental conditions, and even enemy doctrine, influence the choice of techniques and strategies for HG. The underlying issue here is the degree of confidence for a priori knowledge of the problem. These assumptions (introduced via models, a priori probabilities, representation of explicit knowledge), are viewed as another category of input data. For example, knowledge of the measurement error for sensor data may be used in a Chi-square calculation of a gating boundary (namely to exclude data from being likely to be associated with other data). Similarly, knowledge of the motion of a target may allow one to eliminate unlikely report-to-track pairs based on kinematics. Characterization of this input information is part of the HG design process.

An additional factor, which must be considered as an input variable, involves how the input data must be *ingested*. Incoming data records may be ingested (or processed) in one of three basic ways: (1) process the data sequentially, one-at-a-time, as the data is received; (2) process the data sequentially in groups or batches; or (3) process the data after all data are received by the fusion system. These alternatives are sometimes termed (1) single-scan, (2) multiple or N-scan approach, or (3) batch processing, respectively. The availability of the input data may be a given part of the overall problem, or may be a designer's choice. The availability of the data as a single report, multiple reports, or all reports, affects the extent to which evolving models can be developed (e.g., probability trees) to characterize the data, and to compute the likelihood that one or more hypotheses are viable.

3.3.1.2 Output Data Characteristics

The nature of the output data for HG also characterizes the HG problem and affects the applicable solutions. In particular, the output of hypothesis generation is a list of hypotheses that represents feasible interpretations of the available data (especially with respect to existing data and information). This output is sometimes represented as a matrix as shown in Figure 3.7. The matrix maps reports or observations to hypotheses (namely to interpretations of the meaning of the reports or observations). One point must be made clear in this regard. Unlike the hypothesis evaluation and hypothesis selection functions (for which the inputs are given based on the previous processing), the definition of the output hypothesis types for HG is itself a major design choice. The system designer must identify the domain of possible data interpretations. For example, based on his or her-knowledge of the sensing environment, the designer must specify whether an incoming report could be the result of a sensor false alarm, multipath signal propagation, or enemy countermeasure.

Table 3.1

Summary of Input Data Characteristics for HG

Input Data Categories	Data Format/Description	Utility for Hypothesis Generation
Identity (ID) attributes	Data that allows an entity to be classified into one of a finite number of classes or types (e.g., emitter characteristics such as pulse repetition interval, pulse width, or scan type)	Allows differentiation of one observation from another based on identity or distinguishable attributes
Kinematic data	Geographic location, velocity, positional information such as range, range-rate, angles, and their associated uncertainties	Allows link between data and known objects by matching kinematic or positional measurements against predicted positions
Parameter attributes	Continuous nonkinematic information of an object or entity (e.g., frequency, pulse width, and radar cross-section)	Differentiate between objects based on parameter values(e.g., distinguish one emitter from another based on observed frequency or pulse repetition interval)

Table 3.1 (continued)

Summary of Input Data Characteristics for HG

input Data Categories	Data Format/Description	Utility for Hypothesis Generation
A priori sensor or scenario data	Information about sensor performance(e.g., probability of missed detections, false alarms, independence of data elements)	Provides basis for creation or elimination of hypotheses based on knowledge of sensing environment, or another factor
Linguistic data	Textual information	Allows human supplied information to support hypothesis generation
Space-time patterns	Known spatial or temporal relationships (relative position of emitters and platforms, order of battle)	Allows use of a spatial or temporal *template* to determine feasibility of observation-entity pairings
High uncertainty-in-the-uncertainty information	Uncertainty representations such as fuzzy set concepts, evidential reasoning, modified Dempster Shafer	Provides an alternative to probabilistic formulations of uncertainty (e.g., allows specification of the general level of uncertainty, and fuzzy concepts)
Unknown structures and patterns	Clusters of data with unknown patterns (e.g., scatter plots).	Requires assumption of potential inter-data relationships and automated investigation via pattern recognition techniques
Error probability density functions	Statistical descriptions of sensor uncertainties or state vector propagation uncertainties	Information about measurement uncertainty and state vector propagation uncertainty can be incorporated to properly scale distance metrics for gating
Target characteristics	A priori information about target characteristics: size, shape, parameters, target motion, capabilities	Use of target information provides a means to determine likelihood of a hypothesis (e.g., based on target motion and characteristics)
Signal propagation models	Models related to signal propagation in a dynamic environment	Assists in determining possible alternate explanations of data such as multipath, or other propagation effects

In general, there are several categories or classes of hypotheses. These are summarized in Table 3.2 and described briefly as follows.

- *Report-to-report*: Each individual observation or report (comprising a vector of parameters such as location, attributes, and identity information) may be hypothesized to be related to a previously observed report. Thus, the incoming reports may be grouped with other previous reports, thereby hypothesized that they belong to the same physical target or entity. Report-to-report similarity may be based on some measure of observation-to-observation closeness or similarity. Typical examples of such metrics are distance measures or probabilistic measures.

- *Report-to-track*: Incoming reports may be hypothesized to be evidence associated with an existing tracked object. Thus, one may consider that a received observation, O_1, is new evidence associated with tracked object, 1. In this case, the hypothesis must consider the movement of the tracked object, 1, since its previous observation time, t_0, to the time of the current observation, t_i. This requires the ability to update the position of the tracked object (namely to solve the dynamic equations of motion, to move object 1 from the time it was last seen at time, t_0, to the time of the current observation, t_i), and to compute a predicted observation for object 1.

- *Track-to-track*: An incoming report may already be in the form of a state vector (i.e., a smart sensor may compute a state vector based on multiple single-sensor observations). The incoming track may be hypothesized to be related to an existing tracked object. The comparisons are similar to report-to-report hypotheses, except that state vectors are compared, rather than direct observations.

- *Spatio-temporal*: Incoming data reports may be associated over space and time to define complex multicomponent entities.

Table 3.2

Summary of Output Data for HG

Hypothesis Type	Hypothesis	Example
Report-to-report	Observation/report belongs to a previously obtained observation of report of an entity	Two radar observations of R, Az, El at times t_1 and t_2 are of the same target
Report-to-track	New report is new evidence of an existing (known) tracked object	An observed position (r, Az, El) belongs to target having the estimated state vector x
Track-to-track	Same as above, except the sensor/source proves a state vector (track estimate) as input	Two observed state vectors, $x(t1)$ and $x(t2)$ are of the same target
Spatio-temporal	An incoming report belongs to one or more geographically distributed (or historical) reports, previously observed	Report A belongs to multiple existing reports about geographical position B

Table 3.2   (continued)

Summary of Output Data for HG

Hypothesis Type	Hypothesis	Example
Multi-spectral	Two (or more) non-commensurate sensor reports are related to the same physical target or entity	An IR report (r, IR spectra) at time t1 belongs to the same target as a radar report (r, AZ, EL, RCS) at time t_2
Cross-level	Observations of different physical entities belong to the same complex entity	Sensor 1 report on emitter A belongs to the same entity as a Sensor 2 report
Multisite sources	Observations from different sites or sources are hypothesized to be associated with the same physical entity	Report A from source I, belongs with report B from source J
Multiscenes	Two alternate descriptions (scenes) are developed to interpret a collection of data, at a snapshot in time	Two scenes represent feasible interpretations of the same data
2-D set partitioning	Two mutually exclusive lists of data are matched to multiple alternative hypotheses	See text
N-D set partitioning	N mutually exclusive lists of data are matched to multiple hypotheses	See text
Labeled set covering	Further generalization of N-D set partitioning	See text

- *Multispectral*: Non-commensurate sensor reports (i.e., observations from sensors observing the same entity, via different physical phenomena such as infrared and radar) may be hypothesized to be of the same physical entity or target. Multispectral hypotheses may be developed based on common information such as position or velocity.
- *Cross-level*: Observations of different physical entities (such as an emitter or platform), may be hypothesized to be related to the same weapon system or complex entity. Such cross-level association may be based on geographical proximity of the different observations and knowledge of the syntax or makeup of a hypothesized complex entity.
- *Multisite sources*: Information from multiple sites or sources may be hypothesized to be associated with the same physical entity or complex entity.
- *Multiscenes*: Alternate worldviews may be developed to interpret the observational data. For example, in the lower right-hand side of Figure 3.3, only a single interpretation (namely that there are five moving objects which explain the 24 data points) is shown. One approach would be to

postulate this as only one of several alternative hypotheses to explain the data. Thus, multiple scenes may be developed, each representing an alternate hypothesis that explains the data.

- *Two-dimensional set partitioning:* Two mutually exclusive lists of data or reports may be output and matched to multiple alternative hypotheses. For example, two radar-scans (each containing multiple observations) may be ingested and associated with multiple existing data sets and tracks. An example may be seen in Figure 3.3 if two scans are simultaneously ingested [e.g., the scan of data at time t_5 on the left-hand side of Figure 3.3 is processed simultaneously with the data from the right-hand side of Figure 3.3 (also at time t_5)].

- *N-dimensional set partitioning:* This is similar to two dimensional set partitioning, except that N scans are processed (or ingested) at one time.

- *Labeled set covering:* This allows a piece of data to be associated with other data from more than one object within one feasible hypothesis (e.g., a radar report associated with two separate IR object tracks).

The list above shows that there are numerous variations of how to identify possible hypotheses based on factors such as the character and level of the input data, or whether the input data must be processed one at a time or multiple observations at a time. Other factors in defining feasible hypotheses involve the extent to which knowledge is available about target behavior, sensor performance, signal propagation, and the tactical engagement environment (e.g., stealth and countermeasures).

3.3.1.3 Performance Measure Characteristics

The hypothesis generation techniques that are selected to implement a data fusion system should maximize the system performance at a minimum cost of resources (including implementation costs and computer resource utilization costs). A particular challenge is to identify a sufficient number of feasible hypotheses, which truly represent the physical situation being observed, while eliminating unlikely hypotheses. This elimination of unlikely hypotheses is necessary in order to reduce the computational workload required for hypothesis evaluation and selection. Performance measures for hypothesis generation include the following:

- Software life-cycle cost and complexity (i.e., affordability);
- Processing efficiency (e.g., computations/seconds/watt);
- Accuracy of identified hypotheses (e.g., extent to which the identified hypotheses represent physically realistic explanations of the available data);
- Number of hypotheses (sufficient number to adequately explain the data without spurious hypotheses);

- Utility of available information (extent to which the HG function utilizes information about the sensor performance, signal propagation, target characteristics, and enemy doctrine);
- Robustness to errors and incorrect modeling (i.e., graceful degradation);
- Hardware timing/sizing limitations (extent to which HG utilizes an appropriate amount of available hardware resources without jeopardizing other processes);
- Ease of training on sample data (i.e., training set requirements).

3.3.2   Overview of Hypothesis Generation Techniques

The approach to developing a solution for hypothesis generation can be separated into two aspects: hypothesis enumeration and identification of feasible hypotheses. A summary of HG solution techniques is provided in Table 3.3. Hypothesis enumeration involves developing a global set of possible hypotheses based on physical, statistical, or explicit knowledge about the observed environment. Subsequently, hypothesis feasibility assessment provides an initial screening of the total possible hypotheses to define the feasible hypotheses for subsequent processing (namely for HE and HS processing).

3.3.2.1   Hypothesis Enumeration

The first function required to accomplish hypothesis generation is the identification of the set of potentially feasible hypotheses. Ultimately, this is based on human design choices (namely there are no automated techniques to generate the allowable set of hypotheses). This is a creative systems engineering step that must be accomplished by a human designer. However, this identification of hypotheses may be based on a number of models or techniques. These include the following:

1. *Physical models*: Physical models may assist in the definition of potential hypotheses. Examples include: intervisibility models to determine if it is possible for a given sensor to observe a specified target (with specified target-sensor interrelationships and environmental conditions); models for target motion (i.e., to move a target from one location in time to the time, of a received observation via dynamic equations of motion, terrain models); and models of predicted target signature for specified types of sensors;
2. *Syntactic models*: Models may describe how targets or complex entities are constructed. This is analogous to how syntactic rules describe how sentences can be correctly constructed in English. Syntactical models may be developed to identify the necessary components (e.g., emitters, platforms, or sensors) that comprise more complex entities such as a surface-to-air (SAM) missile battalion. The syntax may be represented in a variety of

ways and used as guidelines to assist in identification of possible hypotheses.

3. *Doctrine-based scenarios*: Models or scenarios may also describe anticipated conditions and actions for a tactical battlefield or battle space. Anticipated targets, emitters, relationships among entities, entity behavior (e.g., target motion, emitter operating modes, sequences of actions), engagement scenarios, and other information can be represented. This information can be used to establish possible hypotheses.

4. *Probabilistic models*: Probabilistic models may be developed to describe possible hypotheses (see [3]). These models can be developed based on factors such as the following:

   - A priori probability of the existence of a target or entity;
   - Expected number of false correlations or false alarms;
   - Probability of a track having a specified length;
   - Probability of track splitting;
   - Probability of track initiation (e.g., a new track).

   Probability models may be developed based on physical models, previous experiences in a particular tactical situation, or scenarios. The probability models provide a basis for developing (and evaluating) potential hypotheses.

5. *Ad hoc models*: In the absence of other knowledge, ad hoc methods may be used to enumerate potential hypotheses, including (if all else fails) an exhaustive enumeration of all possible report-to-report or report-to-track pairs.

The result of hypothesis enumeration is the definition or identification of possible hypotheses for subsequent processing. It must be stressed that this step is key to all subsequent processing. Failure to identify realistic possible causes (or interpretations) for received data (e.g., such as countermeasures or signal propagation phenomena) cannot be recovered in subsequent processing (namely the subsequent processing is aimed at reducing the number of hypotheses and ultimately selecting the most likely or feasible hypotheses).

3.3.2.2 Identification of Feasible Hypotheses

The second function required for hypothesis generation involves reducing the set of possible hypotheses to a set of feasible hypotheses. This involves eliminating unlikely report-to-report, or report-to-track pairs (hypotheses) based on physical, statistical, explicit knowledge, or ad hoc factors. The challenge is to reduce the number of possible hypotheses to a limited set of feasible hypotheses, without eliminating any viable alternatives that may be useful in subsequent HE and HS processing. There are a number of automated techniques for performing this initial pruning. These are listed in Table 3.3 and summarized as follows.

1. *Pattern recognition techniques*: A wide number of pattern recognition techniques may be used to identify *natural* patterns in data (e.g., groupings of previously uncorrelated reports, patterns among predicted observations (based on existing tracks) and incoming reports). This concept is illustrated in Figure 3.10, where incoming data is compared to existing data (or predicted data based on existing tracks) to determine the extent to which the incoming data is *close* to existing data. This comparison is performed in *observation space* or *feature space*. Pattern recognition techniques can be trained to establish a partitioning of the observation space. The feature space is partitioned by decision boundaries, which provide the basis for determining if a hypothesis is feasible or not (e.g., if it is feasible for a report to be associated with another group of observations, presumed to be associated with a single physical target or entities). In Figure 3.10, characteristics of emitters [pulse width (PW) and pulse repetition frequency (PRF)] are used to distinguish between different classes of emitters. Pattern recognition techniques are listed as follows.

Table 3.3

Solution Techniques for Hypothesis Generation

Processing Function	Solution Techniques	Description	References
Hypothesis enumeration	Physical models Syntax-based models Doctrine-based models Probabilistic models Ad hoc models	• Models of sensor performance, signal propagation, target motion, intervisibility to identify possible hypotheses • Use of syntactical representations to describe the makeup (component entities, interrelationships, etc.) of complex entities such as military units • Definition of tactical scenarios, enemy doctrine, anticipated targets, sensors, engagements to identify possible hypotheses • Probabilistic models of track initiation, track length, birth/death probabilities to describe possible hypotheses • Ad hoc descriptions of possible hypotheses to explain available data; may be based on exhaustive enumeration of hypotheses (e.g., in a batch processing approach)	[2] [2] [5] [3, 4] [2]
Identification of feasible hypotheses	Pattern recognition Gating	• Use of pattern recognition techniques such as cluster analysis, neural networks, or gestalt methods to identify natural groupings in input	[6, 7]

Table 3.3 (continued)

Solution Techniques for Hypothesis Generation

Processing Function	Solution Techniques	Description	References
	techniques	data	
	Logical templating	• Use of a priori parametric boundaries to identify feasible observation pairings and eliminate unlikely pairs; techniques include kinematic gating, probabilistic gating, and parametric range gates	[3, 8]
	Knowledge-based methods		
		• Use of prespecified logical conditions and parametric factors to specify spatio-temporal conditions, causal relations, and entity aggregations for feasible hypotheses	[9, 10]
		• Establishment of explicit knowledge via rules, scripts, frames, fuzzy relationships, Bayesian belief networks, neural networks	[11, 12]

- Cluster algorithms use search strategies to group observations into natural groups to represent a single target. Popular techniques include the nearest neighbor and k-nearest neighbor technique, hierarchical divisive methods, and hierarchical agglomerative methods. Descriptions of these techniques may be found in [2, 6, 7].
- Neural network techniques involve a multilayer structure of nodes and internodal connections (namely to emulate the structure of neurons in a biological brain). Neural networks act as a nonlinear transformation between an input feature vector and a set of output categories. Thus, a neural network may be used to map an input report or track to a feasible association with existing reports and tracks. There are numerous formulations of neural networks, including variations of the basic architecture (e.g., number of layers, number of nodes per layer, interconnectivity among nodes), the formulation of the nodal activation function, and training procedures. Regardless of these variations, neural networks act as a nonlinear transformation used (for this application) to find feasible hypotheses. A general discussion of neural nets is provided by [13];
- Gestalt techniques attempt to find natural groupings in feature space. However, these techniques emulate how human vision works to "connect the dots" to find feasible tracks within dense target tracking environments. That is, these techniques attempt to find groups of data points (or reports), that correspond to tracks (assuming that the observed targets move in a regular fashion). A discussion of gestalt techniques (originally introduced by Zahn) is provided by [14].

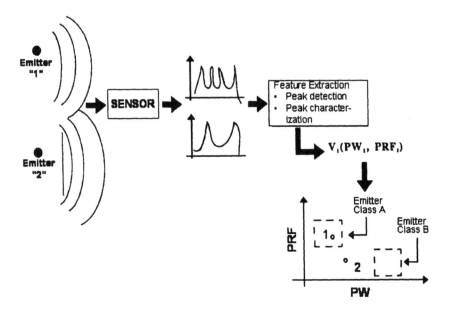

Figure 3.10 Examples of pattern recognition for identification of hypotheses.

2. *Gating techniques*: Gating methods are often used to eliminate unlikely hypotheses based on physical knowledge of the target dynamics or other attributes (e.g., emitter parameter characteristics). The concept of kinematic gating is illustrated in Figure 3.11. In this example, observations related to a single track are shown (namely track 1). Based on a sequential estimation procedure (e.g., a Kalman filter), the last known position of the target is shown at time, t_i. Suppose a new observation, A, is received at time, t_j. Based on knowledge of the maximum velocity of the target and its maneuvering capabilities, a kinematic gate can be created (shown as the wedge shaped figure) to define the feasible region within which the tracked object could move from time t_i to time t_j. Only observations contained within the kinematic gate need be considered as feasible hypotheses (namely the hypothesis that a given observation may be related to track 1). Thus, in this example, the hypothesis that observation A is related to track 1 can be eliminated as a feasible hypothesis.

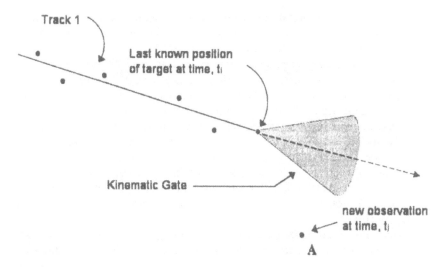

Figure 3.11 Concept of kinematic gating.

Knowledge of a target's maximum speed and maneuver capabilities provides the basis for eliminating consideration of an observation A as belonging to track 1, because the target could not have traveled to point A in the time interval $(t_j - t_i)$. Gates may be created based on various types of knowledge including the following.

- Target kinematic behavior;
- Parametric limits of a target (e.g., size, shape, emitter characteristics such as frequency range, pulse width, pulse repetition intervals), or spectral characteristics such as radar cross-section or infrared features;
- Probabilistic factors or models—in this approach, unlikely observation-observation or observation-to-track pairs are eliminated based on a gating criterion. For example, a Gaussian gate, S_{ij}, may be used. Feasible pairs are those that satisfy the criteria that S_{ij} is less than or equal to S_i, where d is the measurement dimension, dX_{ij} is the residual vector difference of report i and report j (projected to the time of the report), S_{ij} is the residual covariance of the pair, and S_i is the gating threshold selected per report. The error covariance is based on the jointed covariance of the sensor measurement process and the covariance of the predicted target position (e.g., based on a sequential estimation process).

If physical or behavioral information is available, then gating can be an effective way to reduce the combination of possible hypotheses to be subsequently processed via HE and HS processing.

3. *Logical templating techniques*: Logical templating techniques may be applied to extend the concept of gating to include both parametric limits as well as explicit logical conditions regarding temporal relations, causal relations, aggregations of entities to more complex entities, and other factors. Logical templating is especially applicable to those situations involving hypotheses related to complex entities, space-time patterns, complex target behavior [e.g., changing operation modes for an emitter, doctrinal based motion of a target (such as a search pattern)], or other situations in which *explicit* knowledge is available and potentially useful to reduce or eliminate possible hypotheses. A discussion of template based methods is provided by [9, 10];

4. *Knowledge-based methods*: Finally, a wide variety of knowledge-based methods may be used to combine parametric gates and logical conditions to perform automated reasoning to eliminate unlikely hypotheses. The key to these techniques (as is the key to logical templating) is to implement implicit and explicit knowledge to reason about hypotheses. Numerous techniques exist, including the following.

- Expert systems that represent knowledge via production rules, frames, scripts, or nets and provide the capability for automated inferencing to achieve conclusions based on input data and a prespecified knowledge base (see for example [11]);
- Fuzzy inference systems, which are similar to expert systems, except that they utilize fuzzy logic and fuzzy set theory to represent the explicit knowledge [15];
- The use of Bayesian belief networks to represent probabilistic information and conditional probability relationships. Pearl [12] and Neopolitan [16] describe the use of network-like structures to represent chains of probabilistic reasoning, accounting for conditional dependencies. Thus, these provide a basis for formally reasoning about hypotheses to determine feasibility;
- Monotonic logical grids [applied at the Navy Research Laboratory (NRL) during the 1980s], which organize the input data space in sequential logical hierarchical arrangements;
- Smart search trees: [8] describe the use of efficient search trees to reduce the time to identify feasible hypotheses. They define "pseudotracks" obtained by averaging over multiple target tracks to approximate the covariance distribution of the tracks within a cluster of tracked objects.

Whatever method is selected, the intent of the hypothesis feasibility screening is to produce a reduced set of hypotheses for subsequent evaluation by the hypothesis evaluation (HE) and hypothesis selection (HS) processes.

3.4 HYPOTHESIS EVALUATION

In the previous section, we described the hypothesis generation (HG) function, whose purpose is to generate an association matrix, which lists feasible hypotheses to *explain* the data. The association matrix, output from the HG function, identifies the feasible hypotheses and links the observations to hypotheses. In a two-dimensional association matrix for example, the rows of the matrix represent received sensor observations or data, and the columns of the matrix represent alternative hypotheses. In order to quantitatively compare these alternative hypotheses, it is necessary to compute a measure of the *likelihood* or *validity* of the alternate hypotheses. The purpose of the hypothesis evaluation (HE) function is to develop these metrics.

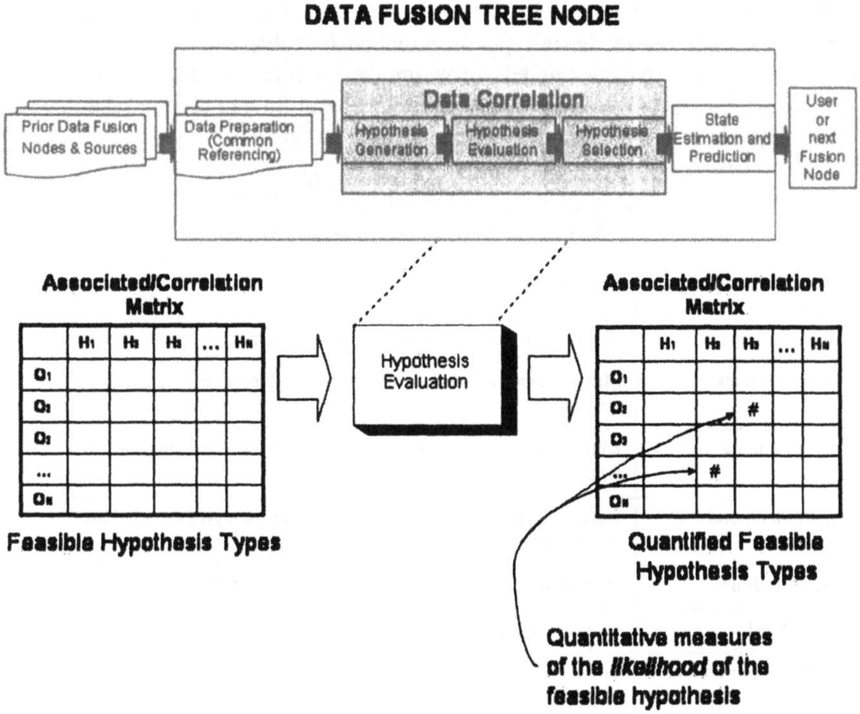

DATA FUSION TREE NODE

Figure 3.12 An overview of the hypothesis evaluation function for correlation.

For each batch of data to be fused (i.e., each fusion node), the hypothesis evaluation function within data correlation scores the feasible correlation (i.e., association) hypotheses in the association matrix. These HE scores are used by the hypothesis selection function (described in Section 3.5) to select the specific hypotheses to be used by the state estimation function. This role for the HE function is shown in Figure 3.12. The feasible hypotheses are input, for example, using the columns of a correlation/association matrix.

Initially, the association matrix input to HE may contain only values of zero and one. A zero in the (i-j) element of the association matrix indicates that the association between an observation, O_i, and a hypothesis, H_j, is not feasible or viable. Conversely, a value of one in the (i-j) element of the association matrix would indicate that the association between observation, O_i, and a hypothesis, H_j, is feasible. The HE function searches for feasible observation/hypotheses combinations, and computes a score to establish degree of feasibility or likelihood. Thus, the HE function, in effect, replaces the ones in the association matrix with a numerical score related to feasibility, probability, distance, or other type of metric.

The available metrics, which can be used to score the feasible associations, depend largely upon the character of the feasible hypothesis types. A summary of typical association hypothesis types is provided in Table 3.4. The remainder of this section characterizes the HE problem (in Section 3.4.1) and summarizes available solutions (in Section 3.4.2). A discussion of how to choose from among these alternate solution techniques is provided in Section 3.5.

Table 3.4

Summary of Association Hypothesis Types

Feasible Hypothesis Types	Description	Examples of Outputs
Report-to-report	Instantaneous measurements	Associated measurements over time
Report-to-track	Measurements and full state estimate	Reports associated with tracks
Track-to-track	Kinematics, attributes, and ID est.	Feasible track associations
Spatio-temporal	Data associated over time and space	Single sensor report associations
Multispectral	Noncommensurate attributes	Radar and IR feasible associations
Cross-level	Emitter, object, site, unit	Feasible weapon system associates
Multisite sources	Internetted and off-board tracks	Angle-only and like-sensor associates
Multiscenes	Alternative feasible world views	Separate scene hypotheses (MHT)
2-D set partitioning	Two mutually exclusive lists	Two radar scans associated (one each)

Table 3.4 (continued)

Summary of Association Hypothesis Types

Feasible Hypothesis Types	Description	Examples of Outputs
N-D set partitioning	N mutually exclusive lists	N radar scans associated (one each)
Labeled set covering	All data may be associated or false	2 IR tracks association with one radar track

3.4.1   Characterizing the Hypothesis Evaluation Problem

The hypothesis evaluation (HE) problem is characterized primarily by the nature of the input data, and to a lesser extent the nature of the output, and the measures of performance developed for HE. The selection of HE techniques is based on these factors. This section provides a brief description of the HE input data characteristics, the output characteristics, and common measures of performance.

3.4.1.1   Input Data Characteristics

The inputs to the HE function are the feasible associations developed by the HG function. These feasible associations include *pointers* or links to the corresponding input data (i.e., to the original sensor data or information from other data fusion nodes). The input data are categorized according to the available data type(s), the measures of uncertainty for the data, and commonality with other data being associated. Input data includes both recent sensed data and a priori source data. Each data type may have a corresponding measure of certainty (albeit, possibly highly ambiguous). Common types of data are listed as follows and summarized in Table 3.5:

- *Identity (ID) attributes* refer to the data that classifies the observed target or entity into one of a finite number of types. Examples include types of emitters, platform type (e.g., tank, aircraft, and enemy battalion), and weapon system level entities.
- *Kinematic attributes* refer to all geographic location, velocity, and higher derivative data, along with their corresponding uncertainties or covariance.
- *Parameter attributes* refer to all continuous, nonkinematic features that characterize an object or entity (e.g., radio frequency (RF), pulse repetition interval (PRI), pulse width (PW), length, or shape).
- *A priori sensor/scenario data* involves information needed to evaluate the a priori confidence of observations, or determine the default assumptions for association hypotheses (e.g., likelihood of missed detections, and false alarms), independent of the given data.

- *Linguistic data* includes language syntax and semantics.
- *Space-time patterns* refer to data for which there are known relationships in the spatial and temporal domains. Scripts/frames are often used to represent these spatio-temporal relationships between entities (e.g., emitters associated with a site/platform, communications sequences, and task force components).
- The *high uncertainty-in-the-uncertainty* refers to data for which the first order description of its uncertainty is not sufficiently accurate to enable hypothesis evaluation with other data. Thus a random set or possibilistic [e.g., fuzzy set, evidential, or modified Dempster-Shafer (DS)] representation is usually applied).
- *Unknown structure/patterns* refers to data for which HE techniques have not been derived or are not available. Thus they will need to be discovered on-line (e.g., pattern matching).
- *Partial data* includes defaults, unknown but existing, nonexistent, no information, and closed versus open world assumptions on knowledge bases. Conflicting data includes partially incorrect, wild points, miss-entries, incompatible, and stale data.
- *Nonparametric data* is data for which little or nothing is known a priori about the error statistics (e.g., positive mean, symmetric distribution, unimodal distribution).
- *Differing dimensions and types* refers to a batch of data that includes overlapping variables for some subsets of data that have different dimensions or different characteristics than that for other subsets.
- *Differing conditionals* refers to pieces of data that are related (i.e., correlated) but for which there is not a common Bayesian conditional breakdown.
- *Differing resolution data* occurs when one source observes more detail than another; however, the coarser source still has information necessary to be fused. (for example, high angular resolution IR or visible data associated with coarser radar data, which includes range information). This feature of input has more impact on HG and HS to allow many-to-one associations.
- *Error probability density function (PDF) information* provides a statistical representation of the error in the input data, which is important for probabilistic scoring of the association confidences.

Each of these input data categories impacts the class of HE techniques that are applicable. There are other inputs that affect the specific evaluation equations and the other functions in each fusion node. These include the following:

- Sensor and source descriptors such as probability of detection and false alarm;

- Data misalignment uncertainties both for kinematics and other observed parameters (e.g., RF and PRF);
- Object and scenario modeling parameters (e.g., motion, size, characteristics, military capabilities, and doctrine) and environmental conditions.

Table 3.5

Hypothesis Evaluation Input Data Characteristics

Input Data Categories	Description	Examples of Inputs
Identity(ID)/discrete attributes	Discrete/integer valued	IFF, class, type of platform emitter
Kinematics	Continuous-geographical	Position, velocity, angle, range
Parameters attributes features	Continuous-nongeographical	RF, PRI, PW, size, intensity, signature
A priori sensor/scenario data	Association hypothesis statistics	P_D, P_{FA}, birth/death statistics, coverage
Linguistic	Syntactic/semantic	Language, message content
Object aggregation in space-time	Scripts/frames/rules	Observable sequences, object aggregations
Possibilistic (fuzzy, evidential)	High uncertainty-in-uncertainty	Free text, a priori and measured object ID
Unknown structure patterns	Analog/discrete signals	Pixel intensities, RF signatures
Nonparametric data	Partially known error statistics	$P(R|H)$ only partially known
Partial and conflicting data	Missing, incompatible, incorrect	Wild points, closed versus open world, stale
Differing dimensions	Multidimensional discrete/continuous	3-D and 2-D evaluated against N-D track
Differing conditionals	Similarity of $P(A|B)$ & $P(C|D)$	Information about location and movement
Differing resolutions	Coarseness of discernability	Sensor resolution differences: radar + IR
Error density functions given	Probabilistic joint error distribution.	Covariance matrices; ID confidence trees

3.4.1.2   Output Data Characteristics

The characteristics of the HE outputs also drive the selection of the HE techniques. In most cases HE outputs an individual score to quantify the feasibility of a hypothesized association. However, in cases where HG outputs full scene (i.e., batched data) feasible associations, then the outputs represent full scene association scores. Table 3.6 describes the HE output categories. These categories are partitioned according to the output variable type: logical, integer, real, N-dimensional, functional, or none. Usually, the HE outputs are real-valued scores reflecting the confidence in the hypothesized association. However, some outputs provide a discrete qualitative confidence level (e.g., low, medium, or high) while others output multiple scores. These multiple scores could include upper/lower bounds for evidential reasoning, four numbers for trapezoidal fuzzy membership functions, one score per data category (e.g., parameter, ID, kinematic, a priori) or full-function scores with higher order statistics (e.g., fuzzy or random set membership functions).

For some batches of data, the hypothesis evaluation function does not compute a numerical score. Instead, only a yes/no decision on the feasible hypothesis is output. *No explicit association* refers to those rare cases where the data association function is not even represented (i.e., the data is only implicitly associated in performing state estimation directly on all of the data). An example in image processing is the estimation of object centroid or other features based upon intensity patterns, without first clustering or associating the pixel intensity data to the object.

Table 3.6

Hypothesis Evaluation Output Data Characteristics

Score Categories	Output Description	Examples of Outputs
Yes/no or pass through association	0/1 logic[no scoring]	High confidence only association
Discrete association levels	Integer score	Low/medium/high confidence levels
Numerical association score	Continuous real-valued score	Association probability/confidence
Multiscores per association	N-D [integer or real per dim]	Separate score for each data group
Confidence function with score	Score uncertainty functional	Fuzzy membership or density function
No explicit association	State estimates directly on data	No association representation

3.4.1.3   Performance Measure Characteristics

The HE techniques are selected to maximize the probability of meeting the data fusion system performance (i.e., both cost and accuracy) requirements. The performance measure characteristics significant to HE technique selection include the following:

- Software life-cycle cost and complexity (i.e., affordability);
- Processing efficiency (e.g., computations/sec/watt);
- Evaluation (score) accuracy and consistency (i.e., sufficient for mission);
- Ease of user adaptability (i.e., operational improvements);
- Ease of training on sample data (i.e., training set requirements);
- Ease of self-coding/self-learning (i.e., learns how to score on its own);
- Robustness to errors/miss-modeling (i.e., graceful degradation);
- Result explanation (i.e., responds to queries to justify score); and
- Hardware timing/sizing limitations (i.e., requires significant processing or memory to meet timeliness requirements).

3.4.2   Overview of Hypothesis Evaluation Techniques

The basic mathematical frameworks for making association decisions in the face of all major types of uncertainties (e.g., randomness, subjective linguistics, probabilities, partial knowledge, and fuzziness) have reached a sufficiently mature stage for proceeding to develop rational data association scoring methods. The scoring solution techniques for HE can be summarized as follows (see Table 3.7).

1. *Probabilistic computational models* attempt to develop statistical based measures of association feasibility. These include the traditional and widely applied *likelihood, a posteriori* [including Bayesian nets (i.e., influence flow diagrams)], Bayes risk, hidden Markov models, and probabilistic finite state machines, especially for fusion levels 2/3, and *chi-square integral of the tail* (e.g., used in statistical hypothesis testing approaches). Other probabilistic models include conditional event algebra, nonparametric methods, and information theoretic methods described as follows.

    - *Conditional event algebra* brings contingent evidence (rules) under the umbrella of probability theory to treat incomplete or linguistic data. This permits the assignment of a probability weight to the result of "firing" (i.e., executing) a rule and deductions on the basis of such partial evidence. Conditional algebra also provides a measure of similarity between differing conditional events [17, 18].

- *Nonparametric methods* are used when the joint probability density functions (pdfs) are unknown except possibly for general properties such as symmetry (leads to Wilcoxon test) or positive mean (leads to sign test based on Neyman-Pearson criterion) [19]. Nonparametric methods are also known as distribution-free methods and they strive to estimate the pdf. There are four major types of methods: histogram, kernel method, k-nearest-neighbor, and the series method [20].

- *Information theoretic methods* are used in cases of high uncertainty where the joint error distributions are not known and nonparametric methods are not sufficient. The score is the amount of information that would result from the hypothesis [21].

Table 3.7

Classification of HE Solution Techniques

Solution Techniques	*Description*	*References*
Ad hoc	**Non-rigorous, application-dependent**	[22, 23]
Probabilistic	**Based in probability (Bayesian) theory**	[12]
Likelihood	P(R|H) for reports(R) given hypothesis (H):new, false, associated	[24]
Bayes a posteriori	P(H|R)=P(R|H) P(H)/P(R)	[25-28]
Nonparametric (NP)	Histogram, kernel, k-nearest neighbor, and series methods	[19, 20]
Chi-square tail-test	Fix $P(d_2\|H_1)$ then test to reject H_1: Gaussian statistics implies integral of the tail of chi	[29]
Conditional event algebra	Measure similarity of data with differing conditionals: P(A|B)&P(C|D)	[30, 31]
Information/entropy	Measure information [negative entropy] in result	[21]
Possibilistic	**Simpler t-norms for uncertainty-in-uncertainty 2nd order data combination**	[29, 32]
Evidetial/modified Dempster-Shafer	Uniform distribution [belief-plausibility]	[33, 34]
Fuzzy sets	Set membership function on score confidence	[35, 36]
Logic/Symbolic	**Utilizes symbolic data with logical inferencing and semantic representation**	[37]
Scripts/frames	Spatio-temporal and procedural knowledge represented	[29]
Expert system using rules	Ease incorporation of user knowledge and commands with blackboard	[38, 39]

Table 3.7   (continued)

Classification of HE Solution Techniques

Solution Techniques	Description	References
Case-based reasoning	To automatically learn from user actions	[40]
Semantic distance	Score similarity of semantics representing partial information (e.g., using power domains, anti-chains, or-sets, partitioned rep's)	[41-43]
Neural Networks	**Efficient computations per watt massively parallel pattern recognition**	[44, 45]
Unsupervised learning	Clustering via Hebbian, competitive, self-organizing maps	[46]
Feed-forward supervised learning	Trained data-driven nonlinear regression	[47]
Recurrent supervised learning	Spatio-temporal pattern recognition	[48, 49]
Unified Computational Models	**Common math framework treating diverse knowledge with high uncertainty**	[50, 51]
Random set theory	Uses ordering relations to express intensity of membership; combines fuzzy, DS, Bayes	[32, 35, 52]

2. *Possibilistic computational models* (e.g., fuzzy sets, evidential treatment of uncertainty with soft (i.e., approximate) knowledge combination rules using hard (i.e., precise) models. Evidential scoring schemes use upper and lower bounds that are equivalent to a uniform distribution for the uncertainty-in-the-uncertainty. By contrast, fuzzy scoring enables the user to define full membership functions to express the *second-order* uncertainty [36]. The knowledge combination rules are approximations (e.g., t-norm and co-norms) to the rigorous second-order Bayesian rules that significantly simplify the second-order correlated random variable issues.

3. *Logic and symbolic methods* are used when explicit knowledge is available. These methods are applied, for example, for cases where expert knowledge is available and can be represented via *rule-based expert systems, scripts/frames, semantic distance, and case-based reasoning*. Several techniques have been developed to handle the increasing levels of uncertainty, incomplete data, inconsistent levels of detail, conflicting, and partial data. Additional techniques to more consistently represent the data includes the following.

- *Rough sets:* This approach combines knowledge with various levels of detail in representation (i.e., coarseness of discernability), which is based upon equivalence relations describing partitions of classes of discernible objects [53]. By comparison using an image processing analogy, rough sets treat the size of the pixels whereas fuzzy sets treat the gray levels.

- *Partitioned representations:* This approach is utilized for efficient maintenance of contingent (possible worlds) and modally or temporally qualified information [43].

- *Mixed power domains:* This method is used to represent partial information about a set of data using a pair of sets, upper (Smyth) and lower (Hoare) power domains [41].

- *Anti-chains and or-sets:* These are used for efficient representation of partial information of all kinds within a common semantic framework providing algebraic and analytic tools [42].

4. *Neural networks (NNs)* allow incorporation of implicit information, such as the structure of patterns in the data. These provide efficient, massively parallel, nonlinear pattern recognition especially in the analog pulse-stream NN chips at a tera connection per second per watt [45]. The inner product NN chips are currently trainable based upon sample solutions (e.g., backpropagation, learning vector quantization, neo-cognitron, counter propagation) to be self-coding [54]. When no training scores are available, unsupervised (e.g., Hebbian, competitive, and Kohonen) learning can be applied. Recurrent inner-product NN chips have feedback, which is useful for spatio-temporal pattern recognition. They also can be hard-coded to score and search association hypotheses [55]. In fact most NN solutions are not just for HE but address the whole data association problem.

5. *Unified computational models combine* diverse knowledge (e.g., both numeric and symbolic) within a general decision framework in which various types of uncertainty coexist. This includes scoring data with high uncertainty-in-the-uncertainty (e.g., Fuzzy Sets and Dempster-Shafer) within a common mathematical framework (e.g., extracting a single random set of expected values from among a set of possible expected values using for example Choquet integrals or entropy [52, 56]). Another unified model is o-theory, which uses operator-belief theory to bridge the gap between fuzzy and evidential theories [51].

6. *Ad hoc methods* may be used for cases where no statistical data or explicit knowledge is available. These are differentiated from approximations to the methods previously described and are applied to achieve working solutions for specific applications. Since ad hoc methods do not have a rigorous basis, ad hoc scores need to be retailored as changes in the problem occur. Also these methods have a hard time achieving scoring consistency as new data types are addressed. Examples of ad hoc methods include figure of

merit or measure of correlation [22], and confidence factors such as those used in Mycin for linguistic rules.

It can be seen that there are many different techniques that are available to quantitatively assess the identified hypotheses or feasible associations. The reason for this wide variation in quantitative methods is due to the wide variation in data and wide variation in assumptions about sensors, targets, and the observing environment. A discussion of how to select from among these techniques is provided in Section 3.6.

3.5 HYPOTHESIS SELECTION TECHNIQUES

The final step in data correlation involves the actual selection of the best set of hypotheses (or possible assignments of observations-to-observations, or observations-to-tracks) that best fit the observed data. The process involves searching through the assignment matrix supplied by the hypothesis generation (HG) function, and using the metrics provided by the hypothesis evaluation (HE) function to compare different interpretations of the data. Selection of the *best* set of hypotheses to *explain* the data constitutes an optimization problem. The concept is illustrated in Figure 3.13. This section provides a discussion of the characterization of the HS problem and describes techniques for hypothesis selection. A detailed discussion of how to select from among these alternative methods is provided in Chapter 10.

Once the initial clustering and gating of the HG process is completed, and the distance/closeness metric selection and fundamental approach to have been completed, the overall correlation process has reached a point where the *most feasible* set of both multisensor measurements and multisource inputs exists (filtered, in essence, by these preprocessing operations). The HE process has also scored this most feasible set. The final task is to allocate or *assign observations* to the appropriate estimation processes. The estimation process develops improved predictions of the states of interest. This is the process called *hypothesis selection*. The hypothesis set comprises all the possible/feasible assignment *patterns* (set permutations) of the inputs to the estimation and/or data management processes. Any single hypothesis is one of the set of feasible assignment patterns, which together comprise a feasible *scene*.

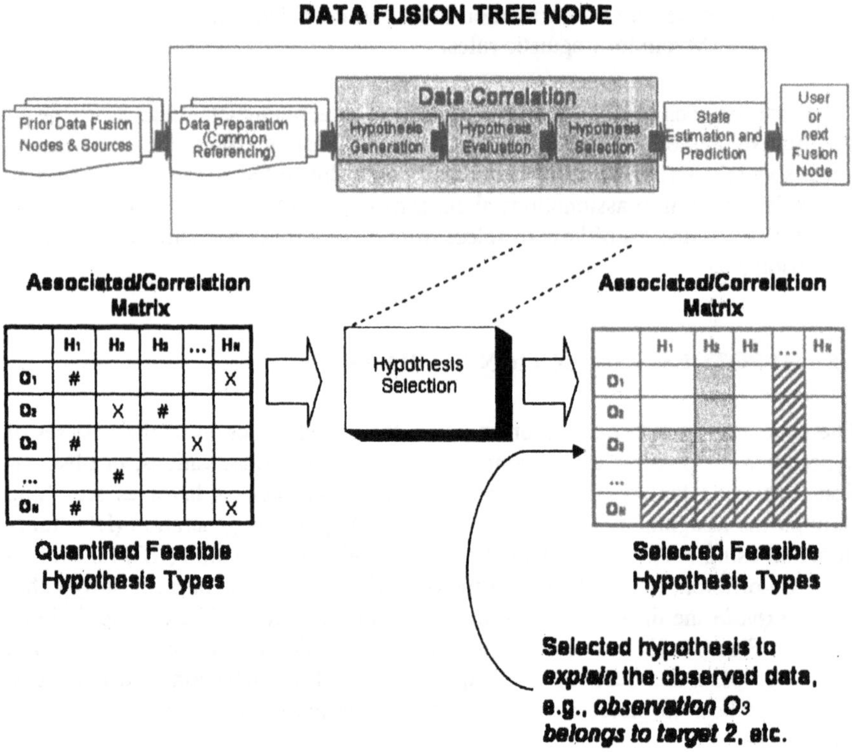

DATA FUSION TREE NODE

Figure 3.13 An overview of the hypothesis selection function for correlation.

While we focus here on position and identity estimation (target tracking and identification), the hypothesis generation—evaluation—selection process is also appropriate to the estimation processes at higher levels of abstraction. These states are related to so-called level 2 and level 3 fusion processes in the JDL data fusion model, which estimates *situational states* or *threat states*. For these cases, the correlation problems are more complex because the problem structures involve many-to-many type correlations. This may involve aggregating individual units, events, and activities to weapon systems and behaviors across a hierarchy. In these cases we are interested in object aggregations and in behaviors (events and activities). As a result, the state estimation processes, unlike the highly numeric methods used for level 1 estimates, are reasoning processes embodied in symbolic computer-based operations.

For the level 1 fusion processing described here, the *input to the hypothesis selection process* is a matrix (or matrices). The typical dimensions are the indexed input data/information/measurement set, and the indexed state estimation processes (e.g., a priori tracks). Other hypotheses may include allowed ambiguity states or any *other* states or conditions, to which the inputs may be assigned (e.g., false alarms). For the basic case involving the association of observations to tracks, the two dimensions of the association matrix are the indexed measurements and the indexed position state estimation processes. In any case, the matrix/matrices are populated with the *costs* of assigning any single measurement to any single estimator (the evaluation metrics determined in the HE step). The cost function may be simply the association scores computed by HE, or a newly developed cost function may be specifically defined for the hypothesis selection step. The usual strategy for defining the optimal assignment (i.e., to *select the optimal hypothesis*) is to find that hypothesis (best assignment pattern) with the lowest total cost of assignment. There are various conditions under which such matrices develop. Two frequently discussed cases in the literature are: (1) when the input systems (e.g., sensors) are initiated or turned on, and (2) when we are maintaining, in a recursive or iterative mode, the dynamic state estimation processes of interest, following the evolving dynamics of the surveillance volume of interest.

When the input systems are *initiated,* the correlation we desire is among an initial set of measurements, in order to initiate the state estimation processes or otherwise process the data. That is, we seek sets of interrelated inputs from which to initialize the various state estimation processes. By enabling any state estimation processes, we in effect assert that a target or object exists that is the *cause* of the inputs. Thus in the beginning, the matrices just discussed have indexed measurement/input values along *both* dimensions of the matrix, and we seek measurement/ input *tuples* (sets of specific inputs from each of the input sources) from which we can then start the state estimator processes. Such operations are typically labeled *measurement-to-measurement* correlation. After initiation, the correlation processing seeks to *maintain* the ongoing state estimation processes, in which case the matrix dimensions are those discussed above. These operations are typically called *measurement-to-track* correlation. In these cases we are in effect correlating *predicted* measurements or inputs (generated by the state estimation routines) with the *actual* values collected or measured by the input sources. The generic forms of these assignment or association matrices are shown in Figure 3.14.

INPUT		1	2	3	N
INPUT	1	C11	C12	C13	C1N
	2	C21	C22	C23	C2N
	3	C31	C32	C33	C3N
.	.				
	.				
	.				
	N				CNN

ESTIMATOR		1	2	3	N
INPUT	1	C11	C12		ETC
	2				
	3				
	.				
	.				
	.				
	N				

a) Input-input or measurement-measurement matrix   b) Input-estimator or measurement-to-track

Figure 3.14 Generic forms of assignment or association matrices (Cij = assignment cost).

These *assignment matrices* may still involve ambiguities in how to best assign the inputs to the state estimators, in spite of the careful preprocessing of the hypothesis generation and evaluation steps. For example, the cost of assigning an input to one or more estimators (or any other hypothesized causal factor) may be reasonable or allowable within the definition of the cost function and its associated thresholds of acceptable costs. If this condition exists across many of the inputs, then the identification of the total lowest-cost assignments of the inputs may be a complex problem, because there will exist many alternative patterns within which to allocate the measurements. The *central problem* to be solved in hypothesis selection is that of defining a way to select that full scene hypothesis, with minimum total cost, out of all those that are feasible/permissible. Often, this involves large combinations of possibilities and leads to a *problem in combinatorial optimization*. In the domain of combinatorial optimization, this problem is called *the assignment problem* and is applicable to many cases other than the measurement assignment problem we discuss here. Because of its wide applicability, this problem has been well-studied by the mathematical and operations research community, as well as by the data fusion community.

3.5.1 Defining the Hypothesis Selection Space

The hypothesis selection (HS) process, is the process most removed from the direct effects of the application domain. It is largely influenced by the nature and structure of: (1) the *basic formulation*, in the sense of deterministic or probabilistic, of how the association/correlation processing will be approached; (2) the *quality and speed* desired from the solution; and, for certain types of solution methods, (3) the *structure and nature* of the *assignment matrix*. The parameters and other characteristics of the problem domain (e.g., factors such as inter-target spacing) *implicitly* influence the correlation (and HS) approach a fusion node designer may take. However, these factors do not *explicitly* influence the design of an HS solution. For example, whether the basic approach to HS is probabilistic or deterministic has a great influence on the nature of the HS processing and the solution to the data assignment problem. The choice of which method to employ is influenced both by aspects of the *operational problem domain* (e.g., few-track and high-clutter problems have proven effectively soluble with a probabilistic framework) and partially upon the existence of those target/clutter characteristics that the designer would make the probabilistic/deterministic choice. However, the features of specific assignment solutions (e.g. computational speed) may also influence the basic approach to assignment processing, and these (implicit) aspects would also influence the selection of an approach. Equally, the rationale for the formulation of the feasible-hypothesis structure as a set-partitioning or set-covering framework by the HG process also significantly influences the HS approach. The set-partitioning approach basically divides the allocable sets of inputs into mutually exclusive sets, whereas the set-covering approach permits shared allocations of specific inputs within its feasible-set definition.

3.5.1.1 Basic Problem Formulation

The HG+HE+HS design processes can have potentially significant coupling effects. Our recommended approach is to *decouple* these process designs so that they can be confidently engineered in isolation. In the case of making a set-wise decision, the HG and HS designers may choose to collaborate on an HS approach by reviewing the specifics of the matrix describing the overall set of feasible hypotheses. In typical applications, it is expected that the data are readily separable, which permits a set-partitioning approach. In other cases, some data may not be so easily separable, leading to the set-covering formulation.

There are also special cases of assignment problem and solution formulation that have been researched in the operations research community but, that have seen limited application in the data fusion domain. The best example of such a special case is that of quadratic assignment. Quadratic assignment is applicable to the case where target behaviors are coupled (e.g., when the assignment to a given target is influenced by another target). Surprisingly, cases and approaches of this type do not appear to have been studied in the data fusion community. However, the operations research and other mathematical research communities have developed a range of solutions for this class of assignment problems.

3.5.1.2 Quality and Speed of Solution

The notions of desired *quality and speed* of an assignment problem solution are among the primary factors influencing the choice of approach. These are influenced by the *dimensionality* of the problem. Dimensionality is directly related to the quantity of data and the number of *causal factors* to which the data is to be assigned. The nature of the dimensionality of the selected approach partitions the assignment solutions into two classes, one in which the solutions are locally optimal and quick, and one in which the solutions are global but suboptimal and slower. Given the dimensionality however, the choices regarding quality (i.e., optimality) and speed of solution are themselves influencing factors on the choice of solution. This is because: (1) for the class of locally optimal solutions (those that only consider a single *batch* or scan-worth of data), all solutions are optimal, and the choice is driven by computing speed; and (2) in the case where *global* but suboptimal solutions are available (such solutions are not truly global but consider multiscan ensembles of data), there are choices regarding both quality (degree of suboptimality) and speed. The notions of local and global optimality also signify that the local solutions do not (usually) have a retrospective characteristic (i.e., they yield a solution optimal for the limited data set considered, which is usually irrevocable). The *global* techniques often permit re-examination of prior assignments over a sliding window of multiscan width, thereby achieving a type of retrospective capability.

3.5.1.3   Structure and Nature of the Assignment Matrix

For the two-dimensional (2D) problem class, the specific structure and characteristics of the assignment matrix can also influence the solution choice. These structural and other characteristics however are frequently the result of engineering choices prior to the HS engineering process. For example, the degree to which the assignment matrix is sparse or populated influences the specific algorithm of choice but sparsity is often the result of various pre-HS processing design choices (e.g., gating techniques) as well as domain parameters (e.g., target spacing and sensor resolution).

3.5.1.4   Input Data Characteristics

Although we have said that the starting point (problem space) for the HS process begins with the assignment matrix, the real starting point for the engineering of an HS solution is reflected in the HG+HE thinking and design choices. These choices lead to the construction of the matrix that specifies the feasible hypotheses of possible allocations, and relationships among the data, estimates, database parameters, and the scores for such feasible relationships. Nominally, the feasible hypotheses are the result of HG process engineering, and the scores result from HE process engineering. For example, consider the case in which the HG designer has chosen a single-scan, set partitioning approach. (This choice may have been influenced by the presence of relatively few targets and, say, good sensor resolution.) The HG designer provides the matrix of possible hypotheses as seen in Figure 3.15.

The HE designer, in turn, selects and computes the hypothesis scores to fill out the matrix with the scores as noted. In this matrix, the 1 codes simply designate the feasibility of an association between a measurement and a track (i.e., a state estimation process). The HG process designer on the basis of the criteria discussed in Section 3.2 generally determines feasibility. However, there are often ambiguous cases [e.g., measurement 1 can feasibly be assigned to either estimator E1 (hypothesis H_a) or to estimator E3 (hypothesis H_c)]. The HE-assigned scores of Sa and Sc measure the viability of each hypothesis respectively.

Input/Estimation Process	Hypothesis Ha	Hypothesis Hb	Hypothesis Hc	Hypothesis Hd	....
Input O1	1		1		
Input O2		1			
Input O3				1	
....					
Estimator E1	1				
Estimator E2		1		1	
Estimator E3			1		
....					
Hypothesis Score	Sa	Sb	Sc	Sd	....

Figure 3.15 Feasible hypothesis matrix.

These matrices provide the framework for the construction of the *assignment* matrices (see Figure 3.16) in which we set up a matrix describing the input-to-estimator relationship more directly.

Input/Estimator	E1	E2	E3	....
O1	Sa		Sc	
O2		Sb		
O3		Sd		
....				

Figure 3.16: Assignment matrix.

While more complex problems lead to more complex setup issues for these matrices, the basic concept remains the same. It can be appreciated that the specifics of the structure and nature of the data in these matrices would affect the specifics of the solution method [e.g., the sparsity of the matrix, (i.e., the degree to which it is populated by the score values)] would clearly affect the relative difficulty of a solution. The central question, as discussed above, is how to find, of all feasible assignment patterns, the lowest cost, optimal result.

3.5.1.5   Output Data Characteristics

The output of the HS process is simply the list of assignments, or an indexing pattern, of all inputs to estimation processes that will employ the assigned data to produce the fused estimate of the surveillance condition or situation. This mapping may not be one-to-one, since the *probabilistic* approach assigns, probabilistically, many inputs to many estimators, as does any *multi-assignment* solution (i.e., in these approaches the data are shared across estimators). Many real-world applications of an assignment solution impose an adjunct-type requirement on the solution—that it *consumes* or assigns all the inputs(i.e., that no inputs are left unassigned). This requirement can obviously affect the nature of the solution. If this requirement is relaxed, then some type of background processing must re-examine the unassigned data to see if new opportunities for possible assignments arise. This induces a retrospective processing and assignment requirement that can be complex if not considered as part of the up-front design process (as is done in the ND approaches).

3.5.1.6   Performance Requirements Measure Characteristics

Throughout this chapter, the correlation process is viewed from a systems perspective. In this view, all subprocesses, such as assignment processing, generally must satisfy or be consistent with the top-level requirements for overall process performance. This view emphasizes the assertion that fusion process (and subprocess) requirements are *derived* requirements that must be traceable to higher levels in the requirements tree. Thus, we repeat here those same type performance requirements as levied on HG and HE:

- Satisfaction of derived operational requirement for assignment processing accuracy and timeliness;
- Software life-cycle cost and complexity;
- Processing efficiency;
- Ease-of-user adaptability;
- Robustness to errors in input;
- Traceability of processing (explanation);
- Satisfaction of derived hardware timing and sizing constraints.

There are, however, specific measures and points of view associated with the details of evaluating assignment problem solution techniques, which are described in Chapter 10. These too must be traceable to the higher-level performance requirements as notionally described above.

3.5.2   OVERVIEW OF HYPOTHESIS SELECTION TECHNIQUES

Methods for hypothesis selection are drawn from the field of mathematical optimization. These methods search the assignment matrix and choose a set of hypotheses that best fit the observed data. The usual descriptions of hypothesis selection (HS) processing and solution techniques, found in the technical literature, are largely confined to level 1 applications. Here, we expand this view to include applications at levels 2 and 3, and to include a more comprehensive view of the *solution space* for assignment problems. These other applications are assigned first, followed by a description of HS techniques for level 1 processing.

3.5.2.1 Assignment Processing at Levels 2 and 3 (Situation and Threat Assessment)

Fusion processing at levels 2 and 3 represents a significant shift from level 1 processing. While bounded by sensor resolution and other constraints, the focus at level 1 is primarily on singular objects, their kinematic characteristics and their identity. At levels 2 and 3, the intent is to *aggregate* the singular level 1 objects into ensembles with meaning (in an operational sense), sometimes called *order of battle* type analysis, and to *understand the behaviors* of both individual and aggregated objects. In what follows, we will briefly discuss one program that has attempted such a formalism. Another way to characterize this processing is as a search for *dependency-constituency* relationships among objects and events and activities. Underlying these processing goals is a need for a symbolic as well as numeric approach to fusion processing (i.e., primarily but not exclusively). Thus, the fusion processes include methods for automated reasoning.

What distinguishes the assignment problem for level 2 and 3 from those at level 1? First, there is the issue of combinatorics, which involves more associations of the many-to-many type, tending to push the solution space in the direction of set covering and ND type approaches. Conceptually this is similar to the dense and closely spaced target problem type at level 1, where the solutions would be of similar nature. Another distinguishing feature however, is that the *distribution of evidence by type* will be much broader than at level 1. This follows by definition since levels 2 and 3 are at higher levels of abstraction, and subsume evidence categories that are descriptive rather than numeric. The impact of this is to push the scoring strategies of HE into the realm of those methods where priors and joint pdfs are (most likely) by and large unknown, and/or into the realm of neural or symbolic approaches.

There are various impacts of this to the assignment or HS function. On one hand, if the scores are essentially single, crisp, numerical values (real or integer), then the factors influencing the choice of a particular solution method would follow the rules for a level 1 solution. In essence, the assignment matrices would have characteristics similar to those for level 1. On the other hand, if the basic scores are *semantically different* than implied by single, crisp values [e.g., the two-valued metrics of the evidential (Dempster-Shafer; D-S) approach or the fuzzy values of a fuzzy set or fuzzy logic approach], then nontraditional methods are required. These assignment techniques range from the mathematically complex to the logic-based types, where the best assignments derive from the application of a reasoning procedure.

3.5.2.2 Assignment Processing at Level 1 (Object)

The basic methods for assignment problem solution for level 1 type problems derive from the broad array of solution techniques, which generally fall into the category of *combinatorial optimization* methods. This term usually signifies optimization problems, which have a finite number of solutions. The following are examples of combinatorial optimization problems: the minimal cost network problem, the transportation problem, and various facilities layout problems (e.g., the so-called traveling salesman problem). When the problems are of relatively small dimensions, an optimal solution can be obtained by exhaustively enumerating all possible combinations of assignments and then selecting the assignments with the lowest total cost or score. Thus, there is a class of solutions that involve *explicit enumeration*, and these yield a globally optimal solution.

For problems of moderate size, many methods exist that avoid exhaustive enumeration yet achieve global optimality. These have been successfully applied to problems of the type we are interested in here. These methods largely derive from the area of mathematical programming including the integer programming approach. One way to constrain the size of a problem is to limit the amount of data considered at any one time. For the dynamic problems of interest here, this is a typical way in which the scope or dimension of a problem is confined (e.g., by limiting the data considered at any one time to a single scan) so that methods can be employed to achieve a *locally* optimal solution. [The term *local* is used here to mean the single-scan-type solutions (i.e., those that consider a small data set, thereby yielding a solution that while it can be called optimal, is applicable to this *local* data set). The term *global* is used to mean the n-scan-type solutions that generate an optimal solution for a wider, *global* data set. The *true* optimal is that solution that treats all the data but for which, in a search space, there may be *local* minima for the objective function.]

In problems of relatively large size, the basic assignment problem formulation leads to an *NP-hard* problem for which reasonable compute-time solutions may not be available. Growth in the magnitude of the problem can result from seeking an approach capable of processing *multiple* scans of data (or data sets from multiple sensors). These more difficult problems search for an optima (namely a minimum or maximum value of the objective function) of an objective function that may exhibit many local optima. The dilemma for any candidate solution is twofold: (1) how to know where it is in relation to the *global* optimum and (2) how to efficiently achieve a near-optimum solution in reasonable compute time. These techniques use the concept of a *neighborhood*, which is a set of feasible solutions that can each be easily reached by a simple operation on the data (called a *move*, such as a pairwise interchange). The efficacy of different techniques for *neighborhood searching* (in the vicinity of a local optimum) is a trade-off issue in their selection. When these simple moves do not achieve a local minimum sufficiently close to the global, other methods, typically employing implicit enumeration are frequently applied (the most frequently cited method being Branch-and-Bound). These concerns for larger problems also lead to the consideration of heuristic solution methods. These methods seek good solutions (knee-of-the-cost/benefit-curve) in a reasonable compute time but offer no guarantees about either feasibility or optimality [57].

Hypothesis selection methods for level 1 fusion processing fall into three basic classes: (1) linear and nonlinear mathematical programming, (2) dynamic programming and branch-and-bound methods as part of the family of methods employing implicit enumeration strategies, and (3) heuristics. While this is a useful generalization, it should be realized that the assignment problems of the type experienced for level 1 data fusion problems arise in many different application areas. Hence, there have been many specific solution types developed over the years, making difficult any attempts for broad generalizations. Nevertheless, this section treats the solution domain in this way, and the following sections expand on each type mentioned.

3.5.2.3   Linear, Nonlinear, and Integer Programming Methods

Linear and nonlinear, and integer programming methods for assignment problems include the extensive array of linear programming methods, including simplex, successive shortest path, relaxation, and primal-dual approaches (e.g., the so-called *auction* algorithms representing near state-of-the-art today); see [58] as a typical reference. These methods often employ sophisticated search methods by which a structured path toward optimality is developed (e.g., steepest descent and coordinate descent methods).

Table 3.8 provides an overview of the conventional methods used for the most frequently structured versions of assignment problems for data fusion level 1 (i.e., deterministic, 2D, set-partitioned problem formulations). These are solutions for the linear case (i.e., linear objective or cost functions) and linear constraints. The linear case is the most frequently discussed case in the literature and the solutions and characteristics cited in Table 3.8 represent a reasonable benchmark in the sense of applied solutions.

Note that there are formulations of the assignment problem that can lead to nonlinear formulations such as the quadratic assignment. Additionally, the problem can be formulated as a *multi-assignment* problem as is done in the set-covering approach. However, these structures usually result in linear multi-assignment formulations if the cost/objective function can be treated as separable. Otherwise the problem will typically result in a nonlinear, quadratic-type formulation (e.g., [59]). Finke [60] is a source from which to extract solutions to the quadratic problem. The key work in multi-assignment structures for relevant fusion type problems is [59], whose authors form a solution analogous to the auction-type approach.

The multitarget tracking problem, as one of the key level 1 data fusion problems, can be cast into an integer programming (IP) formulation (e.g.,[61]). This formulation may involve solely integer variables, in which case it is typically called a *pure* IP problem, or it may involve a combination of continuous and integer variables in which case it is called a *mixed* IP problem. Many applications, including those discussed here, can also be treated with integer variables restricted to just two values, typically 0 and 1. This transformation takes the conventional Bayesian decision problem of finding an assignment of the inputs that minimizes the Bayes risk and converts it to a 0-1 IP problem. In Morefield's original work, the problem becomes a set-packing problem if false alarms are permitted and a set-partitioning problem if they are excluded.

Since we have used several *set*-related terms in this section, some remarks on their distinctions are in order:

- *Set covering:* Conceptually, these problems result from situations where a given set (here the set of all inputs and state conditions) is to be *covered* by a collection of all possible subsets such that the union of elements in the (nonexclusive) subsets are inclusive of the elements in the given set. A cost for each potential subset's inclusion in the solution set is involved, and the optimal solution is the minimum cost solution. These criteria lead to certain properties of the optimization problem setup (see [62]) that cast it as a pure integer-programming problem, which we thus identify as a set-covering problem. Generalizations exist that derive from relaxation of these proper-ties, leading to *weighted* set covering and *generalized* set-covering formula-tions. This type formulation allows sharing of measurements and can lead to the multi-assignment formulation. Solutions are described in [63].

Table 3.8

Frequently-Cited Level 1 Assignment Algorithms

Algorithm and References	Applicability (Problem Space)	Processing Characteristics
Hungarian [64]	Square matrices; optimal pair-wise algorithm	Primal-dual; Steepest descent
Munkres [65]	Square matrices; optimal pair-wise algorithm	
[66]; see also [67]	Rectangular matrices; optimal pair-wise algorithm	B-L faster than squaring–off method of Kaufman
Stephans-Krupa: [68, 69]	Sparse matrices; optimal pair-wise algorithm	
[70]	Sparse matrices; optimal pair-wise algorithm	Appears to be the fastest of the traditional methods; S-K second fastest to JVC; sparse Munkres third fastest; JV is augmenting cycle approach
Auction types [71]	optimal pair-wise algorithm	Primal-dual, coordinate descent; among the fastest algorithms today;

Table 3.8   (continued)

Frequently-Cited Level 1 Assignment Algorithms

Algorithm and References	Applicability (Problem Space)	Processing Characteristics
		parallelizable versions developed; appears much faster than JV algorithm (as does JVC)
Primal simplex/alternative Basis [72]	Applied to relatively large matrices (1,000–4,500) nodes; Optimal pair-wise algorithm	Moderate speed for large problems
Signature methods [73]	Optimal pair-wise algorithm	Dual simplex approach

- *Set packing*: This problem formulation is subtly different than for set covering. Here we seek that collection of subsets (they again each have a cost or value) that maximizes the number of elements that we *pack* into the solution set, while at the same time optimizing the cost or value, but without overlap in the elements. For conventional tracking problems, the goal is to assign the maximum number of (distinct) measurements and achieve minimum cost as well. Morefield's solution is the *classic* application of this problem model.
- *Set partitioning*: This problem combines the set-covering and set-packing criteria (i.e., we are to cover the elements of the given set but to do so without overlap, and at minimum cost). For the conventional tracking application, we explicitly use all measurements but without overlap (i.e., without multi-assignment or sharing). See [74] for a typical solution technique for the set-partitioning case.

Interior point methods: Large-scale assignment problems have also been attacked with what is called the *interior point method* (e.g., [75]). This approach, developed by Narendra Karmarkar [76], is fundamentally different from the classic simplex method for LP. The distinction is that this method proceeds from the interior of the polytope, which is formed from the problem constraints, to find a lowest cost solution, rather than along the extreme points and edges as does the simplex method. While the algorithm exhibits polynomial-type behavior, many of the reported research projects have employed supercomputers for the calculations (e.g., [77]).

Dynamic programming and branch-and-bound methods: Dynamic programming and branch-and-bound methods attack the problem in different ways. The dynamic programming approach could be more explicitly called a solution *strategy* rather than an algorithm, and employs a staged optimization policy for solution development, whereas the branch-and-bound method represents a genre of solution types especially using implicit enumeration as a means to evolve the solution.

Dynamic programming (DP): Dynamic programming is difficult to characterize because it is really a strategy of solution, and the specifics vary considerably in any given application (see [78]). DP first divides the problem into stages or subsets of the problem space (e.g., a set of arcs in a large graph network). Each stage has states associated with it that are the values of the variables of interest, and decisions are made at each stage that are dependent only on the state within the stage (i.e., the processes are assumed memory-less). When a decision is made (e.g., the assignments for one scan of data) a *return* (cost or value) is computed, which is the current optima up to the given stage. By optimizing the return values within a stage, the optimal solution to the *local* or stage-wise problem is determined, and the solution moves to the next stage. DP solutions thus involve recursive equations for the returns between stages. There can be forward recursions and backward recursions depending on the policy or decision flow for problem solution defined by the analyst.

Branch-and-bound techniques: Since we are dealing with problems that have finite solution boundaries, it is natural to consider some type of enumeration procedure for finding an optimal solution. In many cases, however, the number of feasible solutions can still be very large, since the problems tend to grow factorially based on the number of variables and the number of values the variables can assume. However, a practical approach might be produced, if we could define an enumeration strategy that does not explicitly enumerate all feasible solutions, but instead implicitly eliminates large groups of feasible solutions without ever evaluating their cost functions. Such implicit enumeration techniques have been defined and include both the dynamic programming approach and branch-and-bound (B&B). Technical descriptions of the B&B method can be found in many texts on either mathematical programming or combinatorial optimization. A good but dated survey is provided in [79]. Generalized application of B&B to assignment problems can be found in, for example, [80] and [81]. Applications of B&B specifically to the type of problems of interest here can be found in [82, 83].

Modern heuristic techniques for combinatorial problems: Heuristic techniques for solving combinatorial problems can be considered part of the solution family for the combined ND and set-covering problem cases. The distinguishing aspect of Reeves' review [57] is that the heuristics chosen are those that have broad applicability, whereas other such reviews have focused on specialized solutions. They are also distinguished by the way they simulate various naturally occurring processes (e.g., annealing), which have yielded good results when applied to complex problems. The taxonomy of heuristic methods does not appear to be settled; prior reviews classify the methods as greedy construction, neighborhood search, relaxation, partial enumeration, and decomposition/partition. The methods described in [57] are, along with their *natural* analog, seen in Table 3.9.

Table 3.9

Taxonomy of Heuristic Methods

Method	Natural Process Analog
Simulated annealing	Annealing
Tabu search	Intelligent searching
Generic algorithms	Genetic structures
Neural networks	Neural computation

The Reeves book also includes Lagrangian relaxation method, a method that has found frequent use for assignment problems. Lagrangian relaxation (LR) is a derivative of integer programming techniques. Conceptually, the heuristic solutions must concern themselves with both upper and lower bounds (i.e., the estimation of where the current value of the objective function is in relation to the optimal value). Ideally, such solutions should have upper bounds that are as close to the optimal as possible (i.e., small), and Lower Bounds that are correspondingly as high as possible. Figure 3.17, from [57], shows the distinction in heuristics and relaxations in the context of *bound estimation*.

3.5.2.4   The Probabilistic Association Approaches

Finally, we discuss the case of probabilistic association or assignment processing. In this approach, developed by Bar-Shalom [84], the basic idea is that in a given validation region (a subset of the measurements, the construction of which is not explicitly specified as part of the approach), all measurements are probabilistically assigned (a multi-assignment in effect) to the track about which the validation gate is formed. That is, the validation gate is presumed to contain one target-related measurement and the others are assumed to be false alarms or clutter. These non-target measurements are modeled as independently, identically distributed random variables with uniform spatial distributions. In the original formulation this process was done on a scan-at-a-time basis, and led to the construction of the *probabilistic data association filter* (PDAF) (a modified Kalman filter), since the Kalman estimator requires modification for this type of approach. The original formulation of the PDAF concept was for a single target in clutter. There have been many extensions and variations on this approach including multiple target formulations (the *joint* or JPDAF version) and the inclusion of maneuvering effects for targets (the *interacting multiple model* or IMMJPDAF version), among others. This formulation subsumes the specifics of an assignment solution within the overall tracker/state estimation process, and the interaction with HS processing is in the definition of the validation gate or region. That is, the PDAF processor would solicit a set of hypotheses from the HS function that satisfy certain criteria, such as those above a threshold in cost or the top *n,* which would then be passed to PDAF in order to define which measurements would be within what gates for which targets. Then, PDAF would use all these measurements to form, in effect, a pseudomeasurement used for tracking updates. Thus, this approach is often called an *all-neighbor's* approach.

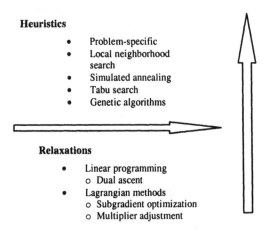

Heuristics

- Problem-specific
- Local neighborhood search
- Simulated annealing
- Tabu search
- Genetic algorithms

Relaxations

- Linear programming
  - o Dual ascent
- Lagrangian methods
  - o Subgradient optimization
  - o Multiplier adjustment

Figure 3.17: Connection between heuristics and relaxation.

REFERENCES

[1] Uhlmann, J. K., "Algorithms for Multiple-Target Tracking," *American Scientist*, vol. 80, 1992, pp. 128-141.

[2] Hall, D., *Mathematical Techniques in Multisensor Data Fusion*, Norwood, MA: Artech House, Inc., 1992.

[3] Blackman, S. S., *Multiple Target Tracking with Radar Applications*, Norwood, MA: Artech House, Inc., 1986.

[4] Bar-Shalom, Y., and E. Tse, "Tracking in a Clustered Environment with Probabilistic Data Association," *Automatica*, vol. II, 1975, pp. 450-460.

[5] Waltz, E., and J. Llinas, *Multi-Sensor Data Fusion*, Norwood, MA: Artech House, Inc., 1990.

[6] Fukunaga, K., *Introduction to Statistical Pattern Recognition*, New York, NY: Academic Press, 1990.

[7] Aldenderfer, M. S., and R. K. Blashfield, *Cluster Analysis*, vol. Sage University Paper 07-044, London, England: Sage Publications, 1984.

[8] Uhlmann, J. K., and M. R. Zuniga, "Results of an Efficient Gating Algorithm for Large-Scale Tracking Scenarios," *Naval Research Reviews*, 1991.

[9] Hall, D., and R. J. Linn, "Comments on the Use of Templating for Multisensor Data Fusion," *Tri-Service Data Fusion Symposium*, Johns Hopkins University, Laurel, MD, 1989, pp. 152-162.

[10] Noble, D. F., "Template-Based Data Fusion for Situation Assessment," *1987 Tri-Service Data Fusion Symposium*, Johns Hopkins University, Laurel, MD, 1987, pp. 226-236.

[11] Jackson, P., *Introduction to Expert Systems*, 3rd. ed., Addison-Wesley Pub. Co., 1999.

[12] Pearl, J., *Probabilistic Reasoning in Intelligent Systems*, San Mateo, CA: Morgan Kaufmann, 1988.

[13] Wasserman, R. B., *Neural Computing: Theory and Practice*, New York, NY: Van Nostrand Reinhold, 1989.

[14] Hall, D., D. Heinze, and G. Guisewite, "Parallel Implementation of a Pattern Matching Expert System," *SCS Eastern Multi-conference*, March, 1989.

[15] Zadeh, L. A., *Fuzzy Sets and Systems*, Amsterdam, Netherlands: North-Holland, 1978.

[16] Neopolitan, R. E., *Probabilistic Reasoning in Expert Systems: Theory and Algorithms*, New York, NY: John Wiley and Sons, 1990.

[17] Calabrese, P. G., "An Algebraic Synthesis of the Foundations of Logic and Probability," *Information Sciences*, 1987, pp. 187-237.

[18] Goodman, I. R., H. T. Nguyen, and E. A. Walker, *Conditional Inference and Logic for Intelligent Systems: A Theory of Measure-Free Conditioning*, Amsterdam, Netherlands: North-Holland, 1991.

[19] Melsa, J., and Cohn, *Decision and Estimation Theory*, New York, NY: McGraw Hill, 1978.

[20] Hand, D. J., *Discrimination and Classification*, New York, NY: John Wiley and Sons, 1981.

[21] Winter, C. L., and A. Stein, "Imagery Exploitation System-Balanced Technology Initiative," *National Symposium on Sensor and Data Fusion (NSSDF)*, San Diego, CA, 1995.

[22] Wright, F. L., and S. Gdowski, "The Art of Multisensor Data Fusion in a Tactical Environment," *IEEE Region 3 Conference*, Colorado Springs, CO, May, 1982.

[23] Kenworthy, M., *Correlation of Intelligence Data*, Colorado Springs, CO: CI/IC Symbolic, 1996.

[24] Bowman, C., "Max Likelihood Track Correlation for Mulitsensor Integration," *18th IEEE Confidence on Decision and Control*, 1979.

[25] Bowman, C. L., "Data Integration (Fusion) Tree Paradigm," *SPIE, Orlando, FL, Proceedings, Signal and Data Processinhg of Small Targets, (Session 1698)*, Orlando, FL, 1992, April 1992.

[26] Bowman, C. L., and A. M. Gross, "Multi-sensor Multi-platform Track Association Using Kinematics and Attributes," *NAECON-85 Proceedings*, May, 1985.

[27] Bowman, C. L., *Intelligent Correlation Agent (INCA) Bayesian Scoring Framework*, N.-.-C.-. NAVELEX, Editor, 1983.

[28] Bowman, C. L., and C. L. Morefield, "Multisensor Fusion of Target Attributes and Kinematic Reports," *3rd ONR/MIT Conference on C31*, 1980.

[29] Bowman, C. L., "Possibilistic Verses Probabilistic Trade-Off for Data Association," *SPIE, Orlando Florida, Proceedings Signal and Data Processing of Small Targets (Session 1954)*, Orlando, FL, April, 1993.

[30] Nguyen, H. T., and I. R. Goodman, "On Modeling of If-Then Rules for Probabilistic Inference," *Journal of Intelligent Systems*, 1995.

[31] Goodman, I. R., and H. T. Nguyen, "A Theory of Conditional Information for Probabilistic Inference in Intelligent Systems," *Information Sciences*, vol. 76, 1994, pp. 13-42.

[32] Goodman, I. R., "PACT: Possibility Approach to Correlation and Tracking," *16th Asilomar Conference*, November, 1982.

[33] Garvey, T. D., and J. D. Lawrence, *Evidential Reasoning: Implementation for Multi-sensor Integration*, San Francisco, CA: SRI, 1983.

[34] Yager, R. R., *Decision Making Under Dempster-Shafer Uncertainties*, New Rochelle, NY: Iona College, 1990.

[35] Nguyen, H. T., and E. A. Walker, "A History and Introduction to the Algebra of Conditional Events and Probability Logic," *IEEE Tranactions on Systems, Man and Cybernetics*, vol. 24, 1994, pp. 1671-1675.

[36] Dubois, D., and H. Prade, *Fuzzy Sets and Systems*, New York, NY: Academic Press, 1980.

[37] Barr, A., and E. Feigenbaum, *The Handbook of Artificial Intelligence*, Stanford, CA: Heuris-Tech Press, 1981.

[38] Bowman, C., L. Gross, and H. Payne, "Data Fusion with Intelligent Assistance: INCA," *16th Asilomar Confidence*, 1982, November, 1982.

[39] Harmon, P., and D. King, *Expert Systems*, New York, NY: John Wiley and Sons, 1985.

[40] Eskridge, T., and J. Barnden, "Application of Connectionism to Analogical Reasoning," *Midwest AI and Cognitive Society Conference*, Carbondale, IL, 1991.

[41] Heckmann, R., "Set Domains," in *European Symposium on Programming*, ed. by N.D. Jones, Berlin, Germany: Springer-Verlan, 1990, p. 177-196.

[42] de Kleer, J., "An Assumption-based TMS," *Artificial Intelligence*, vol. 28, 1986, pp. 127-162.

[43] Dinsmore, J., *Partitional Representations*, New York, NY: Kluwer, 1991.

[44] Bowman, C. L., M. DeYong, and T. Eskridge, "Role for Neural Networks for Avionics," *SPIE, Dual Use Applications of Advanced Imaging and Sensors, Technology Sessions 2566 Proceedings*, July, 1995.

[45] DeYong, M., T. Eskridge, and C. Fields, "Temporal Signal Processing with High-Speed Hybrid Analog-Digital Neural Networks," *Journal of Analog Integrated Circuits and Signal Processing*, vol. 2, 1992, pp. 367-388.

[46] Grossberg, "Competitive Learning: From Interactive Activation to Adaptive Resonance," *Cognitive Science*, vol. 11, 1987, pp. 23-63.

[47] LeCun, Y., et al., "Backpropagation Applied to Hand-Written Zip Code Recognition," *Neural Computation*, vol. 1, 1989, pp. 541.

[48] Elman, J., *Representation and Structure in Connectionist Models*: CLRO, 1989.

[49] Pineda, F., "Generalization of Backpropagation Recurrent and Higher-Order Neural Networks," *IEEE Conference on Neural Information Processing Systems*, Denver, CO, November, 1987.

[50] Goodman, I. R., and H. T. Nguyen, "A Theory of Conditional Information for Probabilistic Inference in Intelligent Systems III: Mathematical Appendix," *Information Sciences*, vol. 75, 1993, pp. 253-277.

[51] Oblow, E., *A Probabilistic-Propositional Framework for the O-Theory Intersectional Rule*, Oak Ridge, TN: Oak Ridge National Laboratory, 1987.

[52] Mahler, R., "Combining Ambiguous Evidence with Respect to Ambitious A Priori Knowledge I: Boolean Logic," *International Journal of Fuzzy Logic and Systems*, 1995.

[53] Pawlak, Z., *Rough Sets*, Boston, MA: Kluwer Academic Publishers, 1991.

[54] Bowman, C. L., "Role for Neural Network in Machine Vision," *Advanced Imaging Association (AIA) Proceedings*, 1995, pp. March, 1995.

[55] Hopfield, J., "Neurons with Graded Response Have Collective Computational Properties," *Proceedings of the National Academy of Sciences (USA)*, vol. 81, 1984.

[56] Nguyen, H. T., and E. A. Walker, "On Decision-Making Using Belief Functions," in *Advances in the Dempster-Shafer Theory of Evidence*, ed. by R.R. Yager, New York, NY: John Wiley & Sons, 1994, p. 331-340.

[57] Reeves, C. R., ed. *Modern Heuristic Techniques for Combinatorial Problems*, New York, NY: Halstead Press, 1993.

[58] Holtzman, A. G., ed. *Mathematical Programming*, New York, NY: Marcel Dekker, 1981.

[59] Tsaknakis, H., et al., "Tracking Closely Spaced Objects Using Multi-Assignment Algorithms," *Data Fusion Systems (DFS 91)*, 1991, Oct. 7-9, 1991.

[60] Finke, G., "Quadratic Assignment Problems," *Annals of Discrete Mathematics*, vol. 31, 1987.

[61] Morefield, C. L., "Application of 0-1 Integer Programming for Multi-Target Tracking Problems," *IEEE Transactions AC*, vol. 22, 1977.

[62] Williams, H. P., *Model Building in Mathematical Programming*, New York, NY: John Wiley & Sons, 1978.

[63] Garfinkel, R. S., and L. Nemhauser, "A Survey of Integer Programming Emphasizing Computation and Relations Among Models," in *Mathematical Programming: Proceedings*, ed. by S.M. Robinson, New York, NY: Academic Press, 1973.

[64] Kuhn, H. W., "The Hungarian Method for the Assignment Problem," *Naval Research Log. Quarterly*, vol. 2, 1955.

[65] Munkres, J., "Algorithms for the Assignment and Transportation Problems," *Journal of SIAM*, vol. 5, 1957.

[66] Bourgeois, F., and J. Lassalle, "An Extension of the Munkres Algorithm for the Assignment Problem to Rectangular Matrices," *Communications of the ACM*, 1971, p. 12.

[67] Silver, R., "An Algorithm for the Assignment Problem," *Communications of the ACM*, vol. 3, 1960.

[68] Salazar, D. L., *Application of Optimization Techniques to the Multi-Target Tracking Problem*, Huntsville, AL: University of Alabama, 1981.

[69] Drummond, O. E., D. A. Castanon, and M. S. Bellovin, *Comparison of 2-D Assignment Algorithms for Sparse, Rectangular, Floating Point Cost Matrices*, 1990.

[70] Jonker, R., and A. Volgenant, "A Shortest Path Augmenting Algorithm for Dense and Sparse Linear Assignment Problems," *Computing*, vol. 38, 1987.

[71] Bertsakas, D. P., and J. Eckstein, "Dual Coordinate Step Methods for Linear Network Flow Problems," *Math Programming Series B*, 1988.

[72] Barr, R. S., "The Alternating Basis Algorithm for Assignment Problems," *Math Programming*, 1977, pp. 13.

[73] Balinski, M. L., "Signature Methods for the Assignment Problem," *Operation Res*, 1985, p. 33.

[74] Marsten, R. E., "An Algorithm for Large Set Partitioning Problems," *Management Science*, vol. 20, 1974.

[75] Goldberg, A. V., "Using Interior-Point Methods for Fast Parallel Algorithms for Bi-Partite Matching and Related Problems," *SIAM Computing*, vol. 21, 1992.

[76] Karmarkar, N., "A New Polynomial Time Algorithm for Linear Programming," *Cominatorica*, vol. 4, 1984.

[77] Astfalk, G., *The Interior-Point Method for Linear Programming*, IEEE Software, 1992.

[78] Cooper, L., and M. W. Cooper, *Introduction to Dynamic Programming*, Oxford, United Kingdom: Pergamon Press, 1981.

[79] Lawler, E. L., and D. E. Wood, "Branch-and-Bound Methods: A Survey," *Operations Research*, vol. 14, 1966.

[80] Ross, G. T., and R. M. Soland, "A Branch and Bound Algorithm for the Generalized Assignment Problem," *Mathematical Programming*, vol. 8, 1975.

[81] Chern, M., "An LC Branch and Bound Algorithm for the Module Assignment Problem," *Information Processing Letters*, 1989.

[82] Brogan, W. L., "Algorithm for Ranked Assignments with Applications to Mulitobject Tracking, I of Guidance, Control, and Dynamics," *AIAA*, 1989, p. 3.

[83] Chen, M., "A Branch-and-Bound Algorithm for Multiple-Target Tracking and its Parallel Implementation," *Proceedings American Control Conference*, Atlanta, GA, June, 1988.

[84] Bar-Shalom, Y., and T. E. Fortmann, *Tracking and Data Association*, San Diego, CA: Academic Press, 1988.

Chapter 4

Level 1 Fusion: Kinematic and Attribute Estimation

Having addressed the data association and correlation problem (discussed in Chapter 3), level one processing seeks to estimate the position, velocity, and attributes of geographically constrained entities. This chapter introduces the problem of kinematic and attribute estimation and develops solutions.

4.1 INTRODUCTION

The problem of fusing multisensor parametric data to yield an improved estimate of the state of the entity was introduced in Chapter 2. Examples of such estimation problems include the following:

1. Using positional data such as line-of-bearing (angles), range or range-rate observations to determine the location of a stationary entity;
2. Combining positional data from multiple sensors to determine the position and velocity of a moving object as a function of time (the tracking problem);
3. Estimating characteristics of an entity, such as size or shape, based on observational data;
4. Estimating the parameters of a model (e.g., the coefficients of a polynomial) which represents or describes observational data.

For these examples, the estimation problem attempts to find the value of a state vector (e.g., position, velocity, and polynomial coefficients) that best fits, in a defined mathematical sense, the observational data. Estimation problems may be dynamic, in which the state vector changes as a function of time, or static in which the state vector is constant in time. This chapter introduces the estimation problem and develops specific solution strategies. A related topic, covariance error analysis, is also introduced. For simplicity, the related data association problem introduced in Chapter 3, is assumed to be solved via the techniques previously described. That is, we assume that the observational data are allocated to distinct groups, each group belonging to a unique entity or object. In practice, data association and estimation techniques must be interleaved to effect an overall solution, especially for multitarget tracking problems.

The history of estimation techniques has been summarized by [1]. The first significant effort to address estimation was Karl Friedrick Gauss' invention of the method of least squares to determine the orbits of planets, asteroids, and comets from redundant data sets. The term redundant data refers to the situation in which more observations are available than the minimum necessary to obtain a solution. In celestial mechanics, techniques for determining orbital elements from minimum data sets are termed initial orbit methods or minimum data set techniques. Gauss utilized the method of least squares in 1795 and published a description of the technique in 1809 [2]. Independently, Legendre invented the least-squares method and published his results in 1806 [3]. The resulting controversy of intellectual propriety prompted Legendre to write to Gauss complaining, "Gauss, who was already so rich in discoveries, might have had the decency not to appropriate the method of least-squares" (quoted in [1] and [4]). Gauss' contribution included not only the invention of the least-squares method, but also the introduction of such modern concepts as the following:

1. Observability: The issue of how many and what types of observations are necessary to develop an estimate of the state vector;
2. Dynamic modeling: The need for accurate equations of motion to describe the evaluation of a state vector in time;
3. A priori estimate: The role of an initial (or starting) value of the state vector in order to obtain a solution;
4. Observation noise: Sets the stage for a probabilistic interpretation of observational noise.

Subsequent historical developments of estimation techniques include Fisher's probabilistic interpretation of the least squares method [5] and definition of the maximum likelihood method, Wiener[6] and Kolmogorov's [7] development of the linear minimum mean square error method, and Kalman's formulation of a discrete-time, recursive, minimum mean square filtering technique (viz the Kalman filter [8]). Numerous texts and papers have been published on the topic of estimation (and in particular on sequential estimation). Blackman [9] provides a detailed description of sequential estimation for target tracking, while Gelb [10] describes sequential estimation from the viewpoint of control theory.

A conceptual view of the estimation processing flow is illustrated in Figure 4.1.

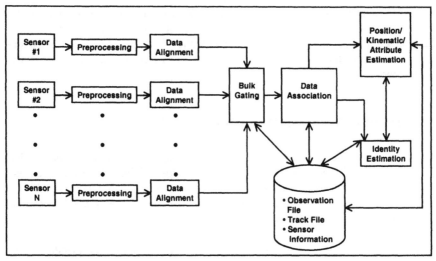

Figure 4.1 Conceptual view of JDL level 1 processing flow.

The situation is illustrated for a positional fusion (e.g., target tracking and identification) problem. A number of sensors observe location parameters such as azimuth, elevation, range, or range rate and attribute parameters such as radar cross-section. The location parameters may be related to the dynamic position and velocity of an entity via observation equations. For each sensor, a data alignment function transforms the "raw" sensor observations into a standard set of units and coordinate reference frame. An association process groups observations into meaningful groups–each group representing observations of a single physical entity or event. The associated observations represent collections of observation-to-observation pairs, or observation-to-track pairs that "belong" together. An estimation process combines the observations to obtain a new or improved estimate of a state vector, $x(t)$, which best fits the observed data. The estimation problem illustrated in Figure 4.1 is the level 1 process within the JDL data fusion process model. It also assumes a centralized architecture in which observations are input to an estimation process for combination. By contrast, an autonomous architectural approach would fuse state vectors as illustrated in Chapter 1, while a hybrid architecture would combine these approaches. The estimation techniques described subsequently in this chapter are applicable to each of these architectures.

In this chapter (and the next) we will treat the location estimation problem and the identity estimation problems as if they are independent. This is done to make the discussion of algorithms easier to follow. However, in implemented systems location estimation and identity estimation processing are often interleaved (since identity information assists in data correlation to improve assignment of data to tracks).

The remainder of this chapter provides a description of techniques for estimation. Section 4.2 provides an overview of design alternatives for estimation, while Section 4.3 introduces batch estimation techniques. Section 4.4 introduces sequential estimation methods, and finally, Section 4.5 describes covariance error analysis methods.

4.2 OVERVIEW OF ESTIMATION TECHNIQUES

Estimation techniques have a rich and extensive history. An enormous amount has been written, and numerous methods have been devised for estimation. This section provides a brief overview of the choices and common techniques available for estimation. Figure 4.2 summarizes the alternatives and issues related to estimation. These include the following:

- System models: what models will be selected to define the problem under consideration? What is to be estimated (namely what is the state vector sought)? How do we predict the state vector in time? How are the observations related to the state vector? What assumptions (if any) can we make about the observation process (e.g., noise and biases)?
- Optimization criteria: How will we define a criteria to specify *best fit*? That is, what equation will be used to specify that a state vector best fits a set of observations?
- Optimization approach: Having defined a criteria for best fit, what method will be used to actually find the unknown value of the state vector that satisfies the criterion?
- Processing approach: Fundamentally, how will the observations be processed (e.g., on a batch mode in which all observations are utilized after they have been received, or sequentially, in which observations are processed one at a time as they are received)?

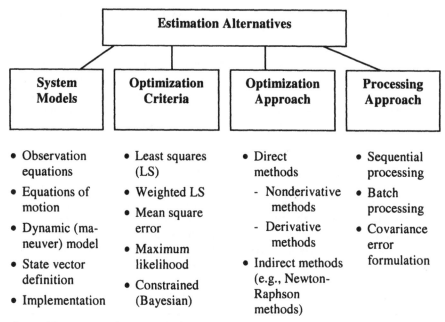

Figure 4.2 Overview of estimation alternatives.

4.2.1 System Models

An estimation problem must be defined by specifying the state vector, observation equations, equations of motion (for dynamic problems), and other choices such as data selection, editing and preconditioning, convergence criteria, and coordinate systems necessary to specify the estimation problem. We will address each of these in turn.

A fundamental choice in estimation is to first specify exactly what parameters are to be estimated (i.e., what is the independent variable or state vector $x(t)$ whose value is sought)? For positional estimation, a typical choice for x is the coordinates necessary to locate a target or entity. Examples include the geodetic latitude and longitude (ϕ, λ) of an object on the surface of the Earth, the three dimensional Cartesian coordinates (x, y, z) of an object with respect to Earth-centered inertial coordinates, or the range and angular direction (r, azimuth, elevation) of an object with respect to a sensor. For nonpositional estimation, the state vector may be selected as model coefficients (e.g., polynomial coefficients) to represent or characterize data. State vectors may also include coefficients that model sensor biases and basic system parameters. For example, at the National Aeronautics and Space Administration (NASA) Goddard Space Flight Center, large-scale computations are often performed to estimate a state vector having several hundred components including; position and velocity of a spacecraft, spherical harmonic coefficients of the Earth's geopotential, sensor biases, coefficients of atmospheric drag, precise location of sensors, and many other parameters [11].

The choice of what to estimate depends upon defining what parameters are necessary to determine the future state of a system, and what parameters may be estimated based on the observed data. The latter issue is termed *observability*. This issue concerns the extent to which it is feasible to determine components of a state vector based on observed data. A weak relationship may exist between a state vector element and the data. Alternatively, two or more components of a state vector may be highly correlated with the result that variation of one component to fit observed data may be indistinguishable from variation in the related state vector component. Hall and Waligora [12] provide an example of the inability to distinguish between camera biases and orientation (attitude) of Landsat satellite data. Deutsch [13] also provides a discussion of observability. The general rule of thumb in selecting a state vector is to choose the minimum set of components necessary to characterize a system under consideration. It is often tempting to choose more, rather than fewer, components of a state vector. An example of this occurs frequently when researchers attempt to represent observational data using a polynomial. Higher order polynomials utilized to fit data invariably result in smaller residuals (i.e., the difference between the actual data and the value of the polynomial that represents the data). Indeed if N observations are available, a polynomial of degree N-1 will fit the data *exactly*. However, such a polynomial may be highly unsuitable and useless for predictive purposes. Hence, a cautious, *less is better* approach is recommended for selecting the components of a state vector.

A second choice required to define the estimation problem is the specification of the observation equations. These equations relate the unknown state vector to predicted observations. Thus, if $x(t)$ is a state vector, and $y_i(t_i)$ is an actual observation, then

$$z_i(t_i) = g(x_j(t_i)) + n \tag{4.1}$$

predicts an observation, $z_i(t_i)$, which would match $y_i(t_i)$ *exactly* if we knew the value of x and we also knew the value of the observational noise, n. The function $g[x_j(t_i)]$ represents the coordinate transformations necessary to predict an observation based on an assumed value of a state vector. If the state vector x varies in time, then the estimation problem is dynamic and requires further specification of an equation of motion that propagates the state vector at time, t_0, to the time of an observation, t_i. For example,

$$x(t_i) = \Phi(t_i, t_0) x(t_0) \tag{4.2}$$

The propagation of the state vector in time (4.2) may involve only a simple truncated Taylor series expansion

$$x(t_0) = x(t_0) + \dot{x}(t_0)\Delta t + \frac{1}{2}\ddot{x}(t_0)\Delta t^2 \qquad (4.3)$$

where $\Delta t = t_i - t_0$, and $\dot{x}$ represents the velocity at time t_0 and $\ddot{x}$ represents the acceleration at time, t_0. In other situations, more complex equations of motion may be required. For example, in astrodynamical problems, the equation of motion may involve second-order, nonlinear, simultaneous differential equations in which the acceleration depends upon the position, velocity, and orientation of a body and the positions of third bodies such as the Sun and Moon. In that case, the solution of the differential equations of motion requires significant computational effort utilizing numerical integration techniques. An example of such a problem may be found in [11, 13]. The selection of the equations of motion (4.2) depends ultimately on the physics underlying the dynamic problem. Tracking problems may require models to predict propulsion, target maneuvering, motion over terrain or through a surrounding media. Selection of an appropriate equation of motion must trade off physical realism and accuracy versus computational resources and the required prediction interval. For observations closely spaced in time (e.g., for a radar tracking an object for a brief interval) a linear model may be sufficient. Otherwise, more complex (and computationally expensive) models must be used. A special difficulty for positional estimators involves maneuvering targets. Both Blackman [9] and Waltz and Llinas [14] discuss these issues.

Closely coupled with the selection of the equations of motion is the choice of coordinate systems in which the prediction is performed. Some coordinate reference frames may be *natural* for defining an equation of motion. For example, Earth-centered (geocentric) inertial Cartesian coordinates provide an especially simple formulation of the equations that describe the motion of a satellite about the Earth (see [13, 15]). By contrast, for the same case, topocentric (Earth surface) noninertial coordinates provide an equation of motion that must introduce artificial acceleration components (namely due to corriolis *forces*) to explain the same motion. However, despite this, the use of a topocentric non-inertial coordinate frame may be advisable from a system viewpoint. Kamen and Sastry [16] provides an example of such a trade-off. Sanza, McClure et al. [17] present the equations of motion and sequential estimation equations for tracking an object in a spherical (r,θ,φ) coordinate reference frame.

There are a variety of other issues associated with defining system models to characterize an estimation problem. Particular implementation issues include the approach as used for data editing and association, techniques to determine convergence, methods for initiating the estimation process (i.e., how to determine a starting estimate), and formulation of the equations. These implementation issues will be addressed for batch estimation and for sequential estimation in Sections 4.3 and 4.4, respectively.

4.2.2   Optimization Criteria

Having established the observation equations that relate a state vector to predicted observations, and equations of motion (for dynamic problems), a key issue involves the definition of *best fit*. We seek to determine a value of a state vector, x(t), which best fits the observed data. There are numerous ways to define *best fit*. Each of these formulations involves a function of the residuals.

$$v_i = [y_i(t_i) - z_i(t_i)] \qquad (4.4)$$

Here, v_i is the vector difference between the *ith* observation, $y_i(t_i)$, at time t_i and the predicted observation, $z_i(t_i)$. The predicted observation, z_i is a function of the state vector $x(t_o)$ via (4.1), and hence v_i, is also a function of $x(t_o)$.

Various functions of v_i have been defined and used for estimation. A summary of several optimization criteria is provided in Table 4.1. In essence, a function is chosen that provides an overall measure of the extent to which the predicted observations match the actual observations. This function of the unknown state vector, x, is sometimes termed a loss (or cost) function because it provides a measure of the penalty (i.e., poor data fit) for an incorrect value of x. The state vector, x, is varied until the loss function is either a minimum or maximum, as appropriate. The solution of the estimation problem then becomes an optimization problem. Such optimization problems have been treated by many texts (see for example [18]).

Table 4.1

Overview of Optimization Criteria

Criteria	Description	Mathematical Formulation	Comments
Least squares (LS)	Minimize the sum of the squares of the residuals	$L(x) = vv^T$	Earliest formulation provided by Gauss—no a priori knowledge assumed
Weighted least squares (WLS)	Minimize the sum of the weighted squares of the residuals	$L(x) = vWv^T$	Yields identical results to MLE when noise is Gaussian and weight matrix

Table 4.1 (continued)

Overview of Optimization Criteria

Criteria	Description	Mathematical Formulation	Comments
			equals inverse covariance matrix
Mean square error (MSE)	Minimize the expected value of the squared error	$$L(x) = \int (x - \overline{x}) w (x - \overline{x})_P (x/y)(x - \overline{x})^T dx$$	Minimum covariance solution
Bayesian weighted least squares (BWLS)	Minimize the sum of the weighted squares of the residuals constrained by a priori knowledge of x	$$L(x) = vWv(x - x_0) P_{\Delta x_0} (x - x)^T$$	Constrains the solution of x to a reasonable value close to the a priori estimate of x
Maximum likelihood estimate (MLE)	Maximize the multivariate probability distribution function	$$L(x) = \prod_{i-1}^{m} t_1(n_1/x) t_2(n_2/x_2)\ldots t_n(n_n/x)$$	Allows specification of the probability distribution for the noise process

Perhaps the most familiar definition of best fit are the least squares and weighted least squares formulations. The weighted least squares expression in vector form may be written as

$$L(x) = vWv^T \tag{4.5}$$

Equivalently,

$$L(x) = [(y_1 - z_1),\ldots,(y_n - z_n)] \begin{pmatrix} w_{11} & \cdots & 0 \\ \vdots & \ddots & \vdots \\ 0 & \cdots & w_{nn} \end{pmatrix} \begin{bmatrix} (y_1 - z_1) \\ \vdots \\ (y_n - z_n) \end{bmatrix} \tag{4.6}$$

or,

$$L(x) = \sum_{i=1}^{n} (y_i - z_i) w_{ij} (y_i - z_i) \qquad (4.7)$$

The loss function $L(x)$ is a scalar function of x that is the sum of the squares of the observation residuals weighted by W. The matrix, W, allows assignment of weights to the individual observations (e.g., corresponding to different sensor characteristics). The least squares or weighted least squares criterion is used when there is no basis to assign probabilities to x and y, and there is limited information about the measurement errors. A special case involves linear least squares, in which the predicted observations are a linear function of the state vector. In that case, $L(x)$ may be solved explicitly with a closed form solution (see [19] for the formulation).

A variation of the weighted least squares objective function is the constrained function,

$$L(x) = vWv^T + (x - x_0) P_{\Delta x_0} (x - x_0)^T \qquad (4.8)$$

This expression, sometimes termed the Bayesian weighted least squares criterion, constrains the weighted least squares solution for x to be close to an a priori value of x (i.e., x_0). In (4.8), the quantity $P_{\Delta xo}$ represents an estimate of the covariance of x given by the symmetric, positive-definite matrix.

$$P_{\Delta x_0} = \begin{pmatrix} \sigma^2_{x1} & \cdots & \sigma_{x1}\sigma_{xn} \\ & \ddots & \\ \sigma_{x1}\sigma_{xn} & & \sigma^2_{xn} \end{pmatrix} \qquad (4.9)$$

In the event that the components of x can be assumed to be independent, then $P_{\Delta xo}$ is a diagonal matrix. The Bayesian criterion is used when there is prior knowledge about the value of x and a priori knowledge of associated uncertainty via $P_{\Delta xo}$. The resulting optimal solution for x lies nearby the a priori value x_0.

The mean square error formulation minimizes the expected (mean) value of the squared error or minimizes

$$L(x) = \int (x - \hat{x})^T W (x - \hat{x}) P(x/y) dx \qquad (4.10)$$

where $P(x|y)$ is the conditional probability of state vector $\hat{x}$ given the observations y. This formulation assumes that x and y are jointly distributed random variables. The quantity $\hat{x}$ is the conditional expectation of x,

$$\hat{x} = \int xP(x/y)dx \tag{4.11}$$

The solution for $\hat{x}$ yields the minimum covariance of x.

The final optimization criterion shown in Table 4.1 to define *best fit* is the maximum likelihood criterion,

$$L(x) = \prod_{i=1}^{m} l_1(n_1/x)l_2(n_2/x)...l(n_n/x) \tag{4.12}$$

L(x) is the multivariate probability distribution to model the observational noise n_i. The function, L(x) is the conditional probability that the observational noise at times t_0, t_1, ..., t_i will have the values n_0, n_1,..., n_i, if x is the actual or true value of the state vector. The maximum likelihood criteria selects the value of x that maximizes the multivariate probability of the observational noise.

If the measurement errors, n_i are normally distributed about a zero mean, then $l_i(n_i/x)$ is given by

$$l_i(n_i/x) = \frac{1}{(2\pi)^{m/2} |M_i^{1/2}|} \exp(-\frac{1}{2} n_i^T M_i^{-1} n_i) \tag{4.13}$$

The quantity, m, refers to the number of components, at each time, t_i, of the observation vector, and M_i is the variance of the observation at time t_i. The maximum likelihood criterion allows us to postulate non-Gaussian distributions for the noise statistics.

Selection of an optimization criterion from among the choices shown in Table 4.1 depends upon the a priori knowledge about the observational process. Clearly, selection of the MLE criterion presumes that the probability distributions of the observational noise are known. Similarly, the MSE criterion presumes knowledge of a conditional probability function, while the Bayesian weighted least squares assumes a priori knowledge of the variance of the state vector. Under the following restricted conditions, the use of these criterion result in an identical solution for x. The conditions are listed as follows:

1. The measurement (observational) errors are Gaussian distributed about a zero mean;
2. The errors n_i at time t_i are stochastically independent of the errors n_j, at time t_j;
3. The weight for the weighted least squares criterion is the inverse covariance of x.

Under these conditions, the WLS solution is identical to the MLE, MSE, and BWLS solutions.

4.2.3   Optimization Approach

Actual solution of the optimization criterion to determine the value of the state vector x may be performed by one of several techniques. The reader is referred to several texts, which present detailed algorithms for optimization [18-21]. References [19, 21] provide computer codes to solve the optimization problem. In this section, we will provide an overview of optimization approaches and give additional detail in Sections 4.3 and 4.4 for batch and sequential estimation, respectively.

Optimization techniques may be categorized into two broad classes as illustrated in Table 4.2

Table 4.2

Overview of Optimization Techniques

Category	Optimization Techniques	Description
Direct Methods	Nonderivative methods Downhill simplex Direction set Derivative methods Conjugate gradient Variable metric(quasi-Newton)	Direct methods find the value of **x** that satisfies the optimization criteria (i.e., find **x** such that the loss function is either a minimum or maximum). Techniques fall into two classes: derivative methods require knowledge of derivative of loss function with respect to **x**, while nonderivative methods require only the ability to compute the loss function
Indirect Methods	Newton-Raphson methods	Indirect methods find the roots of a system of equations involving partial derivatives of the loss function with respect to the state vector, **x** (i.e., the partial derivative of $L(x)$ with respect to **x** set equal to zero); the only successful techniques are multidimensional Newton-Raphson methods

Direct methods treat the optimization criteria, without modification, seeking to determine the value of x that finds an extremum (i.e., minimum or maximum) of the optimization criterion. Geometrically, direct methods are hill-climbing (or valley-seeking) techniques that seek to find the value of x for which $L(x)$ is a maxima or minima. By contrast, indirect methods seek to solve the simultaneous nonlinear equation given by

$$\frac{\partial L(x)}{\partial x} = \begin{bmatrix} \dfrac{\partial L}{\partial x_1} \\ \vdots \\ \dfrac{\partial L}{\partial x_m} \end{bmatrix} \tag{4.14}$$

where m is the number of components of the state vector. Indirect methods require that the optimization criterion be explicitly differentiated with respect to x. The problem is thus transformed from finding the maximum (or minimum) of a nonlinear equation, to one of finding the roots of m simultaneously nonlinear equations given by (4.14).

A summary of techniques for a direct solution of the optimization problem is given in Table 4.3. An excellent reference with detailed algorithms and computer program listings is [19]. The direct techniques may be subdivided into nonderivative techniques [i.e., those methods that require the ability to compute L(x)] and derivative techniques that rely on the ability to compute L(x) as well as derivatives of L(x). Nonderivative techniques include simplex methods such as that described by [22], and direction set methods that seek successive minimization along preferred coordinate directions. Specific techniques include conjugate direction methods and Powell's methods (see [19, 20]).

Derivative methods for direct optimization utilize first or higher order derivatives of L(x) to seek an optimum. Specific methods include conjugate gradient methods such as Fletcher-Reeves method and the Polak-Ribiere method (see [19, 23]). Variable metric methods utilize a generalization of the one-dimensional Newton's approach. Effective techniques include the Davidon-Flecher-Powell method and the Broyden-Flecher-Goldfarb-Shanns method. A well-known, but relatively ineffective gradient technique is the method of steepest descent. Press et al. provides a discussion of trade-offs and relative performance of these methods.

Indirect solutions of the optimization criterion must find the root of simultaneous nonlinear equations (4.14). While intuitively it would seem easier to find the roots of nonlinear equations than the extreme of a function, such is not the case. Press, Flannery et al. [19] describe the difference in these approaches as follows:

"Why is that task (relatively) easy, while multidimensional root finding is often quite hard? Isn't minimization equivalent to finding a zero of an N-dimensional gradient vector, not so different from zeroing an N-dimensional function? No! The components of a gradient vector are not independent, arbitrary functions. Rather, they obey so-called integrability conditions that are highly restrictive. Put crudely, you can always find a minimum by sliding downhill on a single surface. The test of *downhillness* is thus one-dimensional. There is no analogous conceptual procedure for finding a multidimensional root, where *downhill* must mean simultaneously downhill in N separate function spaces, thus allowing a multitude of trade-offs, as to how much progress in one dimension is worth compared with progress in another."

Table 4.3

Summary of Direct Methods for Optimization

Type of method	Class of Techniques	Algorithm/Strategy	References
Nonderivative methods (do not require derivative of L(x))	Downhill simplex techniques	Utilize a geometric simplex (polygonal figure) to map out and bracket extrema of L(x)	[19] [22]
	Direction set methods	Successive minimization along preferred coordinate directions; examples include conjunctive direction techniques and Powell's method	[19] [20]
Derivative methods [requires derivative of L(x)]	Conjunctive gradient methods	Utilize multidimensional derivative (gradient) information to seek an extremum; techniques include Flecher-Reeves and Polak-Ribiere methods	[19] [23]
	Variable metric (quasi-Newton techniques)	Variations of multidimensional Newton's method; techniques include Davidson-Flecher-Powell and Broyden-Flecher-Goldfarb-Shannon methods	[19]

Reference [19] points out that there is only one generally effective technique for finding the roots of (4.14), namely the multiple dimensional Newton-Raphson method. For a one-dimensional state vector, the technique may be summarized as follows. We seek to find x such that

$$\frac{\partial L(x)}{\partial x} = f(x) = 0 \qquad (4.15)$$

Expand f(x) in a Taylor series,

$$f(x + \delta x) = f(x) + \frac{\partial f}{\partial x} \delta x + \left(\frac{\partial^2 f}{\partial x^2} \right) \delta x^2 + \cdots \qquad (4.16)$$

Neglecting second and higher order terms yields a linear equation,

$$f(x + \delta x) = f(x) + \frac{\partial f}{\partial x} \delta x \qquad (4.17)$$

To find the root of (4.17), set f(x + δx) equal to zero and solve for δx,

$$\delta x = f(x) - \left(\frac{f(x)}{\dfrac{df}{dx}} \right) \tag{4.18}$$

In order to find the root of (4.15), we begin with an initial value of x, say x_1. An improved value for x_i is given by

$$x_i = x_i + \delta x_i \tag{4.19}$$

Equation (4.19) is solved iteratively until $\delta x_i < \varepsilon$, where epsilon is an arbitrarily small convergence criterion. A multidimensional description of the Newton-Raphson technique is described by [19].

4.2.4 Processing Approach

In introducing the estimation problem and alternative design choices, we have implicitly ignored the fundamental question of how data will be processed. As illustrated in Figure 4.2, there are two basic alternatives (1) batch processing and (2) sequential processing. Batch processing assumes that all data are available to be considered simultaneously. That is, we assume that all n observations are available, select an optimization criterion, and proceed to find the value of x that best fits the n observations (via one of the techniques described in Section 4.3). The batch approach is commonly used in modeling or curve-fitting problems. Batch estimation is often used in situations in which there is no time-critical element involved. For example, estimating the heliocentric orbit of a new comet or asteroid involves observations over a period of several days with subsequent analysis of the data to establish an ephemeris. Another example would entail modeling in which various functions (e.g., polynomials and log-linear functions) are used to describe data. Batch approaches have a number of advantages, particularly for situations in which there may be difficulty in associating multiple observations as relating to a common target. At least in principle, one approach to find an optimal association of observations-to-tracks or observations-to-observations is simply to exhaustively try all [n(n-1)/2] combinations. While in practice such exhaustive techniques are not used, batch estimation techniques have more flexibility in such approaches than do sequential techniques. Section 4.3 provides more detail on batch estimation including a processing flow and discussion of implementation issues.

An alternative approach to batch estimation is the sequential estimation approach. This approach incrementally updates the estimate of the state vector as each new observation is received. Hence if $x_n(t_o)$ is the estimate of a state vector at time t_o based on n previous observations, then sequential estimation provides the means of obtaining a new estimate for x, (i.e., $x_{n+1}(t_o)$), based on n+1 observations by modifying the estimate $x_n(t_o)$. This new estimate is obtained without revisiting all previous n observations. By contrast in batch estimation, if a value of $x_n(t_o)$ had been obtained utilizing n observations, then to determine $x_{n+1}(t_o)$, all n+1 observations would have to be processed. The Kalman filter is a commonly used approach for sequential estimation.

Sequential estimation techniques provide a number of advantages including the following:

1. Determination of an estimate of the state vector with each new observation;
2. Computationally efficient scalar formulations;
3. Ability to adapt to changing observational conditions (e.g., noise).

Disadvantages of sequential estimation techniques involve potential problems in data association, divergence (in which the sequential estimator ignores new data), and problems in initiating the process. Nevertheless, sequential estimators are commonly used for tracking and positional estimation. Section 4.4 provides more detail on sequential estimation including a process flow and discussion of implementation issues.

4.3   BATCH ESTIMATION

4.3.1   Derivation of Weighted Least Squares Solution

In order to illustrate the formulation and processing flow for batch estimation, consider the weighted least squares solution to a dynamic tracking problem. One or more sensors observe a moving object, reporting a total of n observations, $y_M(t_i)$, related to target position. Note that henceforth in this chapter the subscript M denotes a measured value of a quantity. The observations are assumed to be unbiased with observational noise whose standard deviation at time t_i is σ_i. The (unknown) target position and velocity at time t_i is represented by an one-dimensional vector, $x(t_i)$. Since the target is moving, x is a function of time given by the equations of motion,

$$x(t) = f(x, \dot{x}, t) \tag{4.20}$$

Equation (4.20) represents l simultaneous nonlinear differential equations. This is an initial value problem in differential equations. Given a set of initial conditions (namely specification of x_o, and x_o at time t_o) equation (4.20) can be solved to predict x(t) at an arbitrary time, t. Numerical approaches to solve (4.20) are described in a number of texts such as [20] and [24]. Specific techniques include numerical integration methods such as Runge-Kutta methods, predictor-corrector methods, perturbation methods, or in simple cases, analytical solutions.

An observational model allows the observations to be predicted as a function of the unknown state vector,

$$y(t) = g(x,t) \tag{4.21}$$

The residuals are the differences between the sensor data and the computed measurement at time t_i, given by,

$$v_i = [y_M(t_i) - g(x,t_i)] \tag{4.22}$$

The weighted least squares criteria or loss function for best fit is specified as,

$$L(x) = \sum_{i=1}^{n} \left(\frac{v_i}{\sigma_i} \right)^2 = \sum_{i=1}^{n} \left(\frac{1}{\sigma_i^2} \right) [y_M(t_i) - g(x,t_i)]^2 \tag{4.23}$$

L(x) is a measure of the closeness of fit between the actual data and the predicted data as a function of x. Note that if different sensor types are utilized (e.g., radar and optical tracker), then the form of the function, g(x, t_i) changes for each sensor type. Hence, for example, g(x, t_i) may represent one set of equations for a radar to predict range, azimuth, elevation, and range-rate, and another set of equations for an optical tracker to predict right ascension and declination angles. Further, the observation represented by $y_M(t_i)$ may be a vector quantity; for example

$$y_M(t_i) = \begin{bmatrix} range(t_i) \\ range - rate(t_i) \\ azimuth(t_i) \\ elevation(t_i) \end{bmatrix}$$

for a radar observation, or

$$y_M(t_i) = \begin{bmatrix} right\ ascension(t_i) \\ declination(t_i) \end{bmatrix}$$

for an optical tracker. In that case, $g(x, t_i)$ and υ_i would correspondingly be vector quantities.

In matrix notation, equation (4.23) becomes

$$L(x) = [y_M - g(x)]^T W [y_M - g(x)] \tag{4.24}$$

where,

$$y_M = \begin{bmatrix} y_M(t_1) \\ y_M(t_2) \\ \vdots \\ y_M(t_n) \end{bmatrix}, \text{ and } g(x) = \begin{bmatrix} g(x,t_1) \\ g(x,t_2) \\ \vdots \\ g(x,t_n) \end{bmatrix} \tag{4.25}$$

and,

$$W = \begin{bmatrix} \dfrac{1}{\sigma_1^2} & 0 & \cdots & 0 \\ 0 & \dfrac{1}{\sigma_2^2} & \cdots & 0 \\ 0 & 0 & \cdots & \\ & & & \dfrac{1}{\sigma_n^2} \end{bmatrix} \tag{4.26}$$

The weighted least squares estimate is the value of x_o, denoted $\hat{x}_0$, that minimizes the function $L(x)$.

Using an indirect approach, we seek x_o, such that

$$\frac{\partial L(x)}{\partial x} = 2[y_M - g(x)]^T W \frac{\partial g}{\partial x} = 0 \tag{4.27}$$

Equation (4.27) represents a set of nonlinear equations in l unknowns (the l components of x_o).

The solution to equations (4.27) may be obtained by a multidimensional Newton-Raphson approach as indicated in the previous section. The following derivation explicitly shows the linearized, iterative solution. First, expand the measurement prediction, g(x,t), in a Taylor series about a reference solution $x_{ref}(t)$, namely

$$g(x,t_i) = g(x_{REF},t) + H_i \Delta x_i \qquad (4.28)$$

where,

$$\Delta x_i = x(t_i) - x_{REF}(t_i) \qquad (4.29)$$

and,

$$H_i = \frac{\partial g(x,t_i)}{\partial x} \qquad (4.30)$$

The value x_{REF} is a particular solution to the dynamical equation of motion (4.20).

A further simplification may be obtained if the equation of motion is linearized; that is,

$$\Delta x_i = \Phi(t_i,t_j)\Delta x_j \qquad (4.31)$$

where, $\Phi(t_i, t_j)$ represents a state transition matrix which relates variations about $x_{REF}(t)$ at times t_i, and t_j,

$$\Phi(t_i,t_j) = \frac{\partial x(t_i)}{\partial x(t_j)} \qquad (4.32)$$

Substituting (4.32) into (4.28) yields,

$$g(x,t_i) = g(x_{REF},t_i) + H_i \Phi(t_i,t_j)\Delta x_j \qquad (4.33)$$

Using this expression for g(x, t_i) in (4.27) yields

$$[\Delta y - H(x_j)\Delta x_j]^T WH(x_j) = 0 \qquad (4.34)$$

where

$$\Delta y = y_M - g(x_{REF}) = \begin{bmatrix} y_M(t_1) - g(x_{REF}, t_1) \\ y_M(t_2) - g(x_{REF}, t_2) \\ \vdots \\ y_M(t_n) - g(x_{REF}, t_n) \end{bmatrix} \tag{4.35}$$

and

$$H(x_j) = \begin{bmatrix} H_1 \Phi(t_1, t_j) \\ H_2 \Phi(t_2, t_j) \\ \vdots \\ H_n \Phi(t_n, t_j) \end{bmatrix} \tag{4.36}$$

Solving (4.34) for $\Delta \hat{x}_j$ yields,

$$\Delta \hat{x}_j = [H(x_j)^T W H(x_j)]^{-1} [H(x_j)^T W \Delta y] \tag{4.37}$$

The increment $\Delta \hat{x}_j$ is added to $x_{REF}(t_j)$

$$\hat{x}(t_j) = x_{REF}(t_i) + \Delta \hat{x}_j \tag{4.38}$$

to produce an updated estimate of the state vector. The improved value of $\hat{x}_j(t_j)$ is used as a new reference value of $x(t_j)$. Equations (4.37) and (4.38) are applied iteratively until $\Delta \hat{x}_j$ becomes arbitrarily small.

Table 4.4 illustrates the linearized iterative solution for several optimization criteria including least squares, weighted least squares, Bayesian weighted least squares, and the maximum likelihood optimization criteria. In (4.37), Δy represents the difference between predicted and actual observations, while $H(x_j)$ expresses the relationship between changes in predicted observations and changes in components of the state vector. For static (nondynamic) problems, the state vector $x(t)$ is constant in time, and the state transition matrix reduces to the identity matrix,

$$\Phi(t_i, t_j) = I = \begin{bmatrix} 1 & 0 & \cdots & 0 \\ 0 & 1 & \cdots & 0 \\ & & \ddots & \\ 0 & 0 & \cdots & 1 \end{bmatrix} \tag{4.39}$$

Table 4.4

Summary of Batch Estimation Solutions

Optimiza-tion Criterion	Mathematical Formulation	Linearized Iterative Solution
Least squares	$L(x) = vv^T$	$\Delta\hat{x}_j = [H^T H]^{-1}[H^T v]$
Weighted least squares	$L(x) = vWv^T$	$\Delta\hat{x}_j = [H^T WH]^{-1}[H^T Wv]$
Bayesian weighted least squares	$L(x) = vWv + (x - x_0)P_{\Delta x_0}(x - x_0)^T$	$\Delta\hat{x}_j = [H^T WH + P_{\Delta x}^{-1}][H^T Wv + P_{\Delta x}^{-1}\Delta\hat{x}_{j-1}]$
Maximum likelihood estimate	$L(x) = \displaystyle\prod_{i=1}^{m} l_i(n_i / x)$	$\Delta\hat{x}_j = [H^T M^{-1} H]^T [H^T M^{-1} v]$

4.3.2   Processing Flow

The processing flow to solve the batch estimation problem is illustrated in Table 4.3 for a weighted least squares formulation. The indirect approach discussed in the previous section is used. Inputs to the process include an initial estimate of the state vector, x_0, at an epoch time, t_0, and n observations, y_{Mk} at times, t_k, with associated uncertainties σ_k. The output of the process is an improved estimate of the state vector, $x_{i+1}(t_0)$ at the epoch time, t_0.

The processing flow shown in Figure 4.3 utilizes two nested iterations. An inner iteration (or processing loop letting k= 1,2,...,n), cycles through each observation, y_{Mk}, performing a series of calculations:

1. Retrieve the kth observation, y_{Mk}, and its associated time of observation, t_k, and observational uncertainty, σ_k;
2. Solve the differential equation of motion (4.20) to obtain $x_{ki}(t_k)$. That is, using the current estimate of the state vector, $x_i(t_0)$, update to time t_k;
3. Compute a predicted observation, $g[x_i(t_k),t_k]$ based on sensor models; note that the predicted observation utilizes a model appropriate to the sensor type (e.g., radar or optical);
4. Compute the transition matrix $\Phi(t_k,t_0)$;
5. Calculate the quantity $H_k\Phi(t_k,t_0)$, and the observation residual $\delta y_k=[y_M(t_k)-g(x_i,t_k)$;
6. Update the matrices,

$$A = A + \frac{H_k^T H_k}{\sigma_k^2} \qquad (4.40)$$

and

$$B = B + \frac{H_k^T \delta y_k}{\sigma_k^2} \qquad (4.41)$$

Steps 1-6 are performed for each observation k=1,2,...,n.

An outer processing loop (i=0,1,...) iteratively computes and applies corrections to the state vector until convergence is achieved. For each iteration, we compute:

$$A^{-1} = (H^T W^{-1} H)^{-1} \qquad (4.42)$$

and

$$\Delta \hat{x}_0 = A^{-1} B \qquad (4.43)$$

with

$$x_{i+1}(t_o) = x_i(t_o) + \Delta \hat{x}_0 \qquad (4.44)$$

Hence, the state vector is successively improved until $\Delta \hat{x}_0$, becomes arbitrarily small.

The process described here and illustrated in Figure 4.3 is meant to convey the essence of the batch solution via a linearized iterative approach. It can be seen that the solution may become computationally demanding. For each observation (which may number in the hundreds to thousands), we must solve a nonlinear set of differential equations, perform coordinate transformations to predict an observation, compute the transition matrix, and perform several matrix multiplications. These calculations are performed for all n observations. Moreover, the complete set (for all n observations) of computations is iteratively performed to achieve an estimate of $x(t_o)$. Upwards of 10-30 iterations of Δx_i may be required to achieve convergence, depending upon the initial value chosen for $x(t_o)$.

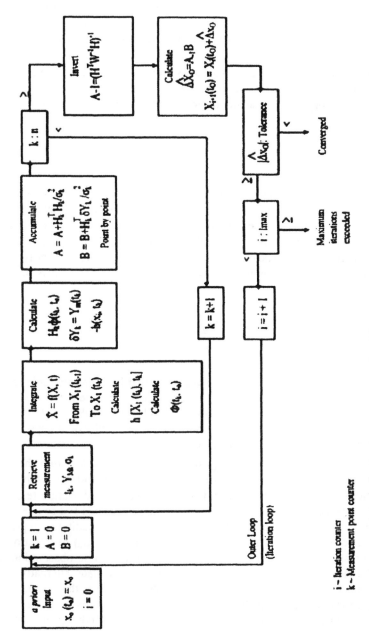

Figure 4.3 Weighted least squares batch process computational flow sequence.

4.3.3   Batch Processing Implementation Issues

The processing flow for batch estimation shown in Figure 4.3 is meant to be illustrative rather than a prescription or flowchart suitable for software implementation. Several implementation issues often arise. We will discuss a few of these issues, including convergence, data editing, the initial estimate of **x**, and observability.

The processing flow in Figure 4.3 shows an outer processing loop in which successive improvements are made to the state vector estimate. The convergence criteria shown in Figure 4.3 tests the magnitude of Δx, and declares convergence when,

$$|\Delta x_0| \le \varepsilon \tag{4.45}$$

The iterations are terminated when the incremental changes to the state vector fall within a sufficiently small increment (for each component of the state vector). This is a logical criterion since we can use physical arguments to establish the values for ε. For example, we might declare that distances within 1m, velocities within 1 cm/sec, and frequencies within 1 Hertz for example are arbitrarily small. Other convergence criteria might equally well be used. An example is the ratio criterion,

$$\left| \frac{\Delta x_0}{x_0} \right| \le \varepsilon \tag{4.46}$$

A number of convergence criteria have been used including multiple logical conditions [e.g., (4.45) or (4.46)]. There is no guarantee that the iterative solution will converge in a finite number of iterations. Thus logical checks should be made to determine how many iterations have been performed with some upper bound (e.g., $k \le 50$) established to terminate the iterations.

In batch estimation it is tempting to perform data editing within the outer processing loop. A common practice is to reject all observations for which the residual ($|v_k|$) exceeds either an a priori limit, or a standard deviation (e.g., 3σ) test. Such a practice is fraught with potential pitfalls. Deutsch [13] discusses some of these pitfalls. Two problems are most notable. Iterative data editing can prolong the outer-loop iteration. It is possible in the ith estimate of $x(t_0)$ to reject one or more observations only to find in the $(i+1)^{st}$ estimate that these observations are acceptable. Hence the iteration for $x(t_0)$ can sometimes oscillate, alternatively rejecting and accepting observations. A second problem involves the case in which all observations are valid and highly accurate. Statistically, there will be some valid observations whose residuals exceed 3σ. Rejecting or editing out such observations corrupts the solution for $x(t_o)$ by rejecting perfectly good data. A rule of thumb is not to reject any data unless there are valid physical reasons for such editing.

Another implementation issue for batch processing involves how to obtain an initial estimate of the state vector, $x_o(t_o)$. Generally, several observations may be used in a minimum data solution to obtain a value of $x(t_o)$. Alternatively, some sensors may provide estimates of the state vector, or an estimate of x may be available from other a priori information. The generation of a starting value for x is very much dependent upon the particular sensors and observing geometry. Sometimes several minimum data sets are used (i.e., observations y_{M1}, y_{M3}, and y_{M5}; observations y_{M2}, y_{M4}, and y_{M6}) with an initial estimate of $x(t_p)$ developed from each data set. Subsequently, the initial estimates are averaged to produce a starting value for the estimation process.

A final implementation issue is the question of observability. We briefly introduced this issue in Section 4.2. Observability is the question of whether improvements in components of $x(t)$ can be obtained based on the observational data. Mathematically this problem is exhibited via an ill-conditioned matrix,

$$\left[H^T W^{-1} H \right] \tag{4.47}$$

Several methods have been introduced to address such an ill-conditioned system. For example, we may require that the determinant of matrix $H^T H$ be non-zero, at each iterative step. Alternatively, nonlinear terms may be introduced to treat the ill-conditioned linear system. While these techniques may be useful, there is no substitute for care in selecting and analyzing the choice of state vector components.

4.4 SEQUENTIAL ESTIMATION

During the early 1960s, a number of technical papers were published describing a sequential approach to estimation. Swerling, Kalman, and Kalman and Bucy [25-27] published papers describing a linearized sequential technique to update an estimate of a state vector. The work was a discrete implementation of earlier continuous formulations by [6, 7] in the late 1940s. An informal history of the development of the Kalman filter, and its application to space flight is provided by [28]. Since then, much work has been performed on discrete estimation. A number of techniques exist. A extensive review of recursive filtering techniques is provided by Sayed and Kailath [29] who compare variants of the Kalman filter including the covariance Kalman filter, the information filter, the square-root covariance filter, the extended square-root information filter, the square-root Chandrasekhar filter, and the explicit Chandrassekhar filter. We will describe one such approach utilizing the Kalman filter. Interestingly, discrete linear estimation techniques are commonly referred to as Kalman filters despite the work of Swerling and Bucy. (Only a few researchers reference the Kalman-Bucy filter or Kalman-Swerling-Bucy filter.)

Several approaches may be used to derive the sequential estimation equations. In particular, the reader is referred to the feedback-control system approach described in [10]. In this section, we will derive the equations for a weighted least squares optimization criterion, for a dynamic tracking problem, identical to that in the previous section. Following a derivation, we will present a processing flow and a discussion of implementation issues.

4.4.1   Derivation of Sequential Weighted Least Squares Solution

Assume that n observations are available from multiple sensors, and that a weighted least squares solution has been obtained; that is,

$$\Delta \hat{x}_n = [H_n W_n^{-1} H_n]^{-1} H_n^T W_n^{-1} \Delta y_n \tag{4.48}$$

where H_n denotes $H(x_n)$ the partial derivatives of the observation components with respect to the state vector (4.36), and Δy_n, denotes the difference between the measured and predicted observation (4.22). Suppose that one additional observation, $y_M(t_{n+1})$ is received. How does the (n+1)st observation affect the estimate of x? Let us utilize the weighted least squares formulation separating the (n+1)st data from the previous n observations. Thus,

$$\Delta \hat{x}_{n+1} = [H_{n+1} W_{n+1}^{-1} H_{n+1}]^{-1} H_{n+1} W_{n+1}^{-1} \Delta y_{n+1} \tag{4.49}$$

where

$$H_{n+1} = \begin{vmatrix} H_n \phi(t_n, t_{n+1}) \\ H_{n+1} \end{vmatrix}; W_{n+1} = \begin{pmatrix} W_n & 0 \\ 0 & \sigma_{n+1}^2 \end{pmatrix}; \Delta y_{n+1} = \begin{vmatrix} \Delta y_n \\ \delta y_{n+1} \end{vmatrix} \tag{4.50}$$

and the (n+1)st residual is,

$$\delta y_{n+1} = y_M(t_{n+1}) - g(x_{REF}, t_{n+1}) \tag{4.51}$$

Substituting (4.50) into (4.49) yields

$$\Delta\hat{x}_{n+1} = \left[[H_n\phi, H_{n+1}] \begin{pmatrix} w_n^{-1} & 0 \\ 0 & \sigma_{n+1}^{-2} \end{pmatrix} \begin{bmatrix} H_n\phi \\ H_{n+1} \end{bmatrix} \right]^{-1} [H_n\phi, H_{n+1}] \begin{pmatrix} w_n^{-1} & 0 \\ 0 & \sigma_{n+1}^{-2} \end{pmatrix} \begin{bmatrix} \Delta y_m \\ \delta y_{n+1} \end{bmatrix}$$

(4.52)

which can be manipulated to obtain

$$\Delta\hat{x}(t_{n+1}/t_n) = \Delta\hat{x}(t_{n+1}/t_n) - K[H_{n+1}\Delta\hat{x}(t_{n+1}/t_n) - \delta y_{n+1}]$$ (4.53)

where,

$$K = P_n(t_{n+1})H_{n+1}^T[H_{n+1}P_n(t_{n+1})H_{n+1}^T + \sigma_{n+1}^2]^{-1}$$ (4.54)

$$P_{n+1}(t_{n+1}) = P_n(t_{n+1}) - KH_{n+1}P_n(t_{n+1})$$ (4.55)

$$\Delta\hat{x}(t_{n+1}/t_n) = \Phi(t_{n+1}, t_n)\Delta\hat{x}(t_n/t_n)$$ (4.56)

$$P_n(t_{n+1}) = \Phi(t_{n+1}, t_n)P_n(t_n)\Phi$$ (4.57)

Equations (4.53)-(4.57) constitute a set of equations for recursive update of a state vector. Thus, given n observations $y_M(t_n)$ (t = 1, 2,..., n), and an associated estimate for x(t), these equations prescribe an update of $y_M(t_n)$ based on a new observation $y_M(t_{n+1})$. Clearly the equations can be applied recursively, replacing the solution for $x_n(t_n)$ by $x_{n+1}(t_{n+1})$ and processing yet another observation, $y_M(t_{n+2})$.

Equation (4.53) is a linear expression that expresses the updated value of $x(t_{n+1})$ as a function of the previous value, $x(t_n)$, a weighting term K, and the observation residual δy_{n+1}. The constant K is called the Kalman gain, which in turn is a function of the uncertainty in the state vector (namely the covariance of x_n given by $P_n(t_{n+1})$, and the uncertainty in the observation, σ_{n+1}). Equation (4.55) updates the uncertainty in the state vector, while (4.56) uses the transition matrix to update the value of x from time, t_n, to time, t_{n+1}. Similarly (4.57) updates the covariance matrix at time t_n to time t_{n+1}. In these equations, the parenthetical expression (t_{n+1}/t_n) denotes that the associated quantity is based on the previous n observations, but is extrapolated to time t_{n+1}. Correspondingly, (t_{n+1}/t_{n+1}) indicates that the associated quantity is valid at time t_{n+1} and also has been updated to include the effect of the (n+1)[st] observation.

The Kalman gain, K, directly scales the magnitude of the correction to the state vector, Δx_n. The gain, K, will be relatively large (and hence will cause a large change in the state vector estimate) under two conditions:

1. When the uncertainty in the prior state vector is large (i.e., when $P_n(t_{n+1})$ is large);
2. When the uncertainty in the $(n+1)^{st}$ observation is small (i.e., when σ_{n+1} is small).

Conversely, the Kalman gain will be small when the state vector is well-known (i.e., when $P_n(t_{n+1})$ is small) and/or when the $(n+1)$st observation is very uncertain (i.e., when σ_{n+1} is large). This result is conceptually pleasing, since we want to significantly improve inaccurate state vector estimates with accurate new observations but do not want to corrupt accurate state vectors with inaccurate data.

There are two main advantages of sequential estimation over batch processing. First, (4.53)-(4.57) can be formulated entirely as scalar equations requiring no matrix inversions. Even when each observation, $y_M(t_n)$ is a vector quantity (e.g., a radar observation comprising range, range-rate, azimuth, and elevation), we can treat each observation component as a separate observation occurring at the same observation time, t_{n+i}. This scalar formulation allows very computationally efficient formulations. The second advantage is that the sequential process allows the option of updating the reference solution $[x_{REF}(t_o)]$ after each observation is processed. This option is generally referred to as the extended Kalman filter. This option provides an operational advantage for dynamic problems such as target tracking because a current estimate of $x(t)$ is available for targeting purposes, or to guide the sensors.

Note that a sequential formulation can be derived for any of the optimization criteria previously described in Section 4.2.

4.4.2   Sequential Estimation Processing Flow

A processing flow for sequential estimation is shown in Figure 4.4. The processing flow is shown for an extended Kalman filter with dynamic noise. Input to the process are an initial estimate of the state vector, $x_0(t_o)$, and associated covariance matrix, $P_0(t_o)$. For each measurement, k=1,2,...,n the following steps are performed.

1. Retrieve the observation yM(tk) and its associated uncertainty, σk;
2. Solve the differential equations of motion (4.20) to propagate the state vector from time tk-1 to tk;
3. Compute the transition matrix $\Phi($ tk, tk-1);
4. Propagate Δx and Pk-1 from time tk-1 to tk using the transition matrix [i.e., (4.56) and (4.57), respectively];

5. Compute a predicted observation, $g(x(tk),tk)$ (4.21), the observation residual, δ_{v_k} , and Hk, the partial derivative of the predicted observation with respect to the state vector;
6. Compute the Kalman gain via (4.54);
7. Update the state vector correction, $\Delta\hat{x}(t_k,t_k)$ [(4.56) and the covariance matrix, Pk(tk) ((4.55)];
8. Update the reference state vector,

$$\hat{x}(t_k) = \hat{x}(t_k) + \Delta\hat{x}(t_k,t_k) \tag{4.58}$$

Steps 1-8 are repeated until all observations have been processed.

Output from the sequential estimation is an updated state vector, $x(t_k)$ and associated covariance matrix, $P_n(t_n)$, based on all n observations.

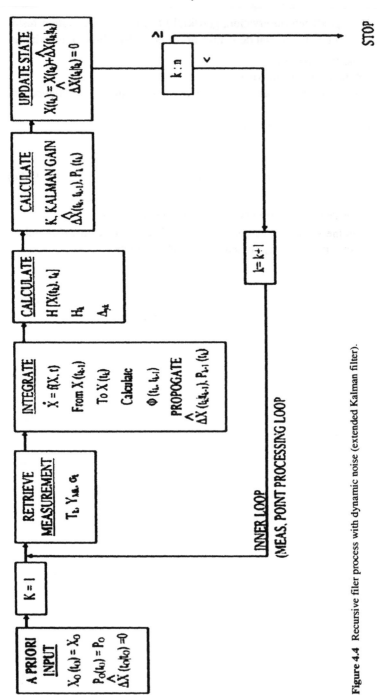

Figure 4.4 Recursive filer process with dynamic noise (extended Kalman filter).

4.4.3   Sequential Processing Implementation Issues

While implementation of a sequential estimation process can be performed in a computationally efficient, scalar formulation, there are a number of other implementation issues. Some of these issues are described below.

One potential problem of a sequential estimation is termed divergence. The problem occurs when the magnitude of the state vector covariance matrix becomes relatively small. This decrease in $P_{\Delta x}$ occurs naturally as more observations are processed, since the knowledge of the state vector increases with the number of observations processed. When $P_{\Delta x}$ becomes relatively small, the Kalman gain becomes correspondingly small [since $P_{\Delta x}$ is a multiplicative factor in (4.54)]. The result is that the estimator *ignores* new data and does not make significant improvements to Δx. While this would seem to be a desirable result of the estimation process, sometimes $P_{\Delta x}$ becomes artificially small, resulting in the filter disregarding valid observations. In order to correct this divergence problem, two techniques are often used: (1) introduction of process noise, and (2) use of a fading memory factor.

Process or dynamic noise is white noise added to the linear state perturbation model, i.e.,

$$\Delta \hat{x}(t_{n+1}) = \Phi(t_{n+1}, t_n)\Delta \hat{x}(t_n) + \eta \tag{4.59}$$

The vector, η, is assumed to have zero mean and covariance Q,

$$\text{cov}(\eta) = Q \tag{4.60}$$

This dynamic noise represents imperfections or random errors in the dynamic state model. The noise vector in turn affects the propagation of the covariance propagation equation, yielding

$$P_n(t_{n+1}) = \Phi(t_{n+1}, t_n)P_n(t_n)\Phi(t_{n+1}, t_n)^T + Q(t_n) \tag{4.61}$$

instead of (4.57). Hence the covariance matrix has a minimum magnitude of $Q(t_n)$. This reduces the divergence problem.

Another technique used to avoid divergence is the use of a fading memory. The concept here is to weight recent data more than older data. This can be accomplished by multiplying the covariance matrix by a memory factor,

$$s = e^{\Delta t / \tau}$$

such that

$$P_n(t_{n+1}) = s\Phi(t_{n+1}, t_n) P_n(t_n) \Phi(t_{n+1}, t_n)^T \qquad (4.62)$$

where Δt is the interval $(t_{n+1}-t_n)$, and τ is an a priori specified memory constant. This is chosen so that $s \geq 1$. The fading memory factor, s, amplifies or scales the entire covariance matrix. By contrast, process noise is additive and establishes a minimum value for P_n. Either of these techniques can effectively combat the divergence problem. Their use should be based on physical or statistical insight into the estimation problem.

A second implementation issue for sequential estimation involves the formulation of the filter equations. The previous section derived a linearized set of equations. Nonlinear formulations can also be developed. Gelb [10] describes a second-order approximation to sequential estimation. Even if a linear approximation is used, there are variations possible in the formulation of the equations. A number of techniques have been developed to ensure numerical stability of the estimation process. Gelb [10] describes a so-called square root formulation and a UDU^T formulation both aimed at increasing the stability of the estimators. It is beyond the scope of this book to describe these formulations in detail. Nevertheless, the reader is referred to work by Tapley (see for example [30]) for a detailed discussion and comparison of results.

The issue of observability is just as relevant to sequential estimation as it is to batch estimation. The state vector may be only weakly related to the observational data, or there may exist a high degree of correlation between the components of the state vector. In either case, the state vector will be indeterminate based on the observational data. As a result, the filter will fail to obtain an accurate estimate of the state parameters.

A final implementation issue addressed here is that of data editing. As in the batch estimation process, indiscriminate editing (rejection) of data is not recommended without a sound physical basis. One technique for editing residuals is to compare the magnitude of the observation residual, $v(t_k)$, with the value $[HPH^T + \sigma_{OBS}]$

$$\frac{|v(t_k)|}{[HPH^T + \sigma_{OBS}^2]} > O_{MAX} \qquad (4.63)$$

If the ratio specified by (4.63) exceeds O_{MAX}, then the observation is rejected and the state vector and covariance matrix are not updated.

4.4.4 The Alpha-Beta Filter

Prior to the publication and popularization of the Kalman filter for sequential estimation, the alpha-beta deterministic filter was applied to tracking moving objects. Variations of the alpha-beta filter have been used since the late 1950s and early 1960s for single-sensor target tracking. The formulation is especially amenable to an analog implementation. References [31-33] provide an overview of the alpha, alpha-beta, and alpha-beta-gamma trackers. Reference [33] in particular compares these filters and demonstrates their performance versus a Kalman filter for a radar target tracking problem. We summarize the alpha-beta filter in this section.

The alpha, alpha-beta, and alpha-beta-gamma filters refer respectively to a first-, second-, and third-order deterministic mean-square-error (MSE) sequential process that relates an MSE estimate of the target state vector to noise-corrupted observations. Kalata [32] derives all three formulations. The formulation for an alpha-beta filter is summarized below.

Consider a one-dimensional target tracking problem. We seek to determine the position and velocity of an object observed at equal time intervals; that is, we seek

$$x = \begin{bmatrix} x \\ \dot{x} \end{bmatrix} \qquad (4.64)$$

given observations, $z(t_k)$, where $t_k = 0, T, 2T, \dots kT$. The equations of motion may be modeled by a truncated Taylor series:

$$x(t_{k+1}) = \Phi(t_{k+1}, t_k) x(t_k) \qquad (4.65)$$

$$\Phi(t_{k+1}, t_k) = \begin{pmatrix} 1 & T \\ 0 & 1 \end{pmatrix} \qquad (4.66)$$

The observations may be related to the (unknown) state vector by the expression

$$z(t_k) = h x(t_k) \qquad (4.67)$$

If the position is directly observed, then the measurement transform h = [1,0].

The alpha-beta filter assumes that the estimated position of a target is the weighted average of the predicted position and the new observation:

$$x_s(t_k) = x_p(t_k) + \alpha[x_m(t_k) - x_p(t_k)] \qquad (4.68)$$

$$\dot{x}_s(t_k) = \dot{x}_s(t_{k-1}) + (\beta/T)[x_m(t_k) - x_p(t_k)] \qquad (4.69)$$

$$x_p(t_{k+1}) = x_s(t_k) + T\dot{x}(t_k) \tag{4.70}$$

where $x_s(t_k)$ is the smoothed (i.e., estimated) position of the object at time t_k, $\dot{x}_s(t_k)$ is the smoothed velocity at t_k, and $x_p(t_{k+1})$ is the predicted position at time t_{k+1}. The measured value of the state vector $x_m(t_k)$ is either observed directly, or available via (4.67). The quantities α and β are smoothing constants whose values are specified a priori.

The performance of the alpha-beta filter clearly depends upon the values of α and β in (4.68) and (4.69). Selection of these smoothing constants depends upon both the dynamics of the object and the observation noise. Different sets of smoothing constants may be used depending upon whether the target is assumed to be moving with constant velocity or accelerating (maneuvering). Typically α and β are related by an equation to reduce the problem to choosing a single parameter [34]. Common choices are

$$\alpha = 2\sqrt{\beta} - \beta \tag{4.71}$$

and

$$\beta = \alpha^2 / (2 - \alpha) \tag{4.72}$$

The relative values of α and β provide a trade-off between noise performance (i.e., the extent to which smoothed values of $x(t_k)$ are impervious to noise) versus an ability of the recursive filter to track objects in spite of abrupt acceleration.

By contrast to this formulation for an α-β filter an alpha filter assumes that the target is moving with constant velocity [hence equation (4.69) reduces to $x_s(t_k) = x(t_{k-1})$]. The α-β-Υ formulation extends the α-β filter to add an acceleration component. The value of the alpha-beta and related filters lies in the simplicity of implementation and minimum computational resources required. This allows multiple targets to be tracked using relatively small computers, hence its popularity in the early days of target tracking. The alpha-beta formulation can be extended to the multiple sensor case (see [31]). However, in that case, compensation must be made for the asynchronous observations. The alpha-beta filter does not perform as well as the more elaborate Kalman filter, but in many instances, allows effective tracking. A performance comparison of these filters is provided by [32].

4.5 COVARIANCE ERROR ESTIMATION

In the discussion of estimation techniques in Sections 4.3 and 4.4, the aim was to estimate a value of the state vector, **x**, given a redundant set of observations, y_i. Batch and sequential approaches were developed to process data to arrive at an optimal value of the state vector. In some situations, however, we are not interested in the specific value of **x**, but rather in the resulting uncertainty. For example, design of a new sensor suite for a tactical platform might trade off the number and accuracy of various sensors to determine the state vector within desired limits. Mission planning is another area in which the state vector accuracy is of primary interest. NASA space flight missions, for example, require careful planning to determine the sensors, observation rates, observational durations, and other factors to establish whether a trajectory can be determined with sufficient accuracy to maintain contact with a vehicle over an extensive time period. This section describes the covariance error analysis technique that allows such analyses to be performed.

We consider the usual observational problem in which one or more sensors observe an object whose (unknown) state vector is x. The observations are related to the state vector by the relation

$$y = f(x, z) + n \qquad (4.73)$$

where x is the state vector, n is unknown observational noise, and z is a vector of model parameters. The vector z comprises a set of parameters whose values are assumed to be constant, but whose values may not be known precisely. The components of x are sometimes referred to as *solve-for* parameters while the components of z are termed *consider* parameters. Examples of such model parameters include coefficients of the Earth's geopotential, atmospheric density coefficients, or parameters related to the observation model (e.g., sensor biases). The batch approach to determining the best estimate of a state vector **x**, minimizes the loss function,

$$L(x) = [y - f(x, z)]^T W [y - f(x, z_0)] + (x - x_0)^T P_{\Delta x_0} (x - x_0) \qquad (4.74)$$

This is a weighted Bayesian loss function. The quantity $P_{\Delta xo}$, the covariance of **x**, and W is an a priori weighting matrix.

The MAP optimal estimate for the state vector, **x**, involves a linearized, iterative, solution of (4.58). The form of the solution is given in Table 4.4. Suppose that we are only interested in the value of $P_{\Delta x}$. Assume that measurement quantities, y_i, are unbiased random variables exhibiting a Gaussian probability distribution, and the observations are stochastically independent (i.e., the error in a measurement at time, t_i, is independent of the error in measurement at any other time, t_k). Under these conditions, the covariance becomes

$$P_{\Delta x} = \Psi \{ F^T WEP_{\Delta z_0} E^T WF + \Psi^{-1} + F^T WEC_{\Delta x_0 \Delta z}^T P_{\Delta x_0}^{-1} + P_{\Delta x_0}^{-1} C_{\Delta x_0 \Delta z} E^T WF \} \Psi^T$$

$$(4.75)$$

where

$$\Psi = (F^T WF + P_{\Delta x_0}^{-1})^{-1} \qquad (4.76)$$

Here, F is the matrix of partial derivatives of the function f with respect to the state vector, x,

$$F = \left(\frac{\partial f}{\partial x} \right) \qquad (4.77)$$

and E is the matrix of partial derivatives of f with respect to the model parameters, z.

$$E = \left(\frac{\partial f}{\partial z} \right) \qquad (4.78)$$

The quantity $P_{\Delta z o}$ is the a priori covariance matrix of the model parameters. The correlation matrix between the a priori state vector, x, and the model parameters, z_0 is represented by $C_{\Delta xo \Delta z}$.

Notice that in (4.75), the observation residuals Δy do not appear. Hence, the covariance matrix of the state vector can be obtained knowing only the following:

- The a priori covariance of the state vector $P_{\Delta xo}$;
- The a priori covariance of the model parameters, $P_{\Delta zo}$;
- The partial derivatives of the observational quantities with respect to the state vector and the model parameters;
- The correlation matrix, $C_{\Delta xo \Delta z}$ (usually unknown and hence neglected).

This formulation allows the uncertainty in the state vector $P_{\Delta x}$ to be computed as a function of the uncertainties in the observations, and uncertainties in the model parameters without any actual observations and without actually estimating x.

Covariance error estimation (i.e., estimating the value of $P_{\Delta x}$ without estimating, x, and without observations), may be used to address a number of issues in a multisensor fusion problem including the following:

1. *Estimation accuracy:* for given observation errors from a sensor suite, what are the corresponding errors in the state vector?
2. *Data span:* How do the errors in the state vector decrease with increasing data?
3. *Data density:* What is the effect of increased numbers of observations?
4. *Data distribution:* To what extent does the distribution of data throughout the observation interval influence the errors in the state vector?
5. *Covariance mapping:* How uniform are the errors in the epoch state vector throughout the data span that was used to determine the epoch Covariance matrix? (Note that this is a separate question from 2.)
6. *Observability and correlation:* What is the relationship between the observable quantities and the state vector parameters? Included in this is the sensitivity of the error in a parameter to added data. Finally, how are the state vector parameters interrelated?
7. *State vector parameters:* To what extent does the selection of state vector parameters influence the resulting achievable accuracy?
8. *Modeling errors:* How do errors in the dynamic model and observation equations influence the errors in the state vector?
9. *A priori covariance effects:* How does the choice of the a priori covariance matrix affect the errors in the state vector parameters?

An example of a covariance error analysis study for orbit/attitude determinations using Landsat data is provided by [35]. Figure 4.5 shows an example of the error analysis results using the research and development goddard trajectory determination system. Numerous such studies have been performed with resulting insight into these types of issues.

A covariance error estimate can also be formulated in a sequential as well as a batch mode. Reference [36] provides a derivation of covariance error analysis in both a sequential and formulation batch. While covariance error analysis can provide valuable insights for sensor design and mission planning, the technique should be used with caution. In particular, the assumption of unbiased, Gaussian distributed, stochastically independent measurements may not be accurate. Effects such as nonwhite noise, observation biases, and cochannel interference may corrupt an actual estimation process. Hence, it is almost axiomatic that the state covariance, $P_{\Delta x}$, predicted by (4.63) will be optimistic (i.e., the actual errors in the estimated state will generally be larger than that predicted by covariance error analysis). Hence, while this type of analysis is useful for studying the relative magnitude of various effects, such as model errors, the absolute error predictions should be viewed with proper skepticism.

4.6   RECENT DEVELOPMENTS IN ESTIMATION

This chapter has provided an introduction to basic target tracking and state estimation techniques. In recent years a number of advanced techniques have been explored. These include probabilistic data association (PDA) and IMMPDAF methods introduced by Y. Bar-Shalom and his students, the particle filter (e.g., Keith Kastella and John Percival), the PHD filter (Ron Mahler), and the covariance intersection method. Each of these techniques is summarized as follows:

- Probablistic data association (PDA) method: Bar-Shalom, Kirubarajan, and Li have introduced several techniques for addressing the challenge of tracking targets with less-than-unity probability of detection in the presence of false alarms (e.g., clutter). The probabilistic data association approach calculates in real-time the probability that each sensor measurement is attributable to the target of interest. The probability information is used in a tracking filter that accounts for the measurement uncertainty. In particular, for each possible observation-to-track pair, the PDA method computes an association probability (namely the probability that the observation is associated with the tracked target of interest). The state update equations account for the uncertainty that the observation is associated with the target represented by the state vector. Bar-Shalom and Kirubarajan [37] provide a detailed description of this technique including probability models that include target models and motion analysis models.

- Interacting multiple models with probabilistic data association filter (IMMPDAF): A special challenge for tracking involves highly maneuvering targets, tracked in the presence of clutter and electronic countermeasures (ECM). Maneuvering targets are difficult to track because it is not possible to observe acceleration directly—we can only observe the results of acceleration, not the instantaneous act of acceleration. Moreover, it is not possible to predict the time and amount of acceleration for targets maneuvering in response to human volition (e.g., a pilot changing course for an aircraft). Thus, state estimates obtained using common tracking algorithms will tend to "lag" behind the true position of the target. That is, the estimation algorithm will tend to predict where the target would have been without the acceleration, and needs to "catch up" with the true state by using the differences between the predicted observations and the actual observations. If the estimation filter is made too sensitive to these observational differences then it may overcorrect the state vector, and if it is not sensitive enough it will never "catch up" with the true target state. The interacting multiple model (IMM) approach uses multiple estimators, each assuming a different model of target maneuverability (e.g., a benign motion model that assumes relatively slow maneuvers and abrupt maneuver models that assume periodic high-g acceleration). These estimators are used simultaneously and the results of each estimator are made available to the other estimators to attempt to develop the best estimate of the target's state at a given time. A discussion of these types of techniques is provided by Lerro and Bar-Shalom [38], and a comparison of interactive multiple model tracking with joint probability of data association (IMMJPDA) versus conventional multiple hypothesis tracking and Kalman filtering using a neural net correlator is provided by Feo, et al. [39]

- The particle filter: A number of researchers have begun to apply so-called particle filters to target tracking [40]. Particle filters use multiple discrete "particles" to represent the belief distribution over the location of a tracked target. This might be the case for a complex target in which observations are received from different elements of the target at different times. Examples of such situations include use of video images or radar tracking. In addition, this approach has been applied to surveillance situations in which multiple targets are tracked within a surveillance region.

- Probability hypothesis density (PHD) tracker: R. Mahler [41] has developed a theoretical foundation for the probability hypothesis density tracker introduced by Stein and Winter. The concept involves a multitarget tracking and evidence accumulation concept called a probability hypothesis surface (PHS). A PHS is the graph of a probability distribution (the probability hypothesis density) that is integrated over a region in target state space to give the expected number of targets in that region. The PHD is the expected value of the point process of a random track-set. This approach allows incorporation of probability models for the "birth" (i.e., the appearance of new targets in a region) and the "death" (i.e., the disappearance of targets in a region). Mahler provides a detailed mathematical description of this approach and provides a comparison to multitarget, Bayesian nonlinear filtering.

- Covariance intersection method: S. Julier and J. K. Uhlman [42] describe the covariance Intersection method. This technique addresses the problem of combining data (or state estimates) from multiple sources that are not independent or have a known degree of correlation. Many of the techniques commonly used in state estimation use the simplifying assumption that the observations are stochastically independent. This assumption simplifies the form of the state estimation equations. The Covariance Intersection method specifically accounts for the dependence of the evidence to be combined. The algorithm takes a complex combination of mean and covariance estimates that are represented in information (inverse covariance) space. This is based on a geometric interpretation of the Kalman filter equations. Julier and Uhlman provide a detailed description of this technique and numerical examples of the algorithm performance.

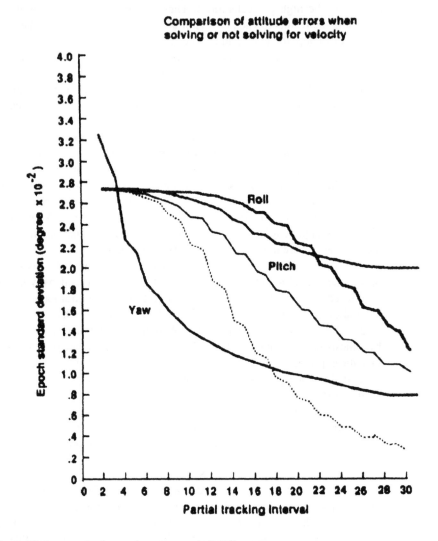

Figure 4.5 An example of a covariance error analysis [35].

REFERENCES

[1] Sorenson, H. W., "Least Squares Estimation: From Gauss to Kalman," *IEEE Spectrum*, 1970, pp. 63-68.

[2] Gauss, K. G., *Theory of Motion of the Heavenly Bodies (reprinted from the 1809 original entitled, Theoria Motus Corporum Coelestium)*, New York, NY: Dover, 1963.

[3] Legendre, *Nouvelles Methods pour la Determination des Orbits des Commetes*, Paris, 1806.

[4] Bell, E. T., *Men of Mathematics*, New York, NY: Simon and Schuster, 1961.

[5] Fischer, R. A., "On the Absolute Criteria for Fitting Frequency Curves," *Messenger of Mathematics*, vol. 41, 1912, pp. 155.

[6] Wiener, N., *The Extrapolation, Interpolation and Smoothing of Stationary Time Series*, New York, NY: John Wiley and Sons, 1949.

[7] Kolmogorov, A. N., "Interpolation and Extrapolation Von Stationaren Zufalliegen Folgen," *Bulletin of the Academy of Sciences, USSR, Ser. Math. S.*, 1941, pp. 3-14.

[8] Kalman, R. E., "New Methods in Wiener Filtering Theory," in *Proceedings of the First Symposium of Engineering Application of Random Function Theory and Probability*, New York: John Wiley and Sons, 1963, pp. 270-388.

[9] Blackman, S. S., *Multiple Target Tracking with Radar Applications*, Norwood, MA: Artech House, Inc., 1986.

[10] Gelb, A., *Applied Optimal Estimation*, Cambridge, MA: MIT Press, 1974.

[11] Zavaleta, E. L., and E. J. Smith, *Goddard Trajectory Determination System User's Guide*, Silver Spring, MD: Computer Sciences Corporation, 1975.

[12] Hall, D., and S. R. Waligora, "Orbit/Attitude Estimation Using Landsat-1 and Landsat-2 Landmark Data," *NASA Goddard Space Flight Center Flight Mechanics/Estimation Theory Symposium*, NASA Goddard Space Flight Center, MD, October 1978.

[13] Deutsch, R., *Orbital Dynamics of Space Vehicles*, Englewood Cliffs, NJ: Prentice-Hall, 1963.

[14] Waltz, E., and J. Llinas, *Multi-sensor Data Fusion*, Norwood, MA: Artech House, Inc., 1990.

[15] Escobol, R. R., *Methods of Orbit Determination*, Melbourne, FL: Krieger Pub. Co., 1976.

[16] Kamen, E. W., and C. R. Sastry, "Multiple Target Tracking Using an Extended Kalman Filter," *SPIE Signal and Data Processing of Small Targets*, Orlando, FL, 1990, pp. 410-424.

[17] Sanza, N. D., M. A. McClure, and J. R. Cloutier, "Spherical Target State Estimators," *American Control Conference*, Baltimore, MD, June, 1994, pp. 1675-1679.

[18] Wilde, D. J., and C. S. Beightler, *Foundations of Optimization*, Englewood Cliffs, NJ, 1967.

[19] Press, W. H., et al., *Numerical Recipes: The Art of Scientific Computing*, New York, NY: Cambridge University Press, 1986.

[20] Brent, R. P., *Algorithms for Minimization without Derivatives*, Englewood Cliffs, NJ: Prentice-Hall, 1973.

[21] Shoup, T. E., *Optimization Methods with Applications for Personal Computers*, Englewood Cliffs, NJ: Prentice-Hall, 1987.

[22] Nelder, J. A., and R. Mead, "A Simplex Method for Function Minimization," *Computer Journal*, vol. 7, 1965, p. 308.

[23] Polak, E., *Computational Methods in Optimization*, New York, NY: Academic Press, 1971.

[24] Henrici, P., *Discrete Variable Methods in Ordinary Differential Equations*, New York, NY: John Wiley and Sons, 1962.

[25] Swerling, P., "First Order Error Propagation in a Stagewise Smoothing Procedure for Satellite Observations," *Journal of Astronautical Science*, vol. 6, 1959, pp. 46-52.

[26] Kalman, R. E., "A New Approach in Linear Filtering and Prediction Problems," *Journal of Basic Engineering*, vol. 82, 1960, pp. 34-45.

[27] Kalman, R. E., and R. S. Bucy, "New Results in Linear Filtering and Prediction Theory," *Journal of Basic Engineering*, vol. 83, 1961, pp. 95-108.

[28] McGee, A., and S. F. Schmidt, *Discovery of the Kalman Filter as a Practical Tool for Aerospace and Industry*, Ames Research Center, CA: NASA, 1985.

[29] Sayed, A. H., and T. Kailath, "A State-Space Approach to Adaptive RLS Filtering," *IEEE Signal Processing Magazine*, vol. 11, 1994, pp. 18-70.

[30] Tapley, D. B., and J. G. Peters, "Sequential Estimation Using a Continuous UDU^T Covariance Factorization," *Journal of Guidance and Control*, vol. 3, 1980, pp. 326-331.

[31] Lefferts, R. E., "Adaptive Correlation Regions for Alpha-Beta Tracking Filters," *IEEE Transactions on Aerospace Electronic Systems*, vol. AES-17, 1981, pp. 738-747.

[32] Kalata, P. R., "The Tracking Index: A Generalized Parameter for a-b and a-b-g Target Trackers," *IEEE Transactions on Aerospace Electronic Systems*, vol. AES-20, 1984, pp. 174-182.

[33] Painter, J. H., D. Kerstetter, and S. Jowers, "Reconciling Steady-State Kalman and Alpha-Beta Filter Design," *IEEE Transactions on Aerospace Electronic Systems*, vol. 26, 1990, pp. 986-991.

[34] Buede, D. M., *Comments on Kalman Filters*, D. Hall, Editor, Washington, D.C., 1991.

[35] Hall, D., and N. V. Kumar, *Error Analysis Studies on Orbit/Attitude Determination Using Landsat-1 and Landsat-2 Data*, Silver Spring, MD: Computer Sciences Corporation, 1979.

[36] Cappellari, J. O., C. E. Velez, and A. J. Fuchs, *Mathematical Theory of the Goddard Trajectory Determination System*, Greenbelt, MD: NASA Goddard Space Flight Center, 1976.

[37] Kirubarajan, T., and Y. Bar-Shalom, "Target Tracking Using Probabilistic Association-Based Techniques with Applications to Sonar, Radar, and EO Sensors," in *Handbook of Multisensor Data Fusion*, ed. by J.L. D. L. Hall: CRC Press, 2002.

[38] Lerro, D., and Y. Bar-Shalom, "Interacting Multiple Model Tracking with Target Amplification Feature," *IEEE Transactions on Aerospace and Electronic Systems*, vol. AES-29, 1993, pp. 494-509.

[39] Feo, M., et al., "IMMJPDA vs. MHT and Kalman Filter with NN Correlation: Performance Comparison," *IEE Proceedings on Radar, Sonar and Navigation*, vol. 144, 1997, pp. 49-56.

[40] Gordon, N., "A Hybrid Bootstrap Filter for Target Tracking in Clutter," *IEEE Transactions on Aerospace and Electronic Systems*, 1997.

[41] Mahler, R., "A Theoretical Foundation for the Stein-Winter Probability Hypothesis Density (Phd) Multi-Target Target Tracking Approach," *Proceedings of the 2000 MSS National Symposium on Sensor and Data Fusion*, October, 2002, pp. 99-118.

[42] Julier, S., and J. K. Uhlmann, "General Decentralized Data Fusion with Covariance Intersection," in *Handbook of Multi-Sensor Data Fusion*, ed. by D.L. Hall and J. Llinas, Boca Raton, FL: CRC Press, 2001.

Chapter 5

Identity Declaration

Identity declaration involves transforming observed attributes of an entity such as size, shape, and spectral characteristics into a label that describes or names the identity of the entity. This is the classic target recognition problem. Techniques for identity declaration are drawn from the literature of pattern recognition. This chapter provides an introduction to the problem and overview of techniques.

5.1 IDENTITY DECLARATION AND PATTERN RECOGNITION

The problem of position and identity fusion was introduced in Chapter 2. Figure 2.2 illustrated the concept of positional fusion by illustrating multiple sensors observing data that could be related to the kinematic state (e.g., position and velocity) of an entity. Chapter 3 addressed the issue of data association (i.e., how to group observations such that they represent a single object or entity). Chapter 4 described techniques by which multiple data could be fused to obtain an estimate of an object's position and velocity. These techniques were also shown to be applicable to fusion of attribute data such as wave length of emitters or other nonkinematic characteristics. In the discussion of positional fusion, alternative architectures were described that allowed the positional fusion to occur at various levels ranging from the fusion of raw data to the fusion of state vectors from multiple sensors.

We turn now to the problem of fusing multiple sensor data to determine an estimate of object identity. Clearly the two problems are related: Positional information can be of value in establishing the identity (and function) of an entity, while identity information can be of value in the estimation problem, particularly for distinguishing among closely spaced targets. In an implemented fusion system these processes would be interleaved. For simplicity, however, we have separated them into distinct problems as illustrated in Figure 5.1.

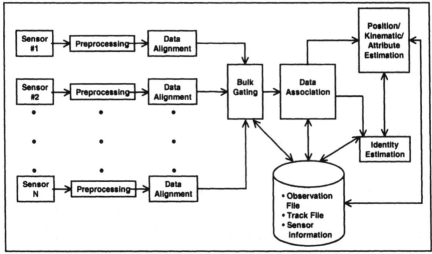

Figure 5.1   The role of identity fusion within level 1 processing.

Identity declaration problems arise in a number of situations. The military applications summarized in Chapter 1 describe instances in which one or more sensors observe emitters, platforms, and weapon systems. Military data fusion applications seek to locate, characterize, and identify the observed entities. A specific example might entail a threat-warning sensor onboard a tactical aircraft that attempts to determine when the aircraft has been illuminated by a weapons guidance device (e.g., laser or missile guidance radar). Another example is the use of radar cross-section data to determine whether an entity is a rocket body, fragment, or reentry vehicle. Identification-friend-foe-neutral (IFFN) equipment uses signature and related data to attempt to identify enemy versus friendly aircraft. More detailed and time-consuming analyses are sometimes performed to *fingerprint* or identify specific emitters or weapons platforms. Steinberg [1] illustrates an example of the types of identity processing onboard a tactical aircraft in a military engagement. Nonmilitary examples of identity declaration problems include identification and isolation of equipment faults in complex systems [2], use of sensor data to monitor manufacturing processes [3], and semi-automated monitoring of human health condition via medical probes and monitors.

The concept of identity fusion was introduced in Chapter 2. Figure 2.5 illustrated multiple sensors observing several objects. A decision level fusion process is shown in which the sensor data (e.g., time series, attribute vector, or imagery) are transformed into feature vectors, y_i, which represent the raw data. Subsequently, each feature vector is transformed into a declaration of object identity using a pattern recognition process. Finally, the declarations of identity from multiple sensors are fused to form a joint declaration of object identity. The first two steps of this process, feature extraction and identity declaration, are illustrated in Figure 5.2 for a single sensor.

Feature extraction transforms the data output from the sensor into a representation or abstraction of that data. Features may be developed for analog, digital, and imagery data. Examples of features extracted from time series data include the following:

- Identification, location, and amplitude of peaks;
- Peak width and duration;
- Average signal amplitude;
- Standard deviation of the variance about a mean;
- Frequency transformation parameters such as Fourier coefficients;
- Location and amplitude of spectral peaks.

Features that may be used to represent imagery data include the following:

- Identification and characterization of image segments;
- Shape, length, width, orientation of objects, and spectral characteristics;

Features are a representation of the data that may be useful in the identification process. For example, humans are able to recognize other humans by an abstraction process, which focuses on distinctive features such as characteristic movements, voice tonality, hair color, or distinct facial features (e.g., a prominent nose). Political cartoonists utilize such features to create very simple but readily identifiable figures. It must be recognized that, while often useful, the extraction of features reduces the information content of sensor data. Section 5.2 discusses this issue in more detail and describes some of the many possible features used to represent data.

The second step, illustrated in Figure 5.2, is an identity declaration process that maps a feature vector into a region in feature space. The right-hand side of Figure 5.2 illustrates that a three-dimensional feature vector, y_i, having components (y_{i1}, y_{i2}, y_{i3}) may be mapped into separable regions in the three-dimensional space. Two regions, A and B, are shown. If these spatial regions represent the identity of two separate entities, then the location of a vector y in region A of the feature space constitutes the identification that y is associated with entity (or target class) A. Hence, the identity declaration process using feature vectors may be seen as analogous to the process used in data association. One difficulty of the identity process occurs when the feature-space regions representing different entities overlap. That results in ambiguity making it impossible to declare a unique identity solely on the basis of the observed features.

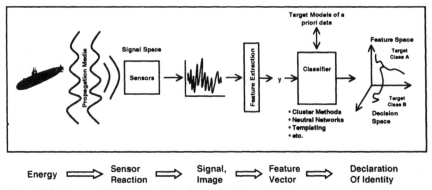

Figure 5.2 Concept of single sensor identity declaration.

The identity declaration process uses pattern recognition techniques such as templating, cluster analysis, adaptive neural networks, or knowledge-based techniques to recognize an entity's identity. These techniques are described in detail in subsequent sections of this chapter. Each of these techniques involves a two-phase operation. In the first phase (the training phase), data from known entities are used to establish the boundaries that partition the feature space into regions, each region establishing identity of an object or entity (i.e., the parametric boundaries for templates or the cluster boundaries for cluster analysis). In the second phase of operation, having established the relationship between feature vectors and identity, the pattern recognition techniques are used in a recognition or classification mode. In such a mode, individual feature vectors are classified by establishing the location of the observed vector with respect to the decision boundaries in feature space. We note that techniques such as neural networks and cluster algorithms "learn" these boundaries in feature space based on examples of data.

A summary of commonly used pattern recognition techniques is shown in Table 5.1. These techniques include: parametric templates, cluster algorithms, neural networks, physical models, knowledge-based approaches, and hybrid methods. Table 5.1 provides a brief description of each technique and establishes a basis for identity declaration and references. These techniques are summarized as follows and described in more detail in Sections 5.3-5.8.

- Parametric templates: Pre-established parametric limits or boundaries are used to define identity classes. For example, a pulsed emitter class may be defined by predefined limits for radio frequency (RF), pulse width (PW), and pulse repetition frequency (PRF). Thus, if $RF_{AMIN} < RF < RF_{AMAX}$; $PW_{AMIN} < PW < PW_{AMAX}$; and $PRF_{AMIN} < PRF < PRF_{AMAX}$, then the observed emission, characterized by the observed vector (RF, PW, PRF) is declared to be a member of emitter class A. Classification is based on the location of a feature vector with respect to template boundaries. Such hard-limited decision methods may not represent the true distribution of entity classes in a given feature space.

- Cluster algorithms: A class of algorithms that use heuristic methods to group data into natural clusters, representing object identity. Classification is based on the proximity of a feature vector with respect to trained clusters in feature space.

- Neural networks: Adaptive neural networks mimic biological nerve connections to perform a nonlinear transformation between an input feature vector and output identity classes.

- Physical models: Physical models may be used to match observed signature data against predicted signatures. Identity classification proceeds by computing the correlation between predicted data and observed data. If the correlation exceeds a pre-established threshold, then the observed data is declared to match the predicted data.

- Knowledge-based approaches: Inference techniques may be used to perform identity declaration. Common techniques map identity to parametric features via syntactical representations (i.e., breaking object identity into recognizable subelements). Knowledge-based inference systems utilize rules, frames, or syntactic logic to establish identity. Knowledge-based methods include Bayes nets and conditioned event algebras.

- Hybrid methods: Hybrid approaches combine the use of *implicit* pattern recognition techniques (e.g., neural nets or cluster algorithms) with *explicit* techniques such as knowledge-based inference methods. Hybrid approaches include the use of syntactical representations of objects or targets (e.g., semantic-based descriptions of targets or entities with pattern recognition of low-level components) and methods in which implicit methods are initially trained using rule-based information.

The decision-level approach to identity fusion illustrated in Figure 2.6 (and 5.1) and described in this section is analogous to state vector fusion for positional estimation. Each sensor performs independent processing to produce an estimate, and these estimates are subsequently combined via a fusion process. Alternative architectures for positional fusion are described in Chapter 1 and illustrated in Figure 1.10. Figure 1.10 illustrated the possibility of fusing raw data (i.e., sensor observations of parametric data related to position), fusing state vectors (preprocessed sensor data to yield single-sensor estimates of position), or a hybrid approach to fuse state vectors with raw data. Similar architectures may be used for identity fusion. Figure 5.3 illustrates three approaches to fusing identity data: (1) decision-level fusion, (2) feature-level fusion, and (3) data-level fusion. Hybrid approaches combining these alternatives may also be utilized. Figure 5.3(a) illustrates decision-level fusion. This is identical to the approach illustrated in Figure 2.4. In this approach each sensor performs a transformation to obtain an independent declaration of identity. The declarations of identity from each sensor are subsequently fused. Techniques for fusing declarations of identity include voting methods, Bayesian inference, Dempster-Shaffer's method, generalized evidence processing theory, and various ad hoc methods. These techniques are described in Chapter 6.

Table 5.1

Summary of Pattern Recognition Techniques

Pattern Recognition Techniques	Description	Basis of Identity Declaration	References
Parametric techniques	A priori boundaries established in feature space to identify unique objects/identities	Location of a feature vector with respect to template boundaries	[4, 5]
Cluster algorithms	Techniques to group data into natural clusters, representing object identity	Proximity of a feature vector with respect to the centroid of a labeled cluster	[6-8]
Neural networks	Emulation of biological nerve connections to produce a nonlinear transformation between feature vectors and identity classes	Mapping between input parameter neural net layer and output identity declaration layer	[9, 10]
Physical models	Use of physical models to predict observed data (e.g., radar cross-section) as a function of object identity	Match of observed versus predicted data (e.g., cross-correlation)	[11]
Knowledge-based methods	Representation of object identity or characteristics via rules, frames, and syntactic representations	Inference techniques determine the extent to which observed features satisfy object representations	[12-14]
Hybrid methods	Combination of implicit and explicit knowledge to characterize an entity	Representation of data characteristics and entity information	[15]

A feature-level fusion approach is illustrated in Figure 5.3(b) In this approach, each sensor observes an object, and a feature extraction is performed to yield a feature vector from each sensor. These feature vectors are then fused and an identity declaration is made based on the joint feature vector. Techniques available to perform the declaration process include cluster analysis, neural networks, templates, and knowledge-based techniques, which are described in the remainder of this chapter. In this feature-level approach, an association process must be used to sort feature vectors into meaningful groups. Location information may be useful to aid this association process, since in principle the feature vectors may be widely different quantities (e.g., Fourier coefficients and time domain characteristics).

A third approach to identity fusion is illustrated in Figure 5.3(c). In this data-level fusion approach, data from commensurate sensors are fused directly, with subsequent feature extraction and identity declaration from the fused data. In order to perform such data-level fusion the sensors must either be identical (e.g., several IR sensors), or commensurate (e.g., an IR sensor and visual imagery sensor). Association is performed on the raw data to ensure that data being fused relates to the same object or entity. Having fused the sensor data, the identification process proceeds identically to the process for a single sensor (e.g., Figure 5.2). For imagery sensors, data level fusion is often referred to as pixel (picture element) level fusion. The extent to which data-level fusion can be performed with diverse data types depends upon the availability of accurate physical models. For example, synthetic aperture radar images may be fused with visible or infrared images if appropriate corrections can be modeled to account for geometrical perspective, image scaling, and related factors.

Which of these alternative approaches is preferred? The relative merits are similar to the relative merits of the positional fusion architectures. In general, better accuracy is obtained by fusing information closer to the source. Thus, data-level fusion is potentially more accurate than feature-level fusion, which in turn is potentially more accurate than decision-level fusion. Clearly, however, data-level fusion is only viable for sensors producing the same type of observations (e.g., images and signal data of the same type). Moreover, data-level fusion requires accurate data alignment among the sensors. The choice of the fusion approach also has implications for system implementation. For example, the communications load to send data, features, or identity declarations to a central fusion process varies widely. Hence, the ultimate choice of processing approach is a system-level decision requiring careful analysis and trade-offs (see, for example [16]). The remainder of this chapter addresses the concept of feature extraction, techniques for identity declaration, and feature-level fusion.

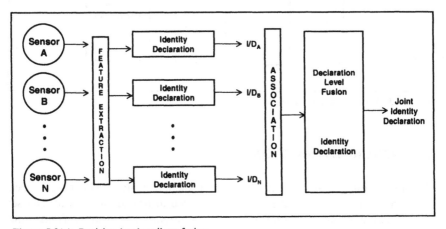

Figure 5.3(a) Decision-level attribute fusion.

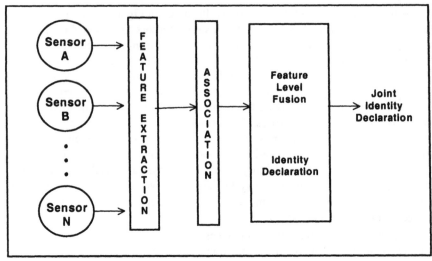

Figure 5.3(b) Feature-level attribute fusion.

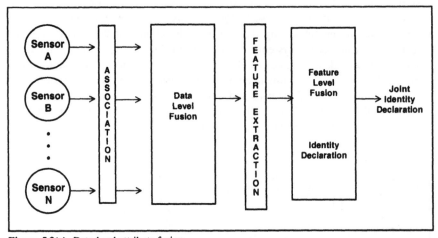

Figure 5.3(c) Data-level attribute fusion.

5.2 FEATURE EXTRACTION

Features are often used in identity declaration and the identity fusion process. Features are an abstraction of the raw data intended to provide a reduced data set, which accurately and concisely represents the original information. Transformation of the raw data into a feature vector is termed feature extraction. This process is similar to, but potentially different than, the concept of data compression. The latter concept seeks a reduced form of the data, which would allow the data to be accurately reconstructed at a subsequent time, with minimum requirement for data storage or transmission. An example of data compression is the use of a differencing operation specifying when transitions occur in a binary time series. Thus, a binary data set:

$$y(t_i) = 0, 0, 1, 1, 1, 0, 0, 0, 0, \ldots$$

$$t_i = t_0, t_1, t_2, \ldots t_i, \ldots$$

may be compressed as,

$$y(t_i) = 0(2), 1(3), 0(4), \text{ etc.}$$

$$t_i = t_0 + I(\Delta t)$$

The parenthetical expression indicates the duration before a transition occurs from 1 to 0 or 0 to 1 in the series. Deming [17] provides an introductory description of the use of data compression in computer science.

By contrast, feature extraction attempts to find characteristics of the data that aid in the identification process. In general, these features may not allow reconstruction of the original data. For example, accurate representation of emitter data received by a collector could require specification of signal amplitude, polarization, frequency, and other information as a function of time. Pertinent features required for emitter identification, however, may only entail average frequency, polarization, and pulse repetition frequency. Other information such as signal amplitude may be of little use for emitter identification. Hence, representation of the data for compression purposes may be significantly different than development of a feature vector for identification purposes.

It is not mandatory that features be extracted for the identification process. Raw signal or imagery data may be used directly. Such identification using raw data is a model-matching process. One example of this is the process used for absolute image registration processing of Landsat and GOES-D weather satellite data. Hall et al. [18] describe a technique to locate reference points in these images. The technique utilizes prestored images of readily recognizable geographic regions (e.g., Haynes point in Maryland). These reference images (termed chips) are compared to new images via a two-dimensional correlation technique. The reference chip is systematically compared to a reference image by computing a correlation coefficient for every (i,j) location within the new image. Because of the large computational requirements, this is generally performed only within a small subarea of the overall image. The center of the reference chip is placed at every (i,j) pixel location in the new image and correlation coefficients are computed. The global maximum of the correlation coefficients is found. If this maximum coefficient exceeds an a priori threshold, then pixel coordinates in the new image are declared to be the precise locations of the reference chip (whose geographic coordinates are known). This allows geographical reference points to be accurately located in Landsat and GOES-D images. This technique allows temporal phenomena such as clouds or undersea shoals to be geographically located with respect to the reference coordinates.

Similarly, objects such as aircraft and tanks could be identified in surveillance imagery by comparing a library of reference photos against each new image. A similar concept may be used for nonimagery data. For example, the radar cross-section (RCS) of an observed object could be compared to prestored RCS signatures in a library. In this case, the identification process becomes one of computing correlation coefficients to determine the correlation of observed data against prestored signature data. If the computed correlated coefficient exceeds an a priori threshold, then the observed object is declared to have the same identity as the prestored data. When prestored data is not available, simulated signature data may be generated via physical models. For some sensor types, such as infrared imagery and radar, modeled data may be much more useful than signature data. The reason for this is that actual data may be appropriate only for very similar conditions to those in which the reference signatures were observed. Thus, variations in observational conditions may not allow accurate comparisons between prestored signatures and new observations.

In practice, identification is rarely done by the direct comparison of observed data against prestored signatures. Such an approach can be time-consuming and computationally expensive. Moreover, direct comparison of data against prestored signatures is not always necessary if representative features can be found. This is analogous to attempting to identify an automobile in a large parking lot by systematically comparing its photographs against photographs of every other automobile in the lot. A far simpler approach would use the license plate number as a feature to identify the automobile. Even if all automobiles in the lot were of the same year, color, and model, the license plate number allows unique (and rapid) identification of one automobile from all others. The use of features to facilitate identity declaration is not only imposed by computation limits. Human cognition is geared towards extraction of key characteristics for recognition purposes. Research indicates that even very young children (less than 6 months old) are able to recognize the "smiley face" symbol as symbolic of a human face.

Successful identification based on features depends upon the selection of features that provide separable regions in feature space. While this is clearly fundamental, often a significant amount of effort in developing an identification process is aimed at peripheral issues such as distance metrics, instead of a concern for the choice of features. Feature selection should be based on a physical knowledge of the objects whose identity or class is sought. In the previous example involving the search for an automobile in a large parking lot (presumably filled with similar automobiles), possible identifying features include color, shape, size, number of doors, tire pressure, and other characteristics. For most automobiles, a poor choice of an identifying feature would be tire pressure. Not only is tire pressure relatively difficult to observe, but it is also ambiguous. Most automobiles have tire pressures ranging from 25 to 35 pounds per square inch. Hence, unless the automobile in question has a flat tire, observed tire pressure does not facilitate identification. Other ambiguous features include the number of doors and the number of tires. People having color-impaired vision often have particular difficulty in distinguishing automobiles. These people sometimes develop identification strategies such as buying a car having a distinctive shape (e.g., a Volkswagen Beetle), or creating a unique feature such as a sticker on the windshield or an ornament on the radio antenna.

Selection of features for identity declaration should be based on physical modeling or experimentation, and features should be selected that are invariant across entity variability and viewing angles. The cluster analysis techniques described in Section 5.4 provide a means of performing numerical experiments to compare the effectiveness of various features. Table 5.2 illustrates a number of candidate features to represent image data [Table 5.2(a)] and one-dimensional (e.g., time series) data [Table 5.2(b)]. Extraction of these features may be performed by a host of image processing and data modeling techniques. The reader is referred to several texts describing image processing including [19-21]. Similarly, a number of texts describe methods for characterizing time series data including [22-24].

Representative features for imagery data include geometrical characteristics of image segments, structural characteristics, statistical features, and spectral information. Man-made objects tend to have distinct, regular boundaries and characteristic spectra. Infrared imagery may be especially useful for detecting rocket plumes or high temperature engines. Remote sensing to detect types of vegetation, existence of diseased plants, or mineral deposits may require other types of features, which combine spectral data with statistical information. Other types of image representations may use pixel-based techniques such as quad-tree representations, or geometric constructs such as fractals.

Features for one-dimensional signal identification are often selected from either a time domain representation of the signal (e.g., amplitude versus time), from the frequency domain (e.g., Fourier coefficients), or from the time-frequency domain (e.g., wavelet coefficients). Time domain features may be viewed as a one-dimensional analog to image features. For limited duty cycle signals (i.e., signals such as voice or pulsed emitters that are *turned on and off*), features can be extracted to describe temporal phenomena such as recurring pulses. Individual pulses may be characterized by pulse shape, rise time, fall time, pulse width, or particular characteristics such as ringing phenomena. Conversely, 100% duty cycle signals (e.g., continuous wave emitters) may be more readily characterized in the frequency domain via Fourier coefficients, existence and location of spectral lines, and features obtained by transformations (e.g., a spectral peak in the power spectral density of a signal raised to the Nth power). Finally, signals that vary their frequency in time (e.g., chirped radar signals) may be more effectively represented in the time-frequency domain.

Table 5.2(a)

Examples of Image Features for Identity Declaration

Feature Type	Representative Features	Description
Geometrical features	Edges Lines Line lengths Line relationships (parallel, perpendicular) Arcs Circles Conic shapes Size Area	Features aimed at representing the geometrical size/shape of objects; man-made objects tend to exhibit regular geometrical shapes with distinct boundaries
Structural features	Surface area Relative orientation Orientation to vertical and horizontal ground plane Juxtaposition of planes, cylinders, and cones	Structural features develop a larger scale and contextual view of image segments
Statistical features	Number of surfaces Area/perimeter Moments Fourier descriptors Wavelet coefficients Mean/variance/moments Kurtosis/skewness Entropy	Statistical features can be used at a local and global image level to characterize image data
Spectral features	Color coefficients Effective blackbody temperature Spectral peaks Spectral signature	Man-made objects tend to have distinct infrared spectral signatures (distinct from non-man-made objects)

Hybrid characterizations may employ both time and frequency information for complex signals. Examples of hybrid characterizations include wavelet representations, use of Wigner-ville representation, and cyclo-stationary techniques (see for example, [25]). Another view of features available from various sensors is illustrated in Table 5.3, which lists a number of common sensors and identifies the signature data available and features that are useful for identification. A special case of signal processing for identification involves processing of human voice data to recognize words, identify the language spoken, or identify the speaker. Commercial voice recognition systems for computer interface typically utilize a comparison between prestored signature data and spoken data. Hence, a voice recognition system is initially trained by requiring a speaker to speak words (typically less than 1,000 words) into a computer. These spoken words are identified by the speaker and used as reference signatures. Subsequent word recognition involves comparison of digitized voice data against the prestored signatures. More sophisticated systems use syntactic rules of sentence structure to identify likely words that would normally follow or proceed recognized words. An enormous amount of research has been performed on the speech recognition problem.

Table 5.2(b)

Examples of Signal Data Features for Identity Declaration

Feature Type	Representative Features	Description
Time domain	Pulse characteristics (e.g., rise time, fall time, and amplitude) Pulse width/duration Moments Ringing and overshoot phenomena Pulse repetition interval Relationship of pulses to ambient noise floor	Selection of time features versus frequency-domain features depends upon the nature of the signal; non-100% duty cycle signals favor time-domain features
Frequency domain	Fourier coefficients Chebysheff coefficients Periodic structures in the frequency domain Spectral lines/peaks Pulse shape and characteristics Forced features (e.g., power spectral density of signal)	Features in the frequency domain are analogous to those in the time domain; 100% duty-cycle signals favor the frequency domain
Hybrid characteristics	Wavelet representations Wigner-ville distributions Cyclostationary represenations	Hybrid features may be useful for signals in which both time and frequency (duration) are important.

Word recognition in human speech is particularly difficult by signal processing techniques. Humans tend to recognize words spoken by fellow humans based on contextual information, meaning, voice inflection, and even body language. Humans are able to understand fellow humans despite differences in pitch, accent, speed, and other factors. Other factors that complicate recognition involve whether a speaker is speaking his or her native tongue or a foreign tongue. Rabiner [26] provides an introduction to speech processing including the use of hidden Markov models and linear predictive coding coefficients.

Selection of features to aid identification requires physical analysis and experimentation. There is no shortcut to feature selection or prescription for finding useful features. Key issues of feature selection include observability, orthogonality, invariance, and predictability. The issue of observability in feature selection is identical to the problem by the same name in parameter estimation described in Chapter 4; namely, is there any connection or mapping between the selected features and object identity? Does a relationship exist? Is it unique to identification classes? Is there ambiguity among classes? Orthogonality seeks to determine whether the selected features are independent (i.e., is the ith component of a feature vector independent of the jth component?). Finally, are the selected features invariant with respect to time, ambient conditions, or state of the observed entity? These are all difficult questions that must be addressed by analysis, modeling, and experimentation [27].

After a feature set is selected, a variety of pattern recognition techniques may be used to perform identity declaration. The next sections will describe some of these methods.

Table 5.3

Features Available from Different Sources

Sensor Type	Signature Format	Useful Features
Thermal imager	• 2-D thermal emission image	• Shape (perimeter/area, aspect ratio, moments), texture, maximum/minimum emission, number and location of hot spots, context
Millimeter-wave	• 1-D reflectivity profile • 1-D or 2-D polarization image • Doppler modulation	• Distribution and range extent of scatterers • Number and location of odd/even bounce scatterers • Propulsion frequencies and linewidths (rpm, blade passing, tread slapping)
Laser radar	• 3-D reflectivity image • Doppler modulation (vibration) • 2-D velocity image	• Size, 3-D shape, scatterer locations • Propulsion, structural, and skin frequencies • Spatial distribution of moving components

Table 5.3   (continued)

Features Available from Different Sources

Sensor Type	Signature Format	Useful Features
Synthetic aperture radar	• 2-D (range-cross range) reflectivity image	• Size, aspect ratio, number and location of scatterers
Television	• 2-D reflectivity (contrast) image	• Shape, texture, internal structure, context
Microwave radar	• Doppler modulation • 2-D (range-cross range) reflectivity image	• Velocity, propulsion frequencies and linewidths • Size, aspect ratio, number and location of scatterers
Acoustic sensor	• Spectrum and time dependence of acoustic emission	• Propulsion frequencies, harmonics, frequency ratios, pump and generator frequencies, peculiar noise sources
Radio frequency interferometer	• Spectrum and time dependence microwave emission	• Frequency, frequency modulation, amplitude modulation, pulse duration, pulse intervals

5.3   PARAMETRIC TEMPLATES

Perhaps the most basic technique for identification is templating. The idea is simple; we assume that the multidimensional feature space can be partitioned into distinct regions, each representing an identification or identity class. An example is illustrated in Figure 5.4 for the identification of pulsed emitters. Two emitters (1 and 2) emit energy pulses, which repeat in time. Figure 5.4 illustrates a sensor observing the emitted energy and providing amplitude versus time data as output. A feature extraction process creates a two-dimensional feature vector [comprising a measure of pulse width (PW) and pulse repetition frequency (PRF)] to characterize the observed data. Such a feature extraction function would utilize automated peak detection and characterization techniques.

The feature vectors are mapped into a two-dimensional feature space whose components are PRF and PW. Suppose that prior measurements or analyses have established two distinct emitter classes of interest, A and B. In Figure 5.4, these emitter classes are shown as having distinct nonoverlapping boundaries (the rectangles shown in Figure 5.4). Identity classification is simply the process of determining whether the observed data (namely y_1 and y_2) fall inside either one of the feature space boundaries. In this specific example, Figure 5.4 illustrates that one of the observed emitters (emitter 1) falls inside the feature space boundaries of emitter class A. Hence, observation 1 would be declared as having the same identity as emitter class A. The second observation, y_2, falls outside the feature space boundaries of both emitter class A and class B. In the absence of other information, observation y_2 would be declared as having an unknown identity.

The process of identity declaration via parametric templates is analogous to the association process described in Chapter 3; namely, the location of a feature vector is compared to prespecified locations in feature space. If an observation, y_1, falls close to or within the boundaries of an identity class, then the observation is declared to have the same identity as the associated identity class. Hence, a similarity measure must be computed, and each observation is compared to a priori classes, much like observation-to-track correlation. Association measures may include any of the measures described in Chapter 3.

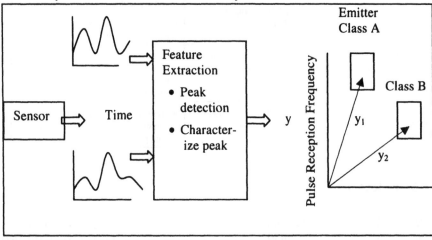

Figure 5.4   Concept of templates for identity classification.

Many types of boundaries may be used to partition a feature space [e.g., geometrical (rectangular or ellipsoidal volumes), statistical (Bayesian), or ad hoc boundaries]. Fukunaga [28] describes a variety of these techniques. A factor that complicates the template method is the problem of overlapping volumes in feature space. Such overlap causes an ambiguity in identification that may not be resolvable. Nevertheless, the same techniques described in Chapter 3 may be applied to attempt to resolve the ambiguities. Several strategies may be applied. A hard decision declaration process will declare a unique identity. A soft decision approach may declare an identity with an associated probability or uncertainty.

Conceptually simple, the templating approach to identity declaration is often used in data fusion systems. The technique can be implemented in a computationally efficient manner. Its effectiveness depends strongly on the choice of features and their associated distribution in feature space.

5.4   CLUSTER ANALYSIS TECHNIQUES

Cluster analysis methods are a set of heuristic algorithms that have become popularized in the biological and social sciences. Cluster analysis is a generic name for a wide variety of procedures that can be used to group data into natural groupings or clusters. These data clusters are interpreted by an analyst as a classification or identification. For example, parameters such as pulse repetition interval (PRI) and radio frequency (RF) might be used to separate radars by function or activity (e.g., tracking versus acquisition versus search). The concept is illustrated in Figure 5.5. Figure 5.5(a) shows a training cycle in which sensor data from known targets are input to the classifier. Feature vectors are extracted and input into a trainable classifier. The classifier forms decision surfaces or boundaries in feature space. Ideally the decision surfaces allow each target or object to be uniquely located in the multidimensional feature space. Having performed the training, then the classifier can be used in a classification cycle as illustrated in Figure 5.5(b). When the classifier is in the classification mode, it will observe an unknown target. The resulting feature vector is mapped into feature space, and its proximity to known identity classes is established. Output from the trained classifier is either the ID corresponding to the region of feature space in which the datum lies or a list of potential identities for the unknown target (prioritized by decreasing similarity).

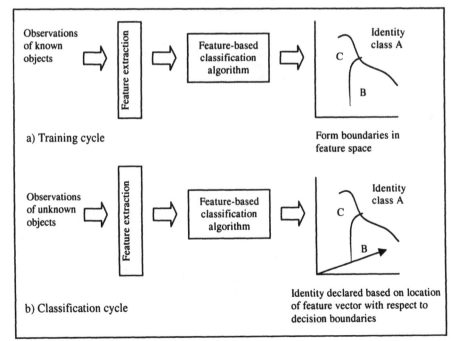

Figure 5.5 Concept of cluster analysis.

Five basic steps are required for any type of cluster algorithm. These include the following:

1.   Selection of the sample data;
2.   Definition of the set of variables (or features) on which to measure the entities in the sample;
3.   Computation of the similarities among the entities;
4.   Use of a cluster analysis method to create groups of similar entities;
5.   Validation of the resulting cluster solution.

All of the above steps are important in cluster analysis. Observational features are identified or created to (it is hoped) allow separation of data into identifiable groups. In the example illustrated in Figure 5.4, the selected features are pulse repetition frequency (PRF) and pulse width (PW). Numerous other features could have been selected with potentially different resulting clusters. Features may also be created by processes such as Fourier transforms (resulting in Fourier coefficients). The ultimate selection of useful features depends on either knowledge of the underlying physical processes or simply by trial and error. The features in turn may be scaled or weighted by a method such as factor analysis, logarithmic transformations, and standardization to unit variance.

Table 5.4

Summary of Cluster Algorithm Families

Hierarchical agglomerative methods	For each observation pair, compute the similarity measure; use linkage rules to cluster the observation pairs that are most similar; continue hierarchically comparing/clustering observation-to-observations; observation-to-clusters, clusters-to-cluster until no additional clustering is feasible via the clustering rules; types of rules include single linkage, complete linkage, and average linkage	[29, 30]
Iterative partitioning methods	1. Begin with an initial partition of data set into a specific number of clusters, compute cluster centroids 2. Allocate each data point to the cluster with nearest centroid 3. Compute new centroids of the clusters 4. Alternate steps 2 and 3 until no data points change clusters	[6]
Hierarchical divisive methods	Logical opposite approach to hierarchical agglomerative methods; initially all data are assigned to a single cluster; the cluster is divided into successively smaller chunks using either a monothetic or polythetic strategy; monothetic clusters are defined by certain variables on which certain scores are necessary for membership; polythetic clusters are groups of entities in which subsets of the variables are sufficient for cluster membership	[31]
Density search methods	Algorithms search the feature space for natural modes in the data that represent volumes of high density; strategies include	[32]

Table 5.4   (continued)

Summary of Cluster Algorithm Families

	a variant of single-linkage clustering and methods based on multivariant probability distributions	
Factor analytic methods	Also known as factor analysis variants, inverse factor analysis, and Q-type factoring; methods form a correlation matrix of similarities among cases; classical factor analysis is then performed on the N x N correlation matrix; data are assigned to clusters based on their factor loadings	[33]
Clumping methods	Clumping methods allow case membership in more than one cluster; cases are iteratively assigned to one or more clusters to optimize the value of a criterion referred to as a cohesion function	[34]
Graph theoretic methods	Use of graph theory to develop a hierarchical tree linking data into clusters (analogous to hierarchical agglomerative methods)	[30, 35]

All cluster algorithms require the definition of a similarity metric or association measure, which provides a numerical value representing the degree of *closeness,* or alternatively of dissimilarity between any two observed feature vectors y_i and y_j (namely y_i represents the observation of a single physical entity). The components of y_i are the observed features such as PRF, PW, and Fourier coefficients. Several types of similarity are described in Chapter 3.

Given a scaled data set and selected similarity metric, a number of algorithms have been developed to search for natural groupings of clusters in feature space. Aldenderfer and Blashfield [7] identify seven families of clustering methods summarized in Table 5.4, which lists each family of clustering methods (i.e., hierarchical agglomerative methods, iterative partitioning, hierarchical divisive, density search, factor analytic, clumping, and graph theoretic methods). For each family of methods, a brief description of the algorithm strategy is provided along with references describing the details of the algorithms. One of the difficulties in studying cluster methods is the plethora of techniques derived in diverse applications such as biology, sociology, and psychology. Many of these algorithms, having different names, are identical mathematical procedures. Everett [31] provides an excellent discussion of cluster methods, and describes specific techniques to implement cluster methods.

In order to illustrate the concept of cluster analysis in more detail we will describe one method in particular, a single-linkage hierarchical agglomerative method. Aldenderfer and Blashfield cite a study they performed to survey applications of clustering techniques. Their review found that two-thirds of the published research using clustering methods selected the hierarchical agglomerative approach. In the training cycle, the approach comprises a number of steps as follows:

1. The attribute and feature vector data are assembled or input to the clustering process.
2. The data are scaled or normalized. A typical practice is to normalize all data to the range (0, 1). This data scaling is performed to avoid the magnitude of any one feature component from dominating the cluster process. Various scaling techniques may be used including standardizing the attribute vectors by inverse standard deviation.
3. For each feature vector, y_i, the similarity coefficient is computed with respect to every other feature vector, y_k; that is, for a Euclidean distance metric,

$$s_{ij} = \left[\frac{1}{n} \sum_{i=1}^{n} (y_{ij} - y_{ik})^2 \right]^{1/2} \qquad (5.1)$$

Where y_{ij} is the ith component of the jth data point, n is the number of components of the feature vector, and s_{jk} is the Euclidean distance (in normalized feature space) between the jth and kth feature vector. Any of the similarity measures described in Chapter 3 can be used in place of equation (5.1). A symmetric resemblence matrix may be formed:

$$s = \begin{bmatrix} 0 & s_{12} & \cdots & s_{1M} \\ s_{21} & 0 & \cdots & s_{2M} \\ \vdots & & & \\ s_{M1} & s_{M2} & \cdots & 0 \end{bmatrix} \qquad (5.2)$$

The s_{ij} are a measure of closeness between the i and j feature vectors. We assume total of M data points.

4. A clustering rule is applied. The single-linkage hierarchical method forms a sequence. A hierarchical tree is created. At the bottom of the tree, each cluster consists of a single observation. At the first level (or iteration), the two closest observations are grouped into a cluster. This hierarchy of potential clusters continues until at the top level all observations are clustered into a single cluster.

The concept is illustrated in Figures 5.6 and 5.7. On the right side of Figure 5.7, there are five data points in a two-dimensioned feature space. On the left-hand side, the corresponding dendrogram (cluster tree) is shown. This chart plots the Euclidean distance between feature vectors. A hierarchical tree is shown. At the bottom level, all observations are considered to be in individual clusters. The cluster method searches the resemblence matrix to find the smallest value of $s_{ij(i \neq j)}$ (i.e., the two closest objects). These objects are clustered into a single cluster. The process continues until all objects are included into a single large cluster.

In the single-linkage method, at each hierarchical iteration (denoted step 0,1,...etc. in Figure 5.7), after two clusters i and j are merged to form cluster k, the similarity between cluster k and any other cluster is specified by the metric,

$$s_{1K} = \min(s_{ij}, s_{j1}) \tag{5.3}$$

That is, a new candidate to cluster membership can be joined to an existing group on the basis of the nearest of any member of the existing group; hence only a *single link* is required between two cases for them to merge. More complex linkage rules can be defined. Aldenderfer and Blashfield describe a general formula for hierarchical agglomeration. The concept of linking two objects together into a cluster is shown on the right-hand side of Figure 5.7 (i.e., linking together data point 3 and data point 4 into a single cluster).

Having developed an M-1 hierarchy of possible clusters (for M data points), the establishment of how many clusters to allow is based on determining the point at which *significant* changes no longer occur in the tree hierarchy. This can be performed manually, or via an ad hoc criterion.

An example of the use of a single-linkage hierarchical agglomerative cluster method for multitarget classification and tracking is described by Balakrishnan and Tapley [36].

Cluster analysis is a valuable tool set for exploring new relationships in data, which may lead to identification paradigms. However, because of the heuristic nature of cluster algorithms, their application is fraught with potential biases. In general, data scaling, selection of a similarity metric, choice of clustering algorithms, and (sometimes) even the order of the input data, may substantially affect the resulting clusters. Aldenderfer and Blushfield provide a dramatic example of these effects. Hence, the use of cluster methods must be judged on their effectiveness and repeatable ability to form meaningful identify clusters. The user must be the sole judge of these results.

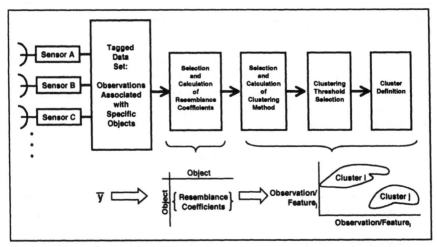

Figure 5.6 Processing flow for cluster analysis.

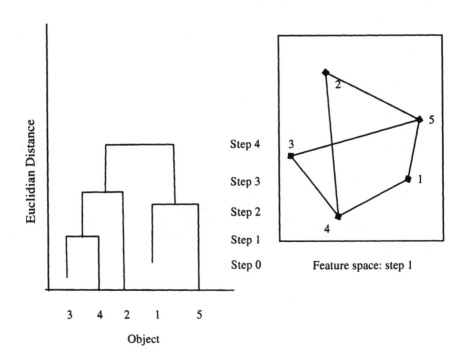

Figure 5.7 Example of hierarchial clustering.

5.5 ADAPTIVE NEURAL NETWORKS

Artificial neural net models, or adaptive neural systems, are hardware or software systems that have many of the characteristics of biological nervous systems. A neural network consists of layers of processing elements, or nodes, which may be interconnected in a variety of ways. Figure 5.8 illustrates a three-layer network with each layer having multiple processing elements. Data vectors are input on the left-hand side of the network and the neural net performs a nonlinear transformation, resulting in an output vector on the right side of the network in Figure 5.8. Such a transformation could produce the type of mapping from data to identity categories as is performed by cluster analysis techniques. Thus neural nets can be used to transform multisensor data into a joint declaration of identity for an entity. Applications of neural nets to data fusion have included Hull-Tech (identifying ships from observed parametric radar data (e.g., PRI and RF) by Prieve and Marchette [37] and sonar applications [9]. G. Willson [38] provides a comparison between neural networks and standard classifiers for radar specific emitter identification. A discussion of other fusion applications is provided by North [39]. Finally, an excellent introductory text is that of Wasserman [10].

The processing concept for a single node of a neural net is shown in the lower half of Figure 5.8. Each node in a layer receives inputs from up to n nodes in the preceding layer. These inputs, x_0, x_1, ..., x_n are combined in the processing element to produce an output, y, which in turn acts as the input to subsequent layers of the net. The combination of x_i to produce y is performed using a nonlinear function of the weighted values of x; that is,

$$y = f\left(\sum_{j=0}^{n} w_j x_j - \theta_i\right) \tag{5.4}$$

The function f may assume a variety of forms including, for example, step functions or sigmoid functions. The argument of f is a dot product of the input vector, **x,** and its internal weighting vector, w, with an additive threshold or bias, or the bias can be considered as a constant element of the input vector, **x**. The weighted sum and bias are operated on by the nonlinear function, f(y). This is sometimes termed the activation function.

A number of variations can be used in formulating a neural network, including the number of layers in a network, number of nodes per layer, interconnectivity among nodes, selection of the function, f, and other details. Lippman [40] and [41] provides a description of these variations and shows a taxonomy of commonly used networks for classification applications. These binary input networks are listed as follows:

- Supervised (Hopfield net, Hamming net);

- Unsupervised (Carpenter/Grossberg classifer) continuous-valued input nets;
- Supervised (perception, multi-layer perception);
- Unsupervised (Kohonen self-organizing feature maps).

The terms *supervised* and *unsupervised* refer to the training mode used. A neural net can be adapted or trained to perform correct classifications (i.e., mapping the input data from known entities into their associated known identity classes) by systematically adjusting the weights, w_i, in the equation for y. This is typically performed using a sample or training data set. Nets trained with supervision (e.g., the Hopfield net and the Perceptron) are provided with information or labels that specify the correct mapping between input data and output classifications. In unsupervised training, the nets are simply trained to form output data clusters without specifying the *correct* input-to-output mapping. Nets may also be trained adaptively, allowing the weights to vary as is done in adaptive signal processing. An example of a neural network trained to recognize letters of the alphabet in shown in Figure 5.9.

A frequently used technique for supervised training of neural networks is the back-propagation (BP) algorithm. BP minimizes an error function usually specified as the mean square difference between the actual network output vector and the desired output. An iterative gradient descent technique is used to systematically adjust the weights, w_i, to produce the desired result. BP requires that the nonlinearity in each node increases monotonically and is differentiable (for the range of its argument). A function that satisfies these conditions is the sigmoid function,

$$f(y) = \frac{1}{\left(1 + e^{-y}\right)} \tag{5.5}$$

A detailed description of BP is provided by Rumdhart et al. [42]. Willson [38] points out that the specific performance of BP depends on a number of factors including computer precision, initial values of w_i and Θ_i, and training data used. Other drawbacks include the possibility that the network will converge to a local vice global minimum. A particular issue is that it is not possible to predict the number of iterations required for convergence. Often thousands of iterations are required for convergence.

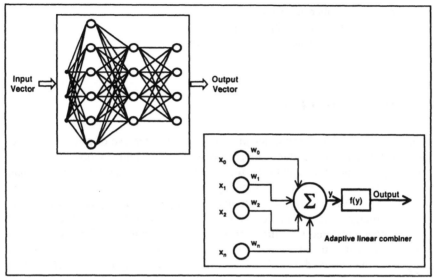

Figure 5.8 Structure of a neural network.

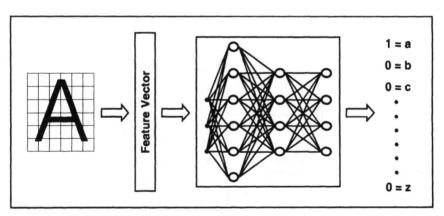

Figure 5.9 Concept of pattern recognition using a neural network.

A number of experiments have been performed with neural nets for fusion of identity data. Preliminary results (i.e., Bowman, [43] and Priebe and Marchette [44]) suggest that neural nets are superior to traditional cluster methods for identity fusion, especially when the input data are noisy, and when there are missing data. However, at this time much groundwork needs to be performed to develop a theoretical framework to understand basic issues such as the following:

1. Selection of a network model;

2. Choice of number of layers and nodes;
3. Development of a training strategy;
4. Incorporation of neural nets with traditional classifiers.

5.6 PHYSICAL MODELS

A direct method of performing identity declaration is the use of physical models of the signal generation and reception process. The technique involves attempting to directly compute the signal (i.e., time domain, signal, data, frequency domain data, or imagery) of an entity based on a physical model. The concept is illustrated in Figure 5.10. A sensor observes an object or target, producing an observed signature or image. The identity declaration process compares the observed data against either prestored signatures (from an a priori signature file) or against a simulated signature based on assumed target characteristics and physical models that predict the observed data. The comparison process may involve a correlation between the predicted and observed data. If the correlation coefficient exceeds an a priori threshold, then an identity match is declared.

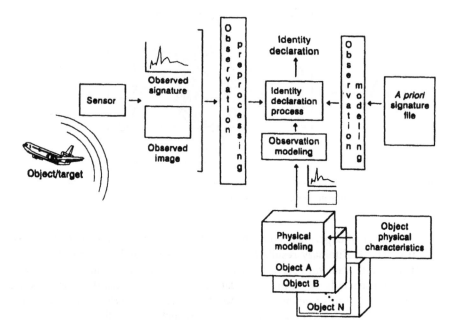

Figure 5.10 Concept of physical models for identity declaration.

Modeling to predict an entity's signature must be based on the physical characteristics of the candidate object. A different physical model may be required for each object or type of object. Examples of physical modeling include the use of black body formulations to predict the infrared spectra of high temperature objects [45] and the use of diffraction models to model the radar cross-section of complex objects [46]. Another example is the use of digital signal simulations to predict the emissions of a transmitter. The complete signal stream, from signal generation (e.g., via Klystron tube to signal receipt) may be simulated. These particular techniques are often used in technical electronic intelligence.

Physical models may become extremely complex, requiring very large software programs. This is one reason why purely physical models are infrequently used. Even when the physical models are relatively simple or when a priori signature data are used, the observation models and preprocessing may become very complex. Suppose for example, that an imagery sensor is used for remote sensing. Further, suppose that we have simple models (e.g., a two-dimensional geometrical drawing) or actual photographs of the entities that we seek to identify by an automatic identity declaration process. In principle the identification process is straightforward; we simply compare the observed imagery against the model or a priori photograph. Unfortunately, the situation may require significant processing in order to match the observed data with the predicted model. Required image preprocessing may entail the following computations:

- Sensor geometric corrections: Complex focal plane arrays and nonlinear optics require solution of integral equations to provide a mapping between the physical geometry of the observed object and its representation in an image plane.
- Detector compensation: Imagery focal plane arrays may exhibit failed or degraded components whose effects require prediction.
- Platform geometry correction: The platform upon which the sensor is located may undergo translation, attitude variations, and other kinematic effects. Corrections must be made to account for their impact in the image plane.
- Modulation transfer function compensation: Models may be utilized to model the optical or observational effects in the sensor. Modulation transfer functions involve two-dimensional convolutions to sharpen an image. These functions may be 3 or 4 dimensional in video processing or processing of synthetic aperture radar date.
- Dynamic range adjustments: An image may be difficult to interpret due to shadows and variations in intensity across the image. Dynamic range adjustments (such as histogram equalizations) may be used to adjust the contrast between adjacent elements in an image
- Automatic exploitation functions: Automated processes may be applied to extract image segments, perform image notations and scaling, dewarp

functions, identify line segments, detect edges, and perform many other functions.

This list of functions is meant to illustrate the complexities that can be encountered in physical modeling to predict images.

The concept of physical modeling for identity declaration is analogous to the estimation techniques described in Chapter 4. Its use is generally limited due to the intensive computational requirements. The technique however, is useful in the investigation of new phenomena, especially in very nonreal time settings (e.g., when an investigator can spend days or months exploring a single observational data set or event). The method is valuable for exploring underlying physical phenomena.

5.7   KNOWLEDGE-BASED METHODS

Knowledge-based techniques such as expert systems and logical templates may be used to perform identity declaration. These methods eschew physical models and instead attempt to emulate the cognitive approach used by humans in performing identity recognition. Expert systems [13, 47] are computer programs that use knowledge representation techniques such as rules or frames, and inference processes to perform reasoning to reach conclusion. Logical template techniques [4] extend the concept of parametric templates described in Section 5.3 to admit logical conditions and relationships. Details of expert systems and logical templates are provided in Chapter 7. The concept of knowledge-based identity declaration is introduced here as an alternative to the methods previously described in this chapter.

Knowledge-based techniques may be based on either raw sensor data or on extracted features. In either case there are two common aspects of knowledge-based methods: (1) techniques to represent knowledge and (2) inference methods to process information to reach a conclusion. Knowledge representation techniques include the following:

1. Rules: (If-then) statements that describe the relationship between evidence (i.e., the existence of observational data) and resulting actions (e.g., declaration of identity and request for additional data);
2. Frames: Complex data structures that identify object attributes, relationships, and characteristics;
3. Scripts: Description of settings and actions that describe real-world situations or events;
4. Cases: Representations of typical examples or cases that may be encountered (i.e., analogous to precedent cases in law);

5. Logical Templates: Descriptions of conditions and constraints about entities, events, and activities;
6. Networks: Graphical or hierarchical constructs to represent relationships such as functional hierarchies, state dependencies, inheritance of properties;
7. Analogical representations: Any special representation scheme that captures particular knowledge (e.g., chemical symbols, musical symbols, and mathematical relations).

These knowledge representation techniques may be used to define complex physical entities. A collection of rules, frames, or scripts is termed a knowledge base.

Knowledge-based systems use inference techniques to process sensor observations against an a priori knowledge base in order to perform identity declaration. Various inference techniques can be used including Boolean logic, decision trees, fuzzy logic, and other methods for processing data to reach valid conclusions. Any of these techniques can incorporate uncertainty in the observational data as well as uncertainty in the knowledge base.

Knowledge representation and reasoning techniques that are especially applicable for identity declarations (or other decisions) are methods that focus on object attributes and their interrelationships. An example is a syntactical approach, based on the structure of human languages. A syntactic approach specifies elements of complex object or entity and interrelationships among those elements. This is analogous to the rules for creating valid sentences (i.e., specification of words and their syntactic elements and rules by which words may be interrelated to form sentences). For example, representation of an entity such as a military tank may identify elements such as wheels, tracks, tank body, and weapon turret. The number and placement of these elements to represent a tank are specified via syntactic rules. The identification process then proceeds to collect data on elementary components of an entity and process the data against a knowledge base to determine if the conditions have been met to establish identity of an object or entity. The concept of knowledge-based identity declaration is illustrated in Figure 5.11.

Several knowledge-based techniques are described in Chapter 7. Details of knowledge representation, inference processes, and uncertainty representation are provided in that chapter. Even prior to that description, however, it is clear that the success of such a technique lies primarily in the makeup of the a priori knowledge base. A knowledge base is created by a technique termed knowledge engineering. Much like systems engineering, this methodology requires skill and remains a bit of an art. Structured methods for knowledge engineering are described in Chapter 7.

When are knowledge-based techniques appropriate to identity declaration? First, even though knowledge-based methods do not use physical models explicitly, they do rely on a thorough understanding of the components and makeup of the entity whose identity is sought. Thus, use of knowledge-based techniques is not a substitute for physical knowledge. The methods simply proceed in a heuristic fashion rather than mathematical modeling. Knowledge-based methods are especially useful when objects can be identified by groups of elemental components, or by interrelationships among components. Further, the knowledge-based methods are applicable as the concept of *entity* becomes more complex. For example, physical models, templates, or neural networks may readily perform identification of emitters. Conversely, identification of military units (e.g., surface-to-air missile (SAM) battalion) requires identification of several components, recognition of interrelationships and functions, and other inferences. Hence these types of identifications may be more properly performed using knowledge-based techniques. Chapter 7 provides a detailed description of such techniques.

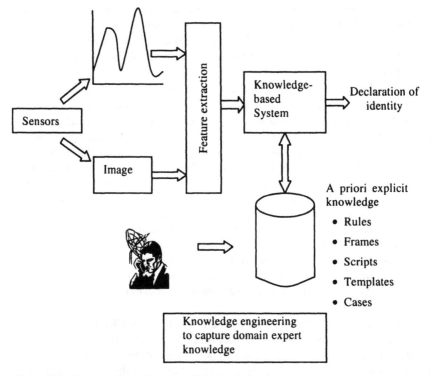

Figure 5.11 Concept of knowledge-based identity declaration.

5.8 HYBRID TECHNIQUES

Hybrid reasoning techniques combine the use of implicit and explicit information. *Implicit* identification methods such as cluster algorithms, neural nets, or parametric templates are trained to represent information inherent in sensor data to transform from a signal, image, or feature vector into a declaration of entity identity or classification. The techniques do not make explicit the transformation between observed data and the classification. Instead these represent implicit models or transformations. By contrast, *explicit* techniques such as knowledge-based systems or physical models use physical information or human-based knowledge to explicitly link observational data to target identity or classification. Each of these methods has limitations. On one hand, implicit methods are very simple to implement and require very limited knowledge of the observed targets or entities. However, they require extensive training data to achieve an accurate transformation. By contrast, explicit techniques require extensive knowledge about the entities being observed or classified (via physical models or information from human experts) and are difficult to implement. However, they allow incorporation of context-based information and "lessons learned" from domain experts. These techniques are not particularly sensitive to newly observed training data.

In recent years, several hybrid methods have been investigated. Figure 5.12 shows the concept of combining a syntactic representation of an aircraft (namely hierarchical specification of components and inter-relationships) with implicit recognition of components (e.g., use of pattern recognition methods to identify components such as engines, wings, and landing gear). In addition, contextual information such as aircraft altitude, speed, and groupings can be represented via rules or logical templates. The hybrid approach combines all of these reasoning methods together to improve the ability to identify complex entities.

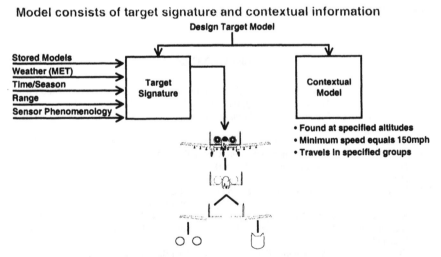

Figure 5.12   Example of syntactic-based hybrid model.

Finally, Figure 5.13 shows an approach developed by Hall and Garga [15]. The approach involves developing an explicit representation, an entity using rules (obtained via a domain expert), and training a neural network to represent these rules. Subsequently, the neural network is allowed to evolve using sample data. As a result, both implicit and explicit information are combined. Experimental results for fault detection in mechanical systems indicate that this hybrid approach is more accurate and more robust that using either the implicit or explicit information alone.

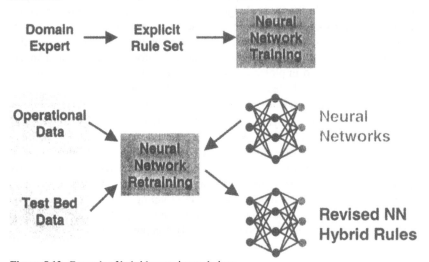

Figure 5.13 Example of hybrid reasoning technique.

REFERENCES

[1]   Steinberg, A. N., "Threat Management System for Combat Aircraft," *Tri-Service Data Fusion Symposium*, Johns Hopkins University, Baltimore, MD, June, 1987, pp. 532-554.

[2]   Rao, B. K. N., ed. *Handbook of Condition Monitoring*, New York, NY: Elsevier Advanced Technology, 1996.

[3]   Koelsch, J. R., "Sensors - The Missing Link," *Manufacturing Engineering*, 1990, pp. 53-56.

[4]   Hall, D., and R. J. Linn, "Comments on the Use of Templating for Multisensor Data Fusion," *Tri-Service Data Fusion Symposium*, Johns Hopkins University, Laurel, MD, 1989, pp. 152-162.

[5]   Noble, D. F., "Template-Based Data Fusion for Situation Assessment," *1987 Tri-Service Data Fusion Symposium*, Johns Hopkins University, Laurel, MD, 1987, pp. 226-236.

[6]   Anderberg, M., *Cluster Analysis for Applications*, New York, NY: Academic Press, 1973.

[7]   Aldenderfer, M. S., and R. K. Blashfield, *Cluster Analysis*, vol. 07-044, London, England: Sage Publications, 1984.

[8]   Breiman, L., *Classification and Regression Trees*, Boca Raton, FL: CRC Press, 1984.

[9]   Widrow, R., and R. Winter, "Neural Nets for Adaptive Filtering and Adaptive Pattern Recognition," *IEEE Computer*, 1988.

[10] Wasserman, R. B., *Neural Computing: Theory and Practice*, New York, NY: Van Nostrand Reinhold, 1989.

[11] Waltz, E., and J. Llinas, *Multi-sensor Data Fusion*, Norwood, MA: Artech House, Inc., 1990.

[12] Winston, P. H., *Artificial Intelligence*, 2nd ed., New York, NY: Addison-Wesley Publishing Company, 1984.

[13] Jackson, P., *Introduction to Expert Systems*, 3rd. ed., New York, NY: Addison-Wesley Pub. Co., 1999.

[14] Russell, S. J., and P. Norvig, *Artificial Intelligence: A Modern Approach*, 2nd ed., New York, NY: Prentice Hall, 2002.

[15] Hall, D., and A. Garga, "Hybrid Reasoning Techniques for Automated Fault Classification," *Society for Machinery Failure Prevention, 51st Meeting*, April, 1997.

[16] Bowman, C., and A. Steinberg, "A Systems Engineering Approach for Implementing Data Fusion Systems," in *Handbook of Multisensor Data Fusion*, ed. by J. Llinas, New York, NY: CRC Press, 2001.

[17] Deming, P., "Saving All the Bits," *American Scientist*, vol. 78, 1990, pp. 402-405.

[18] Hall, D., and S. R. Waligora, "Orbit/Attitude Estimation Using Landsat-1 and Landsat-2 Landmark Data," *NASA Goddard Space Flight Center Flight Mechanics/Estimation Theory Symposium*, NASA Goddard Space Flight Center, MD, October, 1978.

[19] Dougherty, E. R., and C. R. Giardina, *Image Processing, Continuous to Discrete*, Englewood Cliffs, NJ: Prentice-Hall, 1987.

[20] Pennington, K. S., and R. J. Moorhead II, eds. *Image Processing Algorithms and Techniques*, Proceedings of the SPIE, Vol. 1244, 1990.

[21] Russ, J. C., *The Image Processing Handbook*, Heidelberg, Germany: Springer-Verlag, 1998.

[22] Kuut, M., *Digital Signal Processing*, Norwood, MA: Artech House, Inc., 1986.

[23] Roberts, R. A., and C. T. Mullis, *Digital Signal Processing*, Reading, MA: Addison-Wesley Publishing Company, 1987.

[24] Peled, A., and B. Liu, *Digital Signal Processing: Theory, Design, and Implementation*, New York, NY: John Wiley and Sons, 1976.

[25] Daubechies, I., "Orthonormal Bases of Compactly Supported Wavelets," *Communications of Pure and Applied Mathematics*, vol. XLI, 1988, pp. 909-996.

[26] Rabiner, L. R., "A Tutorial on Hidden Markov Models and Selected Application in Speech Recognition," *Proceedings of the IEEE*, vol. 77, 1989, pp. 257-286.

[27] Liu, H., and H. Motoda, eds. *Feature Extraction, Construction and Selection: A Data Mining Perspective*, Kluwer International Series in Engineering and Computer Science, New York, NY: Kluwer Academic Publishers, 1998.

[28] Fukunaga, K., *Introduction to Statistical Pattern Recognition*, New York, NY: Academic Press, 1990.

[29] Sneath, P., and R. Sokal, *Numerical Taxonomy*, San Francisco, CA: W. H. Freeman, 1973.

[30] Dubes, R., and A. Jain, "Clustering Methodologies in Exploratory Data Analysis," *Advances in Computers*, vol. 19, 1980, pp. 113-228.

[31] Everett, B., *Cluster Analysis*, New York, NY: Academic Press, 1980.

[32] Everitt, B., and D. Wishart, *Supplement; CLUSTRAN User Manual, 3rd edition*, Edinburgh, Scotland, 1980.

[33] Skinner, H., "Dimensions and Clusters: A Hybrid Approach to Classification," *Applied Psychological Measurement*, vol. 3, 1979, pp. 327-341.

[34] Cole, A. J., and D. Wishart, "An Improved Algorithm for the Jardine-Sibson Method of Generating Clusters," *Computer Journal*, vol. 13, 1970, pp. 156-163.

[35] Ling, R., "An Exact Probability Distribution of the Connectivity of Random Graphs," *Journal of Mathematical Psychology*, vol. 12, 1975, pp. 90-98.

[36] Balakrishnan, S. W., and D. B. Tapley, "Multi-Target Classification and Estimation Using Clustering Techniques," *Journal of Guidance*, vol. 1, 1990, pp. 121-127.

[37] Prieve, C., and D. Marchette, "An Application of Neural Networks to a Data Fusion Problem," *Tri-Service Data Fusion Conference*, Johns Hopkins University, Laurel, MD, 1987, pp. 226-236.

[38] Willson, G. B., "Radar Classification Using a Neural Network," *Optical Engineering and Photonics in Aerospace Sensing: Applications of Neural Networks*, Orlando, FL, April, 1990, pp. 200-210.

[39] North, R., "Neurocomputing: Its Impact on the Future of Defense Systems," *Defense Computing*, 1988.

[40] Lippman, R. P., "An Introduction to Computing Neural Nets," *IEEE ASSP Magazine*, vol. 3, 1987, pp. 4-22.

[41] Mush, D., and B. Horne, "Progress in Neural Networks: What's New Since Lippman?," *IEEE Signal Processing Magazine*, 1993, pp. 8-39.

[42] Rumelhart, D. E., G. E. Hinton, and R. J. Williams, "Learning Internal Representations by Error Propagation," in *Parallel Distributed Processing: Explorations in the Microstructure of Cognition*, ed. by J.L. McClelland, Cambridge, MA: MIT Press, 1986.

[43] Bowman, C. L., "Artificial Neural Network Adaptive Systems Applied to Multisensor ID," *The Second Tri-Service Data Fusion Symposium*, May, 1988.

[44] Priebe, C., and D. Marchette, "An Adaptive Hull-To-Emitter Correlator," *1988 Tri-Service Data Fusion Conference*, Johns Hopkins University, Baltimore, MD, 1987, pp. 433-436.

[45] Huber, A. J., M. J. Triplett, and J. R. Wolverton, eds. *Imaging Infrared, Scene Simulation, Modeling, and Real Image Tracking*, Proceedings of the SPIE, Vol. 1110, 1989.

[46] Stone, W. R., ed. *Radar Cross Sections of Complex Objects*, New York, NY: IEEE Press, 1990.

[47] Benfer, R. A., E. E. Brent, and L. Furbee, *Expert Systems*, vol. 77, London, England: Sage Publications, 1991.

Chapter 6

Decision-Level Identity Fusion

Decision-level identity fusion involves combined reports or declarations of identity from individual sensors to achieve a joint or fused identity declaration. Techniques range from probabilistic methods such as Bayes method or Dempster-Shafer's method to ad hoc methods such as voting. This chapter introduces the concept of decision-level identity fusion and describes several common techniques.

6.1 INTRODUCTION

Decision-level fusion seeks to process identity declarations from multiple sensors to achieve a joint declaration of identity. Chapter 5 (Figure 5.3) illustrates three possible architectures for identity fusion: (1) a data-level fusion architecture in which data from commensurate sensors are fused prior to feature extraction and identity declaration, (2) a feature-level architecture that develops a joint multisensor feature vector with subsequent identity declaration, and finally, (3) a decision-level fusion architecture in which each sensor performs an identity declaration, followed by a process that fuses the identity declarations to achieve a joint multisensor identity declaration. Techniques for data-level and feature level fusion were introduced in Chapter 5.

This chapter focuses on the concept and techniques for decision-level fusion. Figure 6.1 illustrates the basic concept. Multiple sensors (perhaps of different types) observe a physical entity, situation, or event. Each sensor performs preprocessing including feature extraction and identity declaration to develop a declared identity (or vector of probabilities of various identities) for the observed entity. Examples of decision-level fusion problems include the following:

- Threat-warning systems (TWS) onboard tactical aircraft to identify threats (e.g., illumination by a surface-to-air missile guidance system) [1];
- Multiple-sensor target detection [2];

- Multiple-sensor identification of material flaws in an automated manufacturing system [3];
- Data processing for robotic vision [4].

Figure 6.1 illustrates an identity fusion process in which identity declarations from several sensors are input to a decision-level fusion algorithm. The identity declarations from each sensor must be preprocessed via an association process in order to sort the observations into groups. Each observation group is hypothesized to be associated with a single object or entity. Chapter 3 describes association techniques to sort observations into meaningful groups. Methods to fuse the identity declarations include classical inference, Bayesian inference, Dempster-Shafer's method, Thomopoulos generalized evidence processing theory, and various heuristic approaches such as voting methods. These methods are summarized in this section, and described in more detail in the remainder of this chapter.

The single sensor identity declaration's input to the fusion process may be either hard or soft decisions. Hard decision sensors provide only a single declaration of identity. An example of such a hard decision is a threat-warning sensor on an aircraft that provides a warning light to indicate a threat (or the oil warning light on an automobile). Soft decision sensors provide multiple declarations of identity with an associated probability or confidence factor, as shown in the following.

Identity hypothesis		*Associated probability*
H1	Observed entity is an F-16 aircraft	$P(H_1)=0.53$
H2	Entity is a Boeing 747	$P(H_2)=0.28$
•		
•		
•		
H_n	Entity is a UFO	$P(H_n)=0.03$

Output from the decision-level fusion process is a joint declaration of identity. Ideally this joint declaration is more accurate (and perhaps more specific) than any of the single-source declarations. Figure 1.11 in Chapter 1 (adapted from Waltz [5]) illustrates an example of the improved probability of correct identity declaration by fusing the identity declarations of two IFFN sensors.

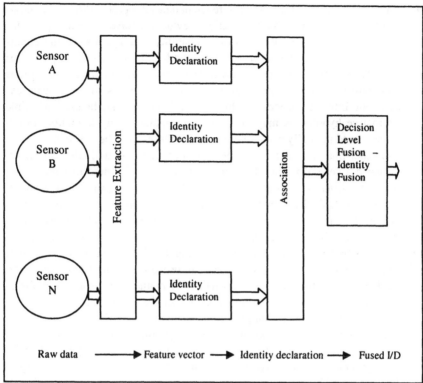

Figure 6.1 General concept of decision-level identity fusion.

Table 6.1 illustrates a summary of decision-level fusion techniques. The table identifies five basic categories of techniques, and for each technique summarizes the kernel process involved, the nature of the required input, the output from the technique, and summary comments. Each technique is introduced below, and described in more detail in Sections 6.2–6.6 of this chapter. A comparison of the techniques and a discussion of implementation issues is provided in Section 6.7.

- Classical inference: Classical inference methods seek to determine the validity of a proposed hypothesis (vice an alternative hypothesis) based on empirical probabilities. Classical inference computes the joint probability given an assumed hypothesis. A test of significance determines the likelihood or probability that data *actually* observed would be observed if in fact the assumed hypothesis is true. In general, classical inference can assess only two hypotheses at a time (a null hypothesis H_0, and its alternative, H_1). Classical inference fails to account for a priori information about the likelihood of a proposed hypothesis.

- Bayesian inference: Bayesian inference updates the probabilities of alternative hypotheses, based on observational evidence. Bayes' formula provides a relationship between the a priori probability of a hypothesis, the conditional probability of an observation given a hypothesis, and the a posteriori probability of the hypothesis. New information is used to update the a priori probability of the hypothesis in a procedure analogous to a scientific experiment. Either empirical or subjective probabilities may be used. Multiple alternative hypotheses may be evaluated simultaneously (although an exhaustive and mutually exclusive set of hypotheses must be defined).

- Dempster-Shafer's method: In 1976, Arthur Dempster and Glen Shafer introduced a generalization of the Bayesian inference method in a book entitled *A Mathematical Theory of Evidence* [6]. Analogous to the Bayesian inference method, the Dempster-Shafer technique updates an a priori mass density function to obtain an a posteriori evidential interval. The evidential interval quantifies the credibility (measure of belief) of a proposition, and its plausibility (lack of evidence refuting the hypotheses). Mass density functions provide the analog to Bayesian probability. The Dempster-Shafer method relaxes the Bayesian restriction on mutually exclusive hypotheses by assigning evidence to propositions rather than hypotheses. In addition, the Dempster-Shafer technique provides for a general level of uncertainty (i.e., does not require an exhaustive set of hypotheses to be defined).

- Generalized evidence processing theory: Thomopoulos [7] introduced a new generalization of Bayesian inference. The generalization is analogous to Dempster-Shafer's method, but develops an optimal way of combining probability masses. The generalized theory seeks to resolve several issues frequently cited about Dempster-Shafer's method.

- Heuristic methods: A variety of nonprobabilistic techniques have been applied for identity fusion. Some of the methods mimic the methods that humans use for joint inferences. These include voting methods, consensus techniques and other methods, which account for the consequences of adopting a hypothesis as true.

Table 6.1

Summary of Techniques for Decision-Level Fusion

Technique	Kernel Process	Inputs	Output
Classical inference	Computes joint probabilities of observation, given an assumed hypothesis	Empirical or classical probability; population distribution for a statistic	Probability of error conditioned on selected hypotheses
Bayesian inference	Updates the a posteriori probability of a hypotheses, given observational evidence	Empirical or subjective probability of observations given hypotheses; a priori	A posteriori probability of hypotheses given evidence

Table 6.1 (continued)

Summary of Techniques for Decision-Level Fusion

Technique	Kernel Process	Inputs	Output
		probability of hypotheses	
Dempster-Shafer's method	Updates the a posteriori belief of propositions given observational evidence; assessment of belief of any proposition being true	Probability mass functions assigned to propositions that may be overlapping; general measure of uncertainty	Credibility intervals with support and plausibility for propositions; general level of uncertainty
Generalized evidence processing theory	Generalization of Bayesian process; uses combination rules to assign evidence and optimize resulting decisions	Same as Dempster-Shafer	Same as Dempster-Shafer, but uses hypotheses not propositions
Heuristic methods	Various methods to combine or choose identity declaration; mimics human group decisions via voting, consensus, or scoring methods	Identity declarations from individual sensors; sensor weights	Joint or consensus declaration of identity

The remainder of this chapter provides details on these techniques.

6.2 CLASSICAL INFERENCE

Statistical inference techniques seek to draw conclusions about an underlying mechanism or distribution, based on an observed sample of data. A probabilistic model is assumed that involves one or more random variables having a defined (but unknown) probability distribution. The probability model provides a connection between observed data and a population. Classical inference typically assumes an empirical probability model (but may be used with analytical or heuristic probability models).

Empirical probability views probability as the limit of long-run frequencies. The concept is illustrated by a coin-tossing experiment. A coin having one side denoted *heads* and the other *tails* is repeatedly tossed. The relative frequency of appearance of heads and tails is observed. Empirical probability assumes that the observed frequency distribution of heads and tails will approximate the probability as the number of trials, K, increases without bound. Thus,

Probability (heads) $\cong$ frequency (heads) (as K—>∞)

and

Probability (tails) $\cong$ frequency (tails) (as K—>∞)

In general, empirical probability assumes that

$$P(E_j) = \lim_{K \to \infty} \left[\frac{K(E_j)}{K} \right] \qquad (6.1)$$

That is, the empirical probability of events, E_j, is equal to the relative frequency of occurrence of E, in K trials, as K approaches infinity. Strictly speaking, empirical probabilities are only defined for repeatable events. It is not possible to ascribe an empirical probability to a single event. Hence, for example, one could not meaningfully specify the probability that *John Jones will win the next election for mayor*, since this is a single event. Prior to the event (i.e., the election), the probability of John Jones' election equals zero. After the event, the probability either remains zero (if John Jones is not elected) or one (if John Jones is elected). Classical inference methods utilize empirical probability and hence are not strictly applicable to nonrepeatable events, unless some model can be developed to compute the requisite probabilities. This formalism then has limitations for data fusion applications involving single events. In the classical approach to identity declaration (see, for example, [8]) we assume that observed data (e.g., radar cross-sections and infrared spectra) are a result of (i.e., are caused by) one of N possible objects whose identity is sought. Hypothesis testing seeks to answer the following question: What is the likelihood that the observed measurements z would be made if in fact the data was caused by observation of object, x? That is, we define two hypotheses:

1. A null hypothesis, H_0: The observed data is caused by an object whose identity is *N*.
2. An alternative hypothesis, H_1: The observed data is not caused by object *N*.

Hypothesis testing proceeds with the following test logic.

1. Assume that the null hypothesis (H_0) is true;
2. Examine the consequences of H_0 being true in the sampling distribution for the statistic (i.e., determine the probability of having observed the data if in fact the hypothesis is true);
3. Perform a hypothesis test, specifically, if the observations have a high probability of being observed if H_0 is true, then declare the data do not contradict H_0;
4. Otherwise, declare that the data tend to contradict H_0.

Notice that classical inference does not declare either hypothesis (H_0 or H_1) to be true or false, but rather establishes the likelihood that the data would be observed if the assumed hypothesis is true.

Two assumptions are required for hypothesis testing. First, it is assumed that an exhaustive and mutually exclusive set of hypotheses can be defined that could explain or cause the observational data. Second, it is assumed that we can compute the probability of an observation, given an assumed hypothesis. An example of classical inference is illustrated by an emitter identification problem. Suppose it was known that two different types of radars exhibited agility in pulse repetition interval (PRI) as illustrated in Figure 6.2, which shows the probability density function versus PRI. Two classes of radars (denoted E_1 and E_2) exhibit overlapping ranges of PRI. The probability that radar class E_2 will emit a pulse repetition interval of ($PRI_N < PRI < PRI_{N+1}$) is given by the cross-hatched area under the probability density function in Figure 6.2. Classical inference seeks to confirm or refute a proposed hypothesis about identity. Thus, if we have observed a pulse repetition interval of PRI_{OBS} we seek to determine whether the radar observed is of class E_1 or E_2. Note that the probability distribution can be caused by target variability and measurement error and can have both random and bias components.

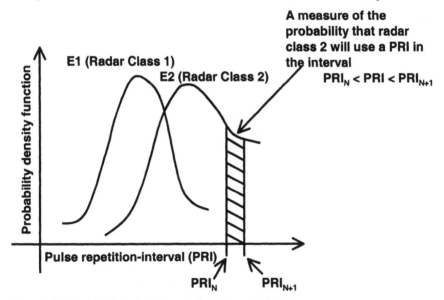

Figure 6.2 Example of classical inference for emitter classification.

Classical inference proceeds by choosing a critical discriminating value of PRI (e.g., PRI_c). If the observed value of PRI is greater than PRI_c (i.e., $PRI_{OBS} > PRI_C$) then we say that the evidence fails to refute the hypothesis that the radar is of class E_2. Alternately if $PRI_{OBS} < PRI_C$ we say that the evidence fails to refute the hypothesis that the radar is of class E_1. Note that classical inference provides information when a hypothesis is given about the identity of an observed entity (i.e., it provides the probability of the sensor observations). In our example, because of the overlap of PRI agility, the decision metric may still result in an erroneous identity declaration. In particular, Figure 6.3 illustrates that there is a finite probability, β, that the observed PRI would be greater than PRI_C for radar of class E_1, and a finite probability, α, that the observed PRI would be less than PRI_C for a radar of class E_2. These misidentification errors are termed type 1 and type 2 errors respectively. The choice of the value of PRI_C is up to the analyst. While it may be changed to minimize either type 1 or type 2 errors, in general there still remains a finite probability of misidentification.

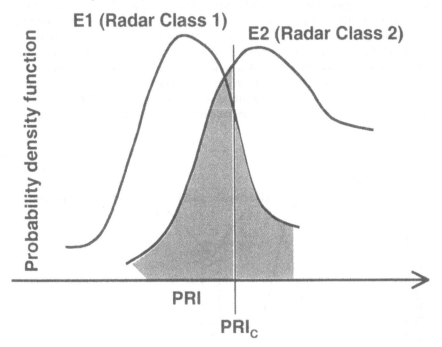

Figure 6.3 Classification error.

In this example we see that the result of classical inference is the probability of an observation (namely evidence) given an assumed hypothesis or cause. Often, however, we seek to determine the probability of the cause (i.e., given the observation, did we in fact see radar of class E_1 or of class E_2).

The classical inference approach may be generalized to include multidimensional data from multiple sensors. However, this requires the a priori knowledge and computation of multidimensional probability density functions. This is a serious disadvantage for realistic applications. Additional disadvantages of classical inference techniques include the following:

1. Only two hypotheses can be assessed at a time (namely the hypothesis H_0 versus an antithetical null hypothesis, H_1);

2. Complexities arise for multivariate data;
3. Classical inference does not take advantage of a priori likelihood assessments.

For these reasons, Bayes modification to classical inference described in Section 6.3 is used in identity fusion more frequently than is classical inference.

In the classical inference approach, it is important to note how the hypothesis acceptance or rejection criterion is established (e.g., in the previous emitter identification problem, how is the value of PRI_C determined). This issue is tantamount to asking how we decide to accept hypothesis H_0, rather than hypothesis H_1. A number of possible decision rules may be employed; they are summarized in Table 6.2 and include the following:

1. Maximum a posteriori: Accepts hypothesis H_0 as being true, if the probability of H_0 given observation y, $(P(H_0 | y))$ is greater than the probability of H_1 given y, $(P(H_1 | y))$.
2. Maximum likelihood: Accepts hypothesis H_0 as true if the $p(y|H_0) > p(y|H_1)$.
3. Neyman-Pearson: Accepts the hypothesis H_0, if the ratio of the likelihood function for H_0 to the likelihood function for H_1 is less than or equal to a constant, c. The constant, c, is chosen to give the desired significance level.
4. Minimax: A cost function is established that quantifies the risk or loss associated with choosing a hypothesis or its alternative (e.g., in our emitter problem one might fear the consequences of incorrectly classifying a threatening emitter type as a neutral or friendly type). The minimax approach selects H_0 such that the maximum possible value of the cost function is minimized.
5. Bayes: A cost function is established that provides a measure of the consequences of choosing hypothesis H_0 versus H_1. A typical cost function is given by

$$C = C_{00}P(H_0)P_a + C_{01}P(H_0)P_b + C_{10}P(H_1) P_cP(H_1) P_d$$

$P(H_0)$ and $P(H_1)$ are the a priori probabilities of hypotheses H_0 and H_1, respectively. P_a through P_d correspond to the probability of a sensor assignment under four possible conditions:

- P_a = Probability the sensor will declare H_0 true if in fact it is true;
- P_b = Probability that the sensor will declare H_0 false if it is true;
- P_c = Probability that the sensor will declare H_1 true if it is true;
- P_d = Probability that the sensor will declare H_1 true if it is false.

Hence, P_a and P_c are detection probabilities, while P_b and P_d, are probability of false alarms. The constants C_{ij} are arbitrarily chosen constants. The Bayes decision criteria selects H_0 over H_1 to minimize the cost function C.

Table 6.2

Decision Rules for Hypothesis Testing

Decision Rule	Criterion	Formulation
Maximum A Posteriori	Based on observational data, which hypothesis is most probable?	$P(H_0/y) > P(H_1/y)$
Maximum Likelihood	Based on observational data, and on the a priori probability of hypothesis, H_0 and H_1, which hypothesis is the most probable?	$P(y/H_0) > P(y/H_1)$
Neyman-Pearson	Based on the observational data, construct a test that exhibits the maximum power to discriminate against the false alarms (i.e., the false acceptance of the alternative hypothesis, H_1)	$\dfrac{L(y/H_0)}{L(y/H_1)} \leq$ constant
Minimax	Choose the a priori assumption that minimizes the maximum cost specified by a cost function	Choose $P(H_0)$ such that the maximum possible value of a cost function is minimized
Bayes' Criteria	Choose the assumption that minimizes a cost function based on detection and false alarm probabilities	$\dfrac{P(y/H_0)}{P(y/H_1)} > \dfrac{P(H_1)(C_{01} - C_{11})}{P(H_0)(C_{01} - C_{00})}$

This summary of decision rules for hypothesis testing is not exhaustive, but merely representative. In particular, mini-max approaches may use complex cost or utility functions. These functions attempt to develop measures of the cost (i.e., liability) or benefit (i.e., utility) of choosing one hypotheses over another. Fishburn [9] provides an excellent history of decision theory using utility functions. Miller [10] provides a review of hypothesis testing. Classical inference techniques have an extensive history and a well-founded mathematical basis. Such techniques may be particularly useful in applications such as identification of defective parts in manufacturing and analysis of faults in system diagnosis and maintenance (see, for example, [11]).

6.3 BAYESIAN INFERENCE

Bayesian inference takes its name from the English clergyman Thomas Bayes. A paper published in 1763 after Bayes death contains the inequality that is known today as Bayes' theorem [12]. The Bayesian inference technique [13] resolves some of the difficulties with classical inference methodology. Bayesian inference updates the likelihood of a hypothesis given a previous likelihood estimate and additional evidence (observations). The technique may be based on either classical probabilities, or subjective probabilities (i.e., it does not necessarily require probability density functions).

Subjective probabilities are commonly used in daily inferences. Humans commonly make judgments based on a subjective assessment of the likelihood of future events. For example, a commuter may choose one route over another based on an assessment of the likelihood of a traffic jam on one road versus another. These assessments of likelihood are subjective, nonrepeatable, and variable from one human to another. Moreover, humans may make a different assessment of the likelihood of an event based on whether it is stated in a positive or negative form. Thus, subjective probabilities suffer a lack of mathematical rigor or physical interpretation found in empirical probabilities. Nevertheless, if used with care, the concept of subjective probability can be useful in a data fusion inference processor. A discussion of the cognitive biases involved in subjective assessment of probabilities is provided by [14].

The Bayesian inference process proceeds as follows. Suppose $H_1, H_2,\ldots,H_j$, represent mutually exclusive and exhaustive hypotheses that can *explain* an event E (i.e., an observation) which has just occurred. Then,

$$P(H_i / E) = \frac{P(E/H_i)P(H_i)}{\sum_i P(E/H_i)P(H_i)} \qquad (6.2)$$

and

$$\sum_i P(H_i) = 1$$

where:

$P(H_i/E)$ = The a posteriori probability of hypothesis H_i being true given the evidence, E;

$P(H_i)$ = The a priori probability of hypothesis H_i being true (without having observed the evidence);

$P(E/H_i)$ = The probability of observing evidence E, given that H_i is true.

This is a pedagogically pleasing formulation in that new evidence, E, is used to improve hypotheses about events. The denominator in (6.2) is a normalization factor that scales probabilities to (0,1).

How would the Bayesian formulation be applied to the example of emitter identification, previously cited in Section 6.2? For that example, we could write:

$$p(E_1 / PRI_{obs}) = \frac{P(PRI_{obs} / E_1)P(E_1)}{\sum_i P(PRI_{obs} / \text{emitter}_i)P(\text{emitter}_i)} \qquad (6.3)$$

and

$$P(E_2 / PRI_{obs}) = \frac{P(PRI_{obs} / E_2)P(E_2)}{\sum_i P(PRI_{obs} / \text{emitter}_i)P(\text{emitter}_i)} \qquad (6.4)$$

In (6.3), $P(E_1| PRI_{OBS})$ is the a posteriori probability that we have seen an emitter of class E_1, given an observed value of PRI (namely PRI_{OBS}). The quantity, $P(PRI_{OBS}|E_1)$ is the probability that an emitter of class E_1 would exhibit a PRI of value, PRI_{OBS}. $P(E_1)$ is the a priori probability that an emitter of class E_1 exists *in nature*. An analogous interpretation may be made for the quantities in (6.4).

The Bayes formulation is pleasing for several reasons. First it provides a determination of the probability of a hypothesis being true, given the evidence. By contrast, classical inference gave us the probability that an observation could be ascribed to an object or event, given an assumed hypothesis. Second, Bayes formulation allows incorporation of a priori knowledge about the likelihood of a hypothesis being true at all. The utility of this latter feature of Bayesian inference is illustrated by the following example: Suppose a patient visits his physician who proceeds to administer a test for a rare blood disease. The medical test has an accuracy of 98%, (i.e., if the patient has the blood disease the test will indicate positive 98% of the time with a 2% false alarm rate). Further suppose that the blood disease exhibits an occurrence in one out of every 10,000 people in the population. If the patient is informed that he has tested positively for the disease, what is the probability that he actually has the indicated disease? The Bayesian formulation would predict the probability as,

$$P(\text{patient has disease}) = \frac{P(\text{test positive/given disease})P(\text{disease})}{P(\text{test is positive})}$$

$$= \frac{(0.98)(0.001)}{(0.98)(0.001) + (0.02)(0.999)}$$

$$= \text{approximately } 0.048 \text{ (i.e., about 5\%)}$$

The final feature of Bayes' formulation is the ability to use subjective probabilities for a priori probabilities for hypotheses, and for the probability of evidence given a hypothesis. This feature allows ready implementation of a Bayesian inference process, especially for multisensor fusion since probability density functions are not required. The reader is cautioned, however, that the output of such a process is only as good as the input data (i.e., the a priori probabilities). Poor input data will bias the computed output hypothesis. Indeed the question of how to specify the a priori hypothesis is an issue that must be addressed in the use of a Bayesian approach. Feller [15] discusses this issue and provides some interesting historical examples. Iversen [16] provides other examples.

Figure 6.4 illustrates the process of using a Bayesian formulation for identity fusion. Multiple sensors observe parametric data (e.g., RCS, PRI, and IR spectra) from an entity whose identity is unknown. Each of the sensors provides an identity declaration, or hypothesis about the object's identity using techniques such as cluster analysis or neural networks. For each sensor, a priori data provide an estimate of the probability that the sensor would declare the object to be of type i given that the object is in fact of type j, [i.e., $P(O_j /\text{Declaration } i)$]. These declarations are then combined via a generalization of (6.2). This provides an updated, joint probability for each possible entity, O_j. Thus,

$$P(O_j / D_1, \ldots, D_n), \quad j = 1, 2, \cdots, M \tag{6.5}$$

is the probability of having observed object j given declaration (evidence) D_1 from sensor 1, declaration D_2 from sensor 2, on so on. We can apply a decision logic and select a joint declaration of identity by choosing the object whose joint probability $P(O_j | D_1 D_2 \ldots)$ has the largest value. The choice of the maximum value of $P(O_j | O_1 \ldots O_n)$ is called a maximum a posteriori probability (MAP) decision rule. The Bayes' formulation provides a means to combine identity declarations from multiple sensors to obtain a joint identity declaration. Required input to the Bayes formulation are the ability to compute $P(E|H_i)$ for each sensor and entity or hypothesis H_i, and the a priori probabilities of the hypotheses $P(H_i)$ being true. When no a priori information exists concerning the relative likelihood of H_i, the principle of indifference is used in which the $P(H_i)$ for all i are set equal.

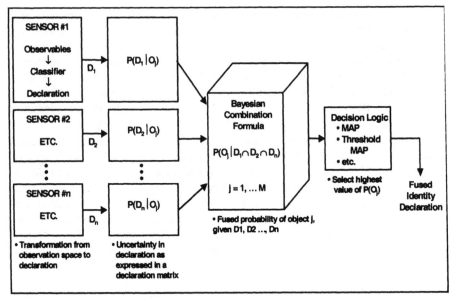

Figure 6.4 Summary of Bayesian fusion.

Disadvantages of Bayesian inference include the following:

1. Difficulty in defining prior likelihood functions: Where do we obtain the values of $P(O_j)$?
2. Complexity when there are multiple potential hypotheses and multiple conditionally dependent events: The combination formula (6.2) becomes much more complex when the observations are not stochastically independent.
3. Requirements that competing hypotheses be mutually exclusive: Cannot assign evidence to object O_i *and* O_j.
4. Lack of an ability to assign general uncertainty.

Peter Denning [17] provides an interesting description of Bayesian inference in a description of a computer program called AUTOCLASS, developed at the NASA Ames Research Center. This program, which is based on Bayesian inference techniques, automatically develops models for large data sets. As an example it automatically classifies stellar objects using data from the infrared astronomical satellite (IRAS).

An IFFN Example

G. Wilson [18] describes an example of the use of Bayesian identity inference provided by an identification-friend-foe-neutral (IFFN) system developed by Ferrante, Inc. of the United Kingdom. This system uses multiple sensors designed to operate onboard an aircraft to perform joint declarations of identity to determine whether observed aircraft are friendly, potential enemies, or neutral. In Wilson's IFFN application, multiple sensors are used to observe a physical phenomena (i.e., infrared or radar cross-section), and based on the data declare the probability of object identity. That is, each sensor constructs a probability matrix of the form,

$$
\begin{bmatrix}
P(D_1/O_1) & P(D_1/O_2) & \cdots & P(D_1/O_M) \\
P(D_2/O_1) & P(D_2/O_2) & \cdots & P(D_2/O_M) \\
& & \ddots & \\
P(D_n/O_1) & P(D_n/O_2) & \cdots & P(D_n/O_M)
\end{bmatrix}
\tag{6.6}
$$

The matrix has N rows corresponding to N possible declarations of identity, and M columns corresponding to M mutually exclusive and exhaustive hypotheses regarding target identity. The (i,j) element of the declaration matrix in (6.6) is the probability that the sensor would declare an object to be object i, if in fact the observed object was object j. M and N are not necessarily equal. Each column in (6.6) must sum to unity, but the sum of each row is not necessarily equal to one.

The values of $P(D_i|O_j)$ for each element in (6.6) may be obtained via physical models or declaration techniques such as those described in Chapter 5. Complex models may be developed for $P(D_i|O_j)$ including effects of target-sensor range, aspect angle, intervening media, and signal-to-noise ratio.

The IFFN Bayesian fusion proceeds as follows. For each sensor, K, observation of an unknown entity results in a declaration, D_K. The row of the sensor declaration matrix provides the probabilities

$$
P(D_k/O_1), P(D_k/O_2), \cdots, P(D_k/O_M)
\tag{6.7}
$$

(i.e., the probability that the sensor will make declaration D_K, given that it has actually observed an entity of type 1, an entity of type 2, etc.). Bayes' rule allows the computation of the probability of the hypotheses given the evidence

$$
P(O_j/D_k) = \frac{P(D_k/O_j)P(O_j)}{\displaystyle\sum_{i=1}^{M} P(D_i/O_j)P(O_j)}
\tag{6.8}
$$

This results in an assessment of the *a posteriori* probabilities of hypotheses 1 through M as predicted by the single sensor observation

$$
\{P(O_1/D_k), P(O_2/D_k), \cdots, P(O_M/D_k)\}
\tag{6.9}
$$

A similar declaration vector may be obtained from each sensor.

Bayes' combination rules allow the sensor declarations to be fused. In general, for each possible object, O_j, the joint (multisensor) probability of having observed object j is,

$$P(O_j / D_1 \& D_2 \cdots D_k) = \frac{P(O_j)P(D_1 / O_j) \cdots P(D_k / O_j)}{\sum_i P(O_i)[P(D_1 / O_i)P(D_2 / O_i) \cdots P(D_n / O_i)]} \tag{6.10}$$

These joint (fused) probabilities are computed for each possible hypothetical object; O_1, O_2, ..., O_M. Decision logic is then applied to choose the object having a maximum probability; that is choose object j such that,

$$P(O_j) \subset \{P(O_1), \cdots, P(O_M)\} \tag{6.11}$$

is a maximum. Wilson [18] has reported initial experiments utilizing this approach to IFFN. A particular difficulty is the definition of the declaration matrix components, $P(O_i|O_j)$. Another issue is the definition of the a priori values of $P(O_j)$. Extensions of Wilson's formulation to other identification applications are readily apparent.

6.4 DEMPSTER-SHAFER'S METHOD

Shafer and Dempster created a generalization of Bayesian theory that allows for a general level of uncertainty (see [19, 20]). The Dempster-Shafer (D-S) method utilizes probability intervals and uncertainty intervals to determine the likelihood of hypotheses based on multiple evidence. In addition, D-S methodology computes a likelihood that any hypothesis is true. The two methods produce identical results when all of the hypotheses considered are mutually exclusive (namely H_i does not overlap H_j for all $i \neq j$) and the set of hypothesis is exhaustive (i.e., no general level of uncertainty).

The succeeding discussion follows the description provided by Lawrence and Garvy [19] and by R. Dillard [21]. Dempster-Shafer's method seeks to model the way humans assign evidence to hypothetical propositions. Dempster-Shafer argues that humans do not assign evidence (i.e., probabilities) to a set of mutually exclusive and exhaustive hypotheses. Instead, they argue that humans assign measures of belief to combinations of hypothesis (i.e., to propositions rather than hypotheses). In the following discussion we distinguish between hypotheses and propositions. A hypothesis is a fundamental statement about nature (i.e., an object is an F-16 aircraft). A proposition may be either a hypothesis or a combination of hypotheses. Propositions may contain overlapping or conflicting hypothesis (e.g., proposition 1 = *the object is an F16 aircraft*, proposition 2 = *the object is an F16 aircraft or Boeing 707 aircraft*, proposition 3 = *the object is a Boeing 707 aircraft*). We see that proposition 1 overlaps proposition 2, proposition 2 overlaps with proposition 3, and proposition 1 conflicts with proposition 3. Consider a set of n mutually exclusive and exhaustive sets of propositions about a subject area; for example,

$$\theta = \{A_1, A_2, \cdots, A_n\} \tag{6.12}$$

This set of elemental propositions is called the frame *of discernment* in D-S terminology. In essence, the frame of discernment is the miniature "world" we are trying to observe and understand. It represents the set of possible causes that would explain observable evidence. Rolling a die provides a simple example of a set of such propositions. Elemental propositions might consist of statements such as the following:

1. The number showing on the die is "1."
2. The number showing on the die is "6."

In this example there are 6 elemental propositions.

If θ denotes the set of n elemental propositions, then there are (2^{n-1}) general propositions that may be developed by Boolean combinations (note the symbol, $\vee$, denotes the Boolean expression, OR) of the original set; that is,

$$2^\theta = \{A_1 \vee A_2, A_1 \vee A_2, \cdots\} \tag{6.13}$$

For our example involving a single die, general propositions include:

1. The number showing on the die is even.
2. The number showing on the die is odd.

One important general proposition is the Boolean disjunction of all of the elementary propositions, denoted $\tilde{\theta}$

$$\tilde{\theta} = A_1 \vee A_2 \vee \cdots \vee A_n \tag{6.14}$$

If evidence is assigned to $\tilde{\theta}$ it is equivalent to a general level of uncertainty. Using the example of the die, if we say that "the number showing on the die is a one or two or three or four or five or six," this is equivalent to saying that "we don't know."

The D-S method assigns evidence to *both* single and general propositions instead of assigning probability to hypotheses (like the Bayesian method). The D-S approach develops the concept of a probability mass, $m(\theta)$, to represent assigned evidence. An observer (or sensor) may assign probability masses such as $m(A_1)$, $m(A_2)$, or $m(A_1 \vee A_2)$. The probability masses are defined such that,

$$m(\theta) \leq 1 \qquad (6.15)$$

where $m(\theta)$ denotes a probability mass assigned either to an elementary proposition (e.g., A_1 or A_2) or to a general proposition in the set 2^θ. Further, the sum of all mass functions assigned to elementary and general propositions is

$$\sum_{i=1}^{n} m(\theta) = 1 \qquad (6.16)$$

How do probability masses relate to probability? The probability of a proposition A_i is given *(induced)* by summing the probability masses for the pertinent elements in θ and 2^θ; that is,

$$\text{Probability } \{A\} = \sum_{A_i \in \theta, 2^\theta} m(\theta, 2^\theta) \qquad (6.17)$$

That is, we sum $m(\theta)$ for the element of θ that contains A_i *exactly* (i.e., there is only one element of θ that contains A_i exactly, since A_i is distinct from A_j in the set of elementary propositions), and in addition, we sum the $m(\theta)$ for those general propositions in 2^θ that contain A_i as an element.

This method of computing the probability of an elementary proposition, provided by (6.17) is one of the features of the D-S approach—namely, that evidence can be assigned not only to mutually exclusive propositions (i.e., the A_i in θ), but also to general propositions that involve overlapping and nonexclusive general propositions. A special aspect of the D-S approach is that one can assign a probability mass to the general proposition, $\tilde{\theta}$ (6.14), which is tantamount to assigning a general level of uncertainty. Since $\tilde{\theta}$ contains every elementary proposition, A_i, $m(\tilde{\theta})$ is an indication of the extent to which an observer (or sensor) is unable to distinguish among any propositions. In particular, if $m(\tilde{\theta})=1$, the sensor is unable to distinguish among any elementary proposition.

Yen [22] suggests that one way to visualize the key difference between Bayesian inference and the Dempster-Shafer method is as illustrated in Figure 6.5. In the Bayesian approach, evidence (e.g., gathered by sensors or other sources) is assigned to one and only one hypothesis. Hence, the left-hand side of Figure 6.5 shows evidence, E_1, assigned to hypothesis H_1, evidence E_2 assigned to hypothesis H_K, and so on. As a result of collecting evidence, and assigning this evidence to hypotheses, it is necessary to determine which hypothesis is "true" (or at least most likely). By contrast, the Dempster-Shafer approach, shown conceptually in the right-hand side of Figure 6.5, shows that evidence from a single source or sensor may be assigned to multiple hypotheses (in effect, to propositions). Thus evidence E_1 is assigned to support hypotheses H_1 and H_2, while evidence E_K is assigned to hypotheses H_2, H_K, and to other hypotheses. This allows the opportunity for evidence from a single source to be assigned to potentially conflicting hypotheses. An issue to be discussed is how to address these conflicting assignments.

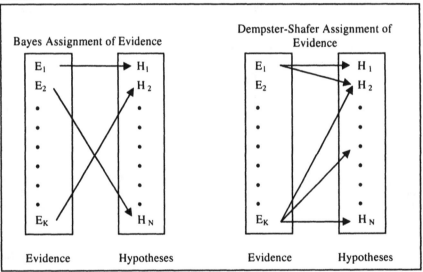

Figure 6.5 Conceptual difference between Bayes & Dempster-Shafer Assignment of Evidence [22].

The D-S approach defines the concept of an evidential interval, denoted [Spt (B_i), Pls (B_i)]. The support for a proposition B_i is defined by

$$Spt(B_i) = \sum_{B_i \in \theta_i \cdot 2^\theta} m(\theta, 2^\theta) \qquad (6.18)$$

This is the sum of the probability masses for a proposition (both within θ and 2^θ). If B_i is a simple proposition (i.e., if $B_i = A_i$), then the Spt (B_i) is simply the probability of A_i. If B_i is a general proposition, (e.g., $B_i = A_1 \vee A_2 \vee A_3$) then the support for B_i is the sum of probability masses contributing to all elements of B_i. For example,

$$Spt(A_1 \vee A_2 \vee A_3) = m(A_1) + m(A_2) + m(A_3) + m(A_1 \vee A_2) + m(A_2 \vee A_3) + m(A_1 \vee A_3) + m(A_1 \vee A_2 \vee A_3)$$
(6.19)

Similarly, the plausibility of a proposition A_i, is defined as the lack of evidence supporting its negation ($\sim A_i$); for example,

$$Pls(A_i) = 1 - Spt(\sim A_i) \qquad (6.20)$$

The output from a D-S process is a set of evidential intervals

$$\begin{bmatrix} Spt(A_1), Pls(A_1) \\ Spt(A_2), Pls(A_2) \\ \vdots \\ Spt(B_k), Pls(B_k) \end{bmatrix} \qquad (6.21)$$

defining the support for (i.e., evidence supporting a proposition), and the corresponding plausibility (i.e., lack of evidence that refutes the proposition). The inputs to the D-S process are probability masses, $m(A_i)$ assigned by an observer or sensor(s). Relative to any mass distribution

$$Spt(A) \leq Pr(A) \leq Pls(A) \qquad (6.22)$$

We note that when probability masses are assigned only to an exhaustive and mutually exclusive set of elementary propositions, then the D-S approach becomes identical to the Bayesian approach.

The Dempster-Shafer Fusion Process

Figure 6.6 illustrates the concept of using a Dempster-Shafer approach to fuse multisensor identity data. Analogous to the Bayesian approach, individual sensors collect parametric data and assign evidence for the identity of an observed entity. These identity declarations are quantified via probability mass functions (vice probabilities in Bayes' approach). Dempster's rules of combination provide a prescription for combining these declarations, resulting in a joint evidential interval, [Spt(H$_i$), Pls(H$_i$)] for each possible hypothesis (i.e., possible identity of observed entities). Here, the support [Spt (H$_i$)] for a hypothesis is a measure of evidence that lends credence to a hypothesis being true, while the plausibility [Pls(H$_i$)] is a measure of the evidence that fails to refute H$_i$. Decision logic is applied to select a hypothesis having the best evidential interval. The result is a declaration of an entity's identity based on the joint results of all the sensors.

The identity fusion process proceeds much like the Bayesian approach described in Section 6.3. Instead of sensor inputs of probabilities of alternate hypotheses (i.e., declarations of identity), the D-S sensors assign probability masses to multiple (nonexclusive, nonexhaustive) propositions. These probability masses are combined via combination rules, and a set of evidential intervals computed. Just as the Bayesian fusion did not prescribe the decision logic to choose a prevailing hypothesis, neither does the D-S approach prescribe the decision logic to choose a proposition as being *true*. The possibility still exists for so-called weak decisions, in which there is insufficient evidence to declare a hypothetical winner from among competing hypotheses.

A useful feature of the D-S approach, however, is the ability to establish a general level of uncertainty. Hence, unlike the Bayesian approach, the D-S method provides a means to explicitly account for unknown possible causes of observational data. This is particularly important in a military hostile observing environment in which observations may be corrupted by enemy countermeasures.

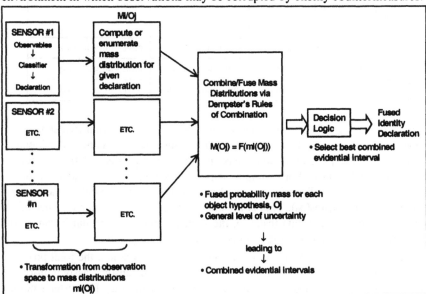

Figure 6.6 Summary of Dempster-Shafer fusion.

Dempster's Rule of Combination

Analogous to the Bayes' formula for combining probabilities, Dempster defined rules of combination. These rules provide a formalism for combining probability masses from independent sources (see for example [20, 21]). Dillard summarizes a number of rules for fusing probability masses from multiple sources. The special case of two sources (sensors) is presented below.

Following Thomopoulos [23], we consider the particular case of the sensors, S_1 and S_2, that assign evidence to three propositions:

- Proposition 1 = u_0 = hypothesis A is true.
- Proposition 2 = u_1 = hypothesis B is true.
- Proposition 3 = u_2 = hypothesis A or B is true.

Sensor S_1 observes parametric data and assigns mass probabilities [$m_1(u_0)$, $m_1(u_1)$, $m_1(u_2)$] to the three propositions. Similarly, sensor S_2 assigns mass probabilities [$m_2(u_0)$, $m_2(u_1)$, $m_2(u_2)$]. Table 6.3 summarizes Dempster's combining rules for this case.

Table 6.3

Dempster's Combination Rules (2 Sources)

S_1 \ S_2	$m_2(u_0)$	$m_2(u_1)$	$m_2(u_2)$
$m_1(u_0)$	$m(u_0)=m_1(u_0)m_2(u_0)$	$k_{10}=m_1(u_0)m_2(u_0)$	$m(u_0)=m_1(u_0)m_2(u_0)$
$m_1(u_1)$	$k_{01}=m_1(u_0)m_2(u_0)$	$m(u_0)=m_1(u_0)m_2(u_0)$	$m(u_0)=m_1(u_0)m_2(u_0)$
$m_1(u_2)$	$m(u_0)=m_1(u_0)m_2(u_0)$	$m(u_0)=m_1(u_0)m_2(u_0)$	$m(u_0)=m_1(u_0)m_2(u_0)$

The elements of the matrix shown in Table 6.3 are the joint two-sensor evidence assigned in accordance with Dempster's rules of combination. For identical propositions, the joint probability mass is simply the product of the masses assigned by each sensor. For example, sensor S_1 assigns mass $m_1(u_0)$ for proposition, u_0, while sensor S_2 assigns mass of $m_2(u_0)$. The joint probability mass is simply, $m(u_0) = m_1(u_0)m_2(u_0)$. This is illustrated in the upper left-hand corner of the matrix in Table 6.3. In this instance, the joint assignment is trivial because both sensors S_1 and S_2 are assigning evidence to the same proposition.

In the case that a proposition overlaps another proposition, the joint probability mass assignment is again straightforward. Suppose sensor 2 assigns $m_2(u_0)$ to proposition u_0, and sensor 1 assigns $m_1(u_2)$ to proposition u_2. Dempster's combination rule assigns a joint mass to proposition, u_0, as follows:

$$m(u_0) = m_1(u_2)\, m_2(u_0)$$

This is illustrated in the lower right corner of the matrix in Table 6.3

How do we treat the assignment of evidence to conflicting propositions? For example, if sensor S_1 assigns evidence $m_1(u_1)$ to proposition u_1 and sensor S_2 assigns evidence $m_2(u_0)$ to proposition u_0, these evidential assignments are in conflict (i.e., in essence, S_1 declares that hypothesis B is true while sensor S_2 declares that hypothesis A is true). Dempster's rules of combination computes a normalizing factor, c, which is the sum of the products of masses assigned to conflicting propositions; for example,

$$c = k_{01} + k_{10} \tag{6.23}$$

for our two-sensor, three-proposition example. Dempster's rule of combination then may be written for two independent sources as

$$m(u_l) = \frac{\displaystyle\sum_{A_i, B_j = u_l} m_1(A_i) m_2(B_j)}{1 - c} \tag{6.24}$$

where

$$c = \sum_{A_i B_m = \phi} m_1(A_k) m_2(B_k) \tag{6.25}$$

Here, ϕ denotes the empty set, and u_l a general proposition defined as a Boolean combination of elemental hypotheses A_i and B_j.

The joint probability mass for u_l is the product sum of all probability masses (from each sensor) over all propositions that contain nonconflicting hypothetical elements A_i and B_j. In our example illustrated in Table 6.3, contributions to mass for hypothesis u_0 include the following:

1. The product mass $m_1(u_0)m_2(u_0)$ when both sensors, S_1 and S_2, assign evidence to proposition u_0 (i.e. the upper left corner of the matrix in Table 6.3).
2. The product mass $m_1(u_0)m_2(u_2)$, when sensor S_1 assigns evidence to proposition u_0, and when sensor S_2 assigns evidence to the nonconflicting proposition u_2 (i.e., the upper right corner of the matrix in Table 6.3)
3. The product mass $m_1(u_2)m_2(u_0)$, when sensor S_2 assigns evidence to proposition, u_0, and sensor S_1 assigns evidence to the non-conflicting proposition u_2 (i.e., the lower right corner of Table 6.3).

These three contributions are summed and "corrected" for the conflicting evidence in which sensor S_2 assigns evidence to proposition u_1, while sensor S_1 assigns evidence to proposition u_0 (i.e., k_{01}) and conversely, k_{10}.

Dillard describes an algorithm for combining probability masses in a more complex situation. Suppose there are two sensors, sensor 1 and sensor 2. Suppose that each sensor outputs a probability mass for an original proposition, A_i, or combinations of these original propositions, i.e.,

$$\text{Sensor 1 Output} = \begin{bmatrix} C_1, m(C_1) \\ C_2, m(C_2) \\ \vdots \\ \tilde{\theta}, m(\tilde{\theta}) \end{bmatrix}, \quad \text{Sensor 2 Output} = \begin{bmatrix} D_1, m(D_1) \\ D_2, m(D_2) \\ \vdots \\ \tilde{\theta}, m(\tilde{\theta}) \end{bmatrix} \tag{6.26}$$

Thus sensor 1 assigns probability masses to u propositions, C_1, C_2, ..., C_u, where the C_k are either original propositions, A_i, or a Boolean combination of original propositions. Similarly, sensor 2 assigns probability masses to propositions, D_1, D_2, ..., D_u. These are either original propositions, A_i, or combinations of original propositions. In this general case the propositions C_i and D_j may overlap, but are not necessarily identical. Table 6.4 summarizes Dillard's algorithm for combining the probability masses including the algorithm steps, and interpretive notes and comments. The algorithm computes both the joint probability mass for propositions, B_k, as well as the general level of uncertainty, $m(\Theta)$. Propositions, B_k, are the conjunction of all input propositions, C_g and D_h. From the joint probability masses, we may compute the evidential interval $[Spt(B_k, Pls(B_k)]$.

Lawrence and Garvey [19] point out that Dempster's rules of combination are both commutative and associative. The order and grouping of combinations do not affect the resulting joint probability masses. Hence, data from sensors may be combined in a hierarchical manner. As a result a variety of parallel implementations might be developed. In addition the same technique may be used to incorporate a priori probability masses assigned to propositions. At this time, work still needs to be done to define efficient implementations. For two or three sensors in a nonparallel implementation, the Dempster-Shafer technique requires approximately twice the computational effort of Bayesian inference.

Table 6.4

Summary of Dillard's Algorithm for Combining Probability Masses

Algorithm Step	Computational/ Logic Test	Interpretation/Notes
1	Set uncommitted = 0 k=0	• Initialize iteration
2	For C_g=C_1, C_2, ..., C_u, θ	• For all sensor evidence

Table 6.4   (continued)

Summary of Dillard's Algorithm for Combining Probability Masses

Algorithm Step	Computational/ Logic Test	Interpretation/Notes

	And $D_h = D_1, D_2, ..., D_u, \theta$
	
	Test:
	
	If $C_g \mid D \leftrightarrow \theta$
	
	   Then, add $m_1(cg^*, m_2(D_h)$ uncommitted
	
	Otherwise
	
	If no B_k such that $B_k = C_g \mid D_h$
	
	   Let $k = k+1$
	
	   $B_k C_g \mid D_h$
	
	   Product-sum$(B_k) = m_1(C_g)^* m_2(D_h)$
	
	Otherwise
	
	   Add $m_1(C_g)^* m_2(D_h)$ to product-sum(B_k)
		• C_g and D_h have no original propositions, A_i, in common

3	For each B_k from step 2 Let $m(B_k) = $ Product-sum$(B_k)/(1$-committed$)$	• $B_k = C_g \mid D_h$ is the Boolean conjunction of proposition C_g and D_h • Note: one of the masses generated is $m(\theta) = (m_1(\theta)^* m_2(\theta) / (1 - \text{uncommitted})$

6.5   GENERALIZED EVIDENCE PROCESSING (GEP) THEORY

The introduction of Dempster-Shafer's (D-S) theory attempted to resolve several issues with Bayesian inference. Specifically, D-S theory provides several features:

1. A mechanism for assigning evidence to overlapping hypotheses (i.e., the ability to assign evidence to hypotheses A_1, A_2 as well as general propositions $(A_1 \lor A_2)$, $(A_2 \lor A_3)$, $(A_1 \lor A_2 \lor A_3)$, and so on. This feature removed the Bayesian requirement to define a set of mutually exclusive and exhaustive hypotheses that could explain observational evidence.

2. D-S theory allows an explicit assignment of evidence to a general level of uncertainty. This allows a sensor to assign evidence to the union of all hypotheses (i.e., $A_1 \lor A_2 \lor A_3 \lor A_u$), in effect declaring *don't know* what hypothesis to declare.

The D-S theory summarized in Section 6.4 combines evidence from multiple sensors via Dempster's rule of combination and allows a priori information (e.g., a priori probability masses for hypotheses) to be updated, much like Bayes' rule updates the a priori probability of a hypothesis based on observation evidence.

Despite these features of D-S theory, there remain criticisms of D-S inference. Critics cite a lack of rigor in defining evidence through independent observation, and some issues in the renormalization techniques used by Dempster's rules of combination. Another generalization of Bayesian theory has been proposed by Thomopoulos and described in [7, 23]. Thomopoulos' formulation is termed the generalized evidence processing (GEP) approach. Fundamentally, GEP addresses the basic assumptions regarding the assignment of evidence to hypotheses or propositions, and the relationship of these assignments to fusion decisions. GEP explicitly separates hypotheses from decisions and meets the choice of different hypothesis tests as different quantization levels of the data.

Consider the two-sensor, three-proposition example described in Section 6.4. Again we consider a situation in which the sensors, S_1 and S_2, observe a phenomenon. Two mutually exclusive hypotheses are considered, H_1 and H_0. A data fusion decision process seeks to make one of three possible decisions based on observed evidence:

- d_0 = hypothesis H_0 is true;
- d_1 = hypothesis H_1 is true;
- d_2 = hypothesis H_0 or H_1 is true.

The GEP method proceeds in a way analogous to the D-S method; that is, each sensor collects evidence and assigns the evidence via probability masses. Unlike the D-S method, however, GEP assigns probability masses and combines those mass data based on the a priori conditional probability of hypotheses H_0 and H_1.

Table 6.5 summarizes the GEP evidence-combining rule for this two-hypothesis, three-decision case. In Table 6.5, the values of the probability masses assigned by sensor S_1 and S_2 are a function of the conditional probabilities for $P(H_0)$ and $P(H_1)$ (i.e., the superscript i for the probability masses indicates that the value is dependent upon $P(H_i)$).

Table 6.5 and the following likelihood ratio threshold rule specify the GEP combination rules:

$$m_1(d_k)m_2(d_m) \rightarrow \text{decision } D_j, \text{if } \frac{m_1^1(d_k)m_2^1(d_m)}{m_1^0(d_k)m_2^0(d_m)} \in F_j \qquad (6.27)$$

where F_j is the decision region that favors decision, d_j. The quantity F_j partitions regions in decision space F_j may be specified to optimize a performance criterion (e.g., minimize false alarms and minimize a risk function).

The combination rules summarized in Table 6.5 along with (6.27) proscribe pair-wise products of probability masses, much like Dempster's combination rules. In D-S, the evidence is combined in accordance with the intersection of propositions (i.e., identity of propositions or overlap of propositions containing mutual hypotheses). In GEP, the combination is based on quantified impact of the resulting decisions. Masses are associated by the threshold rule that permits optimal decisions.

Thomopoulos [7] provides an extension of the combination rules to multiple hypotheses (greater than two) and formulation for various optimization criteria. This formulation appears to be promising and could permit a general fusion and decision process that would optimize criteria such as maximum probability of detection for a given false alarm rate. Additional research remains in the areas of implementation strategies, comparison with D-S and Bayesian formulations, and impacts on fusion system design.

Table 6.5

GEP Evidence Combining Rules [7]

	$m_2^i(d_0)$	$m_2^i(d_1)$	$m_2^i(d_2)$
$m_1^i(d_0)$	$m_1(d_0)m_2^i(d_0)$	$m_1(d_0^i)m_2(d_1)$	$m_1(d_0^i)m_2(d_2)$
$m_1^i(d_1)$	$m_1(d_1)m_2^i(d_0)$	$m_1(d_1^i)m_2(d_1)$	$m_1(d_1^i)m_2(d_2)$
$m_2^i(d_2)$	$m_1(d_1)m_2^i(d_0)$	$m_1(d_2^i)m_2(d_1)$	$m_1(d_2^i)m_2(d_2)$

6.6 HEURISTIC METHODS FOR IDENTITY FUSION

A number of alternative methods are possible for fusing identity data at the decision level. These alternatives are based on techniques used to achieve consensus among human decision makers. The idea is simple: We treat the identity fusion problem as if a group of humans were faced with a decision problem. Each human performs the task of a sensor [i.e., collecting data and choosing or ranking n alternative decisions (hypotheses)]. Thus, the input is a set of observations in either a hard decision format, in which sensors report only a single preference (i.e., declaration of identity), or soft decisions, in which sensors report a ranked list, perhaps with associated confidence factors.

Numerous techniques for group decision-making can be applied to identity fusion. A review of such techniques applied to the problem of selecting research projects is provided by [24]. Applicable techniques for the identity data fusion problem include voting methods, scoring models, ordinal ranking techniques, Q-sort methods, and pair-wise ranking. Each of these is summarized in this section. Table 6.6 provides a summary of these techniques.

Table 6.6

Summary of Heuristic Methods for Identity Fusion

Technique	Description	References
Voting	Treat sensor identity declarations as votes, determine majority, plurality, etc.	[24, 25]
Scoring models	Form weighted sum of ranked identity declarations from sensors; determine maximum weighted score	[26, 27]
Ordinal ranking	Similar to scoring models utilizing ordinal versus scalar quantifications of identity rank	[28]
Q-Sort	Models psychometric process by which a human group achieves consensus	[29, 30]
Pair-wise ranking	Sensors provide preference ranking of I/D versus I+1, I against I+2, and so on; combination rules used to combine pair-wise comparisons of sensor N versus sensor M	[30]

Voting methods address the identity fusion problem by a democratic process. The hard decisions from N sensors are simply counted as votes with a majority or plurality decision rule. Suppose M sensors observe a phenomenon, and each sensor makes an identity declaration from among n alternate hypotheses as illustrated in Figure 6.7 below.

	Hypothesis (Declaration)			
SENSOR	1	2		n
1				
2				
3				
•				
•				
•				
M				
Sum	ϑ_1	ϑ_2	• • •	ϑ_n

Figure 6.7 Identity declaration matrix.

For each possible hypothesis, or declaration of identity, (Hi) we can sum the number of sensors (votes) that declare that hypothesis to be true, namely ϑ_1 in the matrix. The joint declaration of identity is simply the hypothesis Hk such that ϑ_k is a maximum. Decision logic can be applied to insure that the vote is a majority rather than a plurality. Weighted voting schemes (e.g., sensor i has 2 votes while sensor j has 1 vote) may be employed to account for differences in sensor performance. Sensor weights may be computed as a function of target range, signal-to-noise ratio, and other factors. The concept is analogous to the combination methods using Bayesian inference or Dempster-Shafer's method and is illustrated in Figure 6.8.

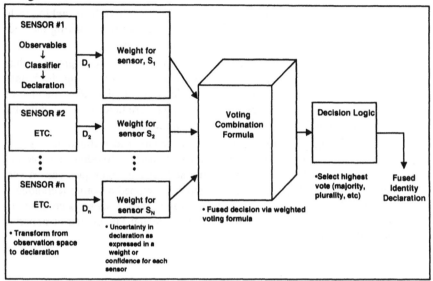

Figure 6.8 Concept of identity fusion using voting.

Scoring models provide another mechanism to establish a joint declaration of identity. These models use a weighted sum to specify the merit of each candidate hypothesis based on a ranking or scoring by each sensor. That is, each sensor, k, assigns a rank or value (i.e., likelihood or confidence factor), Υ_{ik}, for all n possible hypotheses, H_i. Thus Υ_{ik} is the value assigned to the ith hypothesis by the kth sensor. Typically a scale of $[0<\Upsilon<1]$ or $[0<\Upsilon<10]$ is used, where a high score denotes high confidence by a sensor that the hypothesis is true, and zero or a low value of Υ denotes lack of belief by a sensor that the hypothesis is true.

A scoring model simply computes the sum

$$S(\gamma_i) = \left[\frac{1}{c}\right] \sum_{j=1}^{M} w_i \gamma_{ji} \qquad (6.28)$$

M is the total number of sensors, w_j is an a priori weight assigned to the jth hypothesis, Υ_{ji} is the score assigned to the jth hypothesis by the ith sensor, and c is a normalization constant. $S(\Upsilon_i)$ denotes the total score assigned to the ith hypothesis by M sensors. Decision logic is applied to select the hypothesis having the greatest score. Scoring models tend to be robust and consistent (in terms of choosing the correct hypothesis) with more complex optimization or utility models (see for example [26]). Note that product formulations of scoring models tend to be unduly influenced by individual sensor performance (see [24]). Scoring models, like voting models, are computationally efficient and simple to formulate.

Other decision models summarized in Table 6.6 include ordinal ranking techniques described by [28], Q-Sort models [29, 30], and pair-wise ranking methods [30]. These techniques model methods used to achieve consensus.

These latter techniques involve an iterative approach in which a group evaluation occurs followed by re-evaluation based on the disparity among the human evaluators. At this time additional research would be required to adapt them to an automated sensor data fusion problem. However, they are noteworthy as potential heuristic methods to achieve a joint declaration of identity.

A review of voting and related techniques for target identification was conducted by Llinas [25]. A summary of these techniques is provided in Table 6.7. Based on his review, Llinas notes that voting and related techniques can be very sophisticated. Miller and Hall [31], for example, developed a voting technique in which automated "critics" assess the potential value of each sensor's contribution and adapt the voting population based on their potential knowledge or expertise related to each specific vote.

Table 6.7

Summary of Techniques Reviewed by Llinas [25]

Method	Description	Reference
Generalized voting	Multiple expressions for combining classification "votes" from a sensor or algorithm; potential expressions include representation of unanimity, majority, and thresholded majority	[32]
Rank-based classifier fusion	Fusion of declarations from sensors, in which sensors provide a hierarchical ranking (or preference) related to observed target identity; methods include class set reduction, class set reordering, and logistical regression	[33]
Source reliability methods	Methods that assess the source reliability and seek to optimize the overall decision accounting for class-wise reliability; use of a Bayesian minimum cost approach; this is analogous to the decision-making of a main expert to whom the decisions of local experts are forwarded (Llinas)	[31, 34]

6.7   IMPLEMENTATION AND TRADE-OFFS

This chapter has identified a number of techniques to perform identity fusion at the decision level. These include classical inference, Bayesian inference, Dempster-Shafer's method, Thomopoulos' generalized evidence processing (GEP) theory, and various heuristic techniques such as voting and scoring models. What are the trade-offs involved in the utilization of these methods? Key issues include inference performance (i.e., the accuracy of resulting inferences based on input data), required computer resources, requirement for a priori information, and general utility. A brief discussion of these issues is described in this section.

6.7.1   Inference Accuracy and Performance

The performance of various identity fusion techniques has not been studied in a systematic way. Several authors have performed comparisons under limited circumstances. For example, Buede and Martin [35] compare the performance of Dempster-Shafer's method and Bayesian fusion for an IFFN problem involving an aircraft having four sensors [e.g., radar, and infrared search and track (IRST), IFF, and passive electronic support measures (ESM)]. They performed some numerical experiments to compare relative inference accuracy (i.e., the probability of correct identification of a threat) as a function of engagement time. These results are reported in and in [2, 35].

Figure 6.9 illustrates an example of the results with a graph of the probability of correct inference (for Bayesian inference) and assigned belief (for Dempster-Shafer) versus time in seconds. The scenario assumes an engagement between a friendly aircraft having four onboard sensors, and multiple enemy aircraft. Figure 6.9 shows the simulation results of fusing data from ESM and IFF sensors. They suggest that the Bayesian fusion process achieves a greater accuracy than the D-S technique. Moreover, Buede and Martin report that this accuracy advantage is achieved with a factor of four fewer floating point operations than that required for the D-S algorithms.

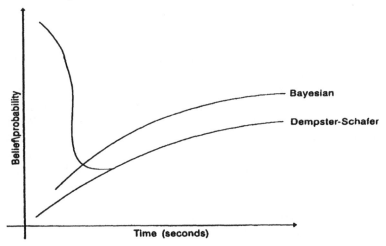

Figure 6.9 Comparison of Decision-Level Techniques [35].

Thomopoulos [7] provides a comparison between his GEP theory and Bayesian inference for several cases involving 2, 3, 4, and 5 sensors. The GEP approach provides approximately a 5% improvement in inference accuracy over the Bayesian approach.

Unfortunately, numerical comparisons of D-S, GEP, and Bayesian techniques are very limited. Proponents of D-S methodology argue that the D-S technique is potentially more robust than Bayesian inference for situations involving many potential causes of observational data. An example would involve a hostile situation in which an enemy employs active countermeasures (e.g., ECM) to degrade the sensing environment. Clearly, much research needs to be performed in this area. Studies need to be performed to assess the comparative inference accuracy for these alternative techniques.

6.7.2   Computer Resource Requirements

The computer resource requirements for various methods are a function of the number of sensors, number of alternative hypotheses, and the specific implementation. Conceptually, the techniques may be ranked by increasing required computer resources as follows: heuristic methods, Bayesian inference, D-S method, and the GEP method. Buede and Martin [35] report that the D-S method requires as much as four times the computational operations as that required for the Bayesian method. A similar factor-of-two result has been reported by Llinas and Hall [36]. Numerical results are not available for other comparisons. One factor that mitigates the additional computational resources required for GEP and D-S theories is the potential for parallel implementations. Clearly, however, additional work is required in this area.

6.7.3   A Priori Data Requirements

Perhaps the most fundamental issue in using any of the approaches for identity fusion described in this chapter is the question of a priori data. Bayesian inference and GEP theory requires a priori probabilities of hypotheses, while the D-S approach requires the a priori assignment of probability masses to propositions. How the a priori is data defined is a key issue that has no simple answer. Possible sources of data include models, statistical databases, historical data, and subjective data. Lack of a reliable source of a priori data impugns the use of any technique. One might argue that the use of a complex inference scheme based on corrupted input data is no better than the use of a simple technique that requires little a priori information.

These issues of accuracy, computational requirements, and source of a priori data all affect the utility of an inference technique. Chapter 11 presents a formal evaluation method that prescribes a mechanism for selecting and evaluating algorithms.

REFERENCES

[1] Steinberg, A. N., "Threat Management System for Combat Aircraft," *1987 Tri-Service Data Fusion Symposium*, Johns Hopkins University, Baltimore, MD, June, 1987, pp. 532-554.

[2] Waltz, E., and J. Llinas, *Multi-sensor Data Fusion*, Norwood, MA: Artech House, Inc., 1990.

[3] Koelsch, J. R., "Sensors - The Missing Link," *Manufacturing Engineering*, 1990, pp. 53-56.

[4] Luo, R. C., and L. Ming-Hsiumn, "Robot Multi-Sensor Fusion and Integration: Optimum Estimation of Fused Sensor Data," *1988 IEEE International Conference of Robots and Automation*, Philadelphia, PA, 1988, April, 1988, pp. 24-29.

[5] Waltz, E. L., "Data Fusion for C^3I," in *Command, Control, Communications Intelligence (C^3I) Handbook*, Palo Alto, CA: EW Communications, Inc., 1986, p. 217-226.

[6] Shafer, G., *A Mathematical Theory of Evidence*, Princeton, NJ: Princeton University Press, 1976.

[7] Thomopoulos, S. C. A., "Theories in Distributed Decision Fusion: Comparison and Generalization," *SPIE 1990 Conference on Sensor Fusion III: 3-D Perception and Recognition*, Boston, MA, November 5-9, 1990, pp. 623-634.

[8] Henkei, R. E., *Tests of Significance*, Beverly Hills, CA: Sage, 1976.

[9] Fishburn, P. C., *Nonlinear Preference and Utility Theory*, Baltimore, MD: The Johns Hopkins Universty Press, 1988.

[10] Miller, K. S., *Hypothesis Testing with Complex Distributions*, Huntington, NY: R. E. Krieger Publishing Company, 1980.

[11] Lipson, C., and N. J. Sheth, *Statistical Design and Analysis of Engineering Experiments*, New York, NY: McGraw-Hill, 1973.

[12] Bayes, T., "Essay Towards Solving a Problem in the Doctrine of Chances," *Philosophical Transactions of the Royal Society*, vol. 53, 1763, pp. 370-418.

[13] Berger, J., *Statistical Decision Theory: Foundations, Concepts and Methods*, New York, NY: Springer-Verlag, 1980.

[14] Plous, S., *The Psychology of Judgement and Decision Making*, New York, NY: McGraw Hill, 1993.

[15] Feller, W., *An Introduction to Probability Theory and its Applications*, vol. 1, New York, NY: John Wiley & Sons, 1957.

[16] Iverson, G. R., *Bayesian Statistical Inference*, Beverly Hills, CA: Sage Publishing Company, 1984.

[17] Denning, P. J., "Bayesian Learning," *American Scientist*, vol. 77, 1989, pp. 216-218.

[18] Wilson, G. B., "Some Aspects of Data Fusion," *1985 Intelligence Conference in Advances in C^2*, United Kingdom, 1985.

[19] Lawrence, J. D., and T. D. Garvey, "Evidential Reasoning: A Developing Concept," *Proceedings of the International Conference on Cybernetics and Society, IEEE*, 1982.

[20] Dempster, A. P., "A Generalization of Bayesian Inference," *Journal of the Royal Statistical Society*, vol. 30, 1968, pp. 205-247.

[21] Dillard, R. A., "Tactical Inferencing with the Dempster-Shafer Theory of Evidence," *The Asilomar Conference of Circuits, Systems, and Computers, 1983*, Naval Post Graduate School, Santa Clara, CA, 1983, pp. 312-316.

[22] Yen, J., *Discussion of Dempster-Shafer's Theory*, D. Hall, Editor, State College, PA, 2003.

[23] Thomopoulos, S. C. A., "Sensor Integration and Data Fusion," *Journal of Robotics Systems*, vol. 7, 1990, pp. 337-372.

[24] Hall, D., and A. Nauda, "An Interactive Approach for Selecting R&D Projects," *IEEE Transactions on Engineering Management*, vol. 37, 1990, pp. 126-133.

[25] Llinas, J., "Fusion-Based Methods for Target Identification in the Absence of Quantiative Classifier Confidence," *SPIE Signal Processing, Sensor Fusion, and Target Recognition VI Conference*, Orlando, FL, April, 1997.

[26] Moore, J. R., and N. R. Baker, "Computational Analysis of Scoring Models for R&D Project Selection," *Management Science*, vol. 16, 1969, pp. 212-232.

[27] Dean, B. V., and M. Nishry, "Scoring and Profitability Models for Evolution and Selecting Engineering Projects," *Journal of Operational Research*, vol. 13, 1965, pp. 550-569.

[28] Cook, W. D., and L. M. Seifford, "R&D ProjectSelection in a Multidimensional Environment: A Practical Approach," *Journal of the Operational Research Society*, vol. 33, 1982, pp. 397-405.

[29] Helin, A. F., and W. E. Souder, "Experimental Test of a Q-Sort Procedure for Prioritizing R&D Projects," *IEEE Transactions on Engineering Management*, vol. EM-21, 1974, pp. 159-164.

[30] Souder, W. E., "System for Using R&D Project Evaluation Methods," *Research Management*, vol. 21, 1978, pp. 29-37.

[31] Miller, D., and D. Hall, "The Use of Automated Critics to Improve the Fusion of Marginal Sensors in ATR and IFFN Applications," *National Symposium on Sensor Data Fusion*, Johns Hopkins Applied Physics Laboratory, Baltimore, MD, May, 1999.

[32] Xu, L., A. Krzyzak, and C. Y. Suen, "Methods of Combining Multiple Classifiers and their Application to Handwriting Recognition," *IEEE Transactions SMC*, vol. 22, 1992.

[33] Ho, T. K., and e. al, "Decision Combination in Multiple Classifier Systems," *IEEE Trans on PAMI*, vol. 16, 1994.

[34] Jeon, B., and Landgrebe, "Decision Fusion with Reliabilities in Multisource Data Classification," *IEEE International Conference on SMC*, Chicago, IL, 1992, October, 1992.

[35] Buede, D. M., and J. W. Martin, "Comparison of Bayesian and Dempster-Shafer fusion," *1989 Tri-Service Data Fusion Symposium*, John Hopkins University, Baltimore, MD, May, 1989, pp. 81-101.

[36] Llinas, J., and D. Hall, *Lecture Notes on Data Fusion*, Torrence, CA: Technology Training Corporation, 1986.

Chapter 7

Knowledge-Based Approaches

A key challenge for multisensor data fusion systems is to perform automated reasoning to perform situation assessment, consequence prediction, and analysis (e.g., for level 2 and level 3 fusion). This chapter describes techniques for automated reasoning or knowledge-based approaches to perform such reasoning. The techniques are drawn from the broad discipline of artificial intelligence and require the ability to represent knowledge (including uncertainty) and to perform reasoning to achieve intelligent results.

7.1 BRIEF INTRODUCTION TO ARTIFICIAL INTELLIGENCE

In previous chapters, techniques and methods were described to fuse multisensor data in order to infer the location, kinematics, attributes, or identity of physical entities such as platforms, emitters, geographical features, mineral deposits, or any of a variety of basic objects. The particular inferences sought are dependent upon the specific application to which fusion algorithms are applied. An example of some applications and corresponding inferences are summarized in Table 7.1. These applications include the following:

1. Tactical military situation assessment aimed at locating, identifying, and determining the behavior of military entities;
2. Military threat assessment designed to identify and locate enemy weapons, units, and to identify potential enemy activities that pose a threat to friendly forces;
3. Automated monitoring of complex equipment (e.g., a nuclear power plant) that monitors equipment status to determine equipment health and identify equipment faults;
4. Medical diagnosis and advisory systems that identify anomalous biological conditions and seek to analyze and determine the cause of patient symptoms;

5. Multisensor remote sensing systems that locate, identify, and monitor natural resources such as crops, forests, mineral deposits, and terrain features.

Basic inferences for each of these applications are summarized in the second column of Table 7.1. Chapters 2-6 of this book describe a number of methods to perform these inferences based on multiple sensor data. Techniques were described for data association, estimation, and identity determination. The techniques to perform these inferences were primarily oriented toward numerical methods; that is, the methods involved process parametric data via numerical algorithms to obtain estimates of parameters or identity.

For each of the applications shown in Table 7.1 there are higher levels of inferences that are typically made by human analysts. For example, military situation analysts seek to determine relationships among objects, identify and characterize complex entities (i.e., aggregates of physical emitters, platforms, and weapons), and interpret the meaning of an order-of-battle database in the context of time, space, prevailing level of hostilities, proximity to political boundaries, and the natural environment (including weather and terrain). Similarly, a maintenance specialist might look at data about equipment faults and performance to diagnose major problems, identify fundamental design or manufacturing problems, and develop recommendations for preventative maintenance and corrective actions. Other examples of these higher-level inferences are illustrated in the third column of Table 7.1. These higher-level inferences traditionally require forms of reasoning and cognition applied by humans. Types of reasoning include the following:

- Pattern recognition utilizing uncertain, incomplete, and conflicting data;
- Spatial and temporal reasoning;
- Establishing cause-effect relationships;
- Prediction of future events/activities
- Planning;
- Induction (establishing general concepts based on examples of specific cases or instances);
- Deduction (reasoning from general principles to specific conclusions about a particular case);
- Abduction (establishing parallels or analogies);
- Learning.

Table 7.1

Examples of Data Fusion Inferences

Application	Basic Inferences	Higher-Level Inferences	Types of Reasoning
Tactical situation assessment	• Location of low-level entities and objects	• Identity of complex entities (aggregates of objects and entities)	• Estimation • Spatial and temporal reasoning

Table 7.1 (continued)

Examples of Data Fusion Inferences

Application	Basic Inferences	Higher-Level Inferences	Types of Reasoning
(level 2)	• Identification of objects	• Relationships among objects • Contextual interpretation	• Functional relationships • Hierarchical relations • Contextual reasoning
Threat assessment (level 3)	• Location and identity of low-level entities and objects • Identification of weapons • Prediction of positions	• Identification of threats • Prediction of intent • Analysis of threat implications • Estimation of capability	• Pattern matching • Prediction • Development of scenarios • Interpretation • Causal reasoning
Complex equipment diagnosis	• Estimation of equipment state parameters • Identification of abnormal conditions	• Establish cause-effect relationships • Development of hierarchical relations • Analysis of process • Recommendation of diagnostic tests • Recommendation of maintenance	• Analysis of hierarchy • Cause-effect analysis • Deduction and induction • Abduction • Case-based reasoning
Medical diagnosis	• Determination of key biological parameters • Assessment of symptoms • Identification of abnormal parameters • Location of injuries	• Analysis of relationships among symptoms • Link symptoms to potential diseases • Recognition of pathological indicators • Location/characterization of injuries	• Pattern recognition • Deduction and induction • Cause-effect analysis • Abduction • Case-based reasoning
Remote sensing	• Location and ID of crops, minerals and geological features • Identification of features and objects	• Identity of unusual phenomena (e.g., crop disease) • Determine relationships among geographical features • Interpretation of data	• Pattern recognition • Spatial and temporal reasoning • Contextual analysis • Case-based reasoning

Human analysts use these and many other cognitive skills to develop complex or higher level inferences based on multisensor data and other information sources. Can computers emulate these capabilities? This fundamental question is precisely the target of research in artificial intelligence (AI). AI may be viewed as the study of how computers or machines can be made to emulate functions normally associated with human or animal behavior, as shown in Table 7.2.

Table 7.2

Relation between Human/Animal Characteristics and AI Areas of Study

Human/Animal Behavior	AI Area of Study
Purposeful/autonomous action	Robotics
Cognition/reasoning	Expert (knowledge-based systems)
Vision	Computer vision
Language understanding and communication	Natural Language processing
Learning	Computer learning

For virtually every aspect of human behavior, there is a corresponding area of study in the field of AI. Numerous texts are available that provide an overview of AI (see for example,[1-7]). Problems in AI are studied by a variety of researchers in more traditional fields (e.g., computer science, psychology, mathematics, biology, biophysics, and many other areas). AI may alternatively be viewed as a laboratory that provides insight into how humans operate. For example, the attempt to make computers understand human language provides insight into how humans do (and do *not)* learn their native language.

One complaint of AI researchers is that AI goals are a moving, ever-receding target. In the early 1970s, for example, it was relatively difficult for computers to perform symbolic computations (e.g., a computer could readily numerically integrate a function such as e^x, via techniques such as the Runge-Kutta methods). However, it was not easy to symbolically integrate such a simple function. Subsequently, a number of symbolic manipulation programs were developed such as formula manipulation compiler (FORMAC) and MACSYMA, developed by Stanford. MACSYMA, in particular, reached a state of sophistication that allowed a computer to perform symbolic mathematical operations normally performed only by college level engineering students (e.g., analytical differentiation and integration in calculus, algebraic manipulations, and analytical solution of differential equations). Current commercial software includes packages such as Mathematica^{TM} and Mathcad. Because of the success of such programs, it has become routine for computers to perform symbolic mathematics. Thus, symbolic computation is not currently considered a feat of AI.

Similarly, speech understanding was considered to be a challenging problem in AI during the mid 1980s. Researchers at Carnegie Mellon University and Stanford developed architectures such as the Blackboard systems. Specific systems such as HEARSAY-1 and HEARSAY-2 explored methods to ingest spoken speech and develop systems that approximated the ability of humans to hear and understand conversations. These programs used a combination of knowledge about words, phonetics, syntax, sementics, and context information to improve speech recognition and understanding. While the general problem of speech understanding (i.e., the problem of semantic understanding) has still not been solved, commercial software such as IBM's package, ViaVoice, provides good capability (with training by individual speakers). Here again, the technology has progressed so that people are not surprised at the ability of computers to "understand" speech.

Despite such successes of AI researchers, true human-level capabilities have not yet been achieved. The goal of a truly *intelligent* machine remains elusive. A lively and entertaining perspective on this pursuit of machine intelligence has been provided by Raymond Kurzweil [1]. Kurzweil describes the history of AI and provides a detailed chronology from 3,000 B.C. to the present. In addition, he predicts some milestones into the next century. Additional thoughts on AI challenges are provided by the article, "AI's Greatest Trends and Controversies" at the Web site; http://www.computer.org/intelligent/articles/AI_controversies.htm.

With these advances in intelligent computing, it is natural that elements of AI research are applied to problems in multisensor data fusion. Table 7.3 shows a summary of their applicability.

Table 7.3

Summary of AI Applicability to Data Fusion

AI Area	Current Status and Limitations	Data Fusion Applications
Knowledge-based systems	• Several thousand reported expert systems • Numerous tools for development of rule-based systems, blackboard systems, and intelligent agents • Limited experience with fielded systems • Limited work with cooperating expert systems–new trends towards use of agent based architectures	• Prototype expert systems for situation assessment, correlation, use of expert systems to guide algorithm selection and use in fusion systems, countermeasure assessment • Smart semi-autonomous weapon systems • Intelligent sensors • Mission planning
Natural language processing	• Increasingly sophisticated commercial tools • Special hardware chips for fuzzy logic • Continuing research script-based and knowledge-based syntactic and semantic processing	• Automated message processing • Intelligent search engines for information retrieval with adaptation to individual users • Speech recognition and conversational (dialog-type) human-computer interactions • Language understanding • Speaker and keyword recognition
Machine learning	• Numerous experiments with self learning systems and mathematical induction • Focus on learning using neural networks and Bayesian Belief Nets	• Adaptive software for situation assessment and entity recognition • Battle management software that learns from experience

Table 7.3 (continued)

Summary of AI Applicability to Data Fusion

AI Area	Current Status and Limitations	Data Fusion Applications
	• Reflexive rule-based modification systems	
Cognitive models	• Extensive research in recent years in new cognitive models such as naturalistic decision making and the recognition-primed decision model • Studies of decision-making under stress; group decision-making; computer-aided cognition • Uncertainty schema (e.g., fuzzy sets, Dempster-Shafer)	• Coordination of battle management • Mathematical models to mitigate human biases • Modeling of human analysts • Utilization of agent-based technology to model human team cognition
Computer vision	• Prototype systems for image interpretation • Computer vision recognition of multiple objects • Gestalt image representation and mapping to semantic terms	• Autonomous mobile sensors • Target motion and tracking • Terrain following • Optical avoidance and path planning • Scene analysis
Robotics	• Routine use of robotics for manufacturing • Robotic arms for hostile environment manipulations • Use of robots for disaster areas • RoboCup simulations of robots for competitions	• Autonomous land vehicles • Robotic air and underwater vehicles • Cooperating teams of semi-autonomous mobile sensors

Three areas in particular are especially important: (1) expert systems or knowledge-based systems; (2) natural language processing (NLP); and (3) computer vision/pattern recognition. In this chapter we will focus on the concept and utility of expert systems. In a survey, Hall and Linn [8] found over 50 reported data fusion systems. Of these, 24 systems utilized the technology of expert systems to perform higher level inferences than those obtainable by analytic techniques. The recent survey by [9] identified over 75 data fusion systems, many of which utilize automated reasoning methods. The advent of expert systems for data fusion has been facilitated by the introduction of numerous commercial tools called expert system shells that provide a ready mechanism for development.

The remainder of this chapter provides an introduction to expert systems and to related technologies—knowledge representation, search techniques, and blackboard architecture concepts. Section 7.3 describes issues in developing expert systems, and Section 7.4 introduces a related method, logical templating. Finally,

Section 7.5 introduces the method of bayesian belief networks, and Section 7.6 describes the use of intelligent agent software.

7.2 OVERVIEW OF EXPERT SYSTEMS

In recent years a large number of expert, or knowledge-based, systems have been successfully developed. Applications have spanned the commercial arena, industrial uses, and military applications. Well-known expert systems include: Digital Electronic Corporation's XCON, which is used to configure VAX minicomputers; PROSPECTOR, which was designed for mineral exploration by the Stanford Research Institute, and MYCIN, the expert system designed to perform medical diagnoses. These systems have evolved from experimental prototypes used to illustrate principles of expert system research to tools used in daily industrial work. By mid 1987, over 1,025 expert systems had been reported in the technical literature. The systems have come to be relied upon to meet the needs of industrial production and sales, and are beginning to be used in military operations. A number of systems have been developed for aerospace applications such as fault diagnosis for orbital refueling operations, flight management, satellite control, and space station automation. This section provides an introduction to expert systems. More detailed discussions can be found in [10-13].

7.2.1 Expert System Concept

Figure 7.1 illustrates the concept of an expert system computer program. The structure of the program comprises four logical parts:

- A knowledge base that contains facts, algorithms, and a representation of heuristics;
- A global database that contains dynamic data;
- A control structure, or inference engine;
- A man-machine interface.

The control structure uses input data and facts and attempts to search through the knowledge base to reach conclusions.

The knowledge base for an expert system contains the information that comprises the system's *expertise.* Knowledge may be presented by several mechanisms:

- Production rules that have the form: If (evidence exists for X) then do Y, where Y may involve performing a computation or updating a database;
- Semantic nets are graphical representations that describe relationships between general objects and more specific instances of the objects (e.g., the

general object, *aircraft,* and specific instances of aircraft, such as B-1, 747, and the Concord);

- Frames are representations of objects and classes of objects via structured data records;
- Scripts use the concept of a play (e.g., acts, scenes, settings, and actions) to describe situations and events.

Each of these types of knowledge representation has advantages and disadvantages for representing information for specific applications.

One of two basic strategies is used for the inference process in an expert system. Backward chaining uses a first-order predicate calculus approach that seeks to determine if data can be found to support or verify an assumed conclusion (i.e., the system assumes that a result is true and seeks to verify or refute the assertion). The backward chaining approach was introduced in the PROLOG language. Alternatively, the forward chaining approach begins with data and seeks to determine if these data allow an inference to be drawn (i.e., the opposite approach to backward chaining). The forward chaining approach is used in production systems. Historically, the forward chaining approach was introduced after the introduction of backward chaining.

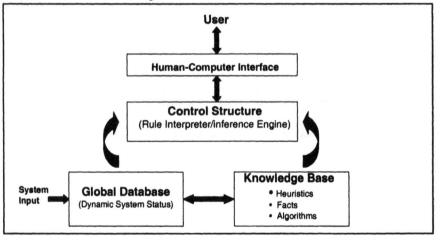

Figure 7.1 Concept of an expert system.

The primary differences between expert systems and ordinary computer programs are twofold. First, an expert system deliberately separates the control process from algorithms in the knowledge base. Ordinary computer programs tend to embed the program control (e.g., logic) with the problem-specific algorithms (knowledge base). This is an efficient design when all of the logical paths are known and fixed, but it breaks down when the complete set of branches is unknown or very large (potentially infinite). One advantage of an expert system is that the knowledge base can be readily modified without reprogramming the program. Cases can be added or paths modified as experience grows, building up a complex base.

A second difference between ordinary computer programs and expert systems is implied by the dynamic and nonexhaustive search through the knowledge base. The expert system program searches for a supportable inference, but does not exhaustively search the knowledge base. Hence, an expert system obtains an answer, but not necessarily the only answer. Indeed, the conclusion may differ, depending on the order in which the data is input. Certainly the order of the chain of reasoning depends upon the data order. This aspect of expert systems makes it difficult to evaluate them.

7.2.2   The Inference Process

The inference process utilized by expert systems is shown conceptually in Figure 7.2. The iterative process begins with an initial data set (i.e., input data to the expert system) and the knowledge base comprising rules, frames, scripts, or semantic nets. (Note that in the following discussion we will restrict knowledge representation to rules, although the discussion is generally applicable to any knowledge representation technique or combination of techniques). The inference process uses the a priori data set, and searches the complete set of rules to identify applicable rules. An applicable rule is one whose antecedent (left-hand side) conditions are satisfied. In general, multiple rules may be found to be applicable. The process of finding applicable rules involves both a search process and pattern-matching capability.

Having identified one (or more rules) that is applicable, the control structure must select one rule for execution (i.e., *instantiation*). The choice of which rule to select is critical since continued progress towards conclusion depends upon which rule is executed. Note that unlike a conventional computer program, we cannot simply execute all applicable rules. The reason is that as soon as even one rule is processed, the dynamic database is modified. As a result, the applicable rule set has changed. Hence, the control structure must research the original complete knowledge base to find a new set of applicable rules. The inferential process continues iteratively until one of two events occurs: (1) a conclusion is reached with a final rule telling the control structure to quit; or (2) no applicable rules can be found via the search process. In the latter case, the expert system must give up its search and declare that no conclusion can be reached or request additional data.

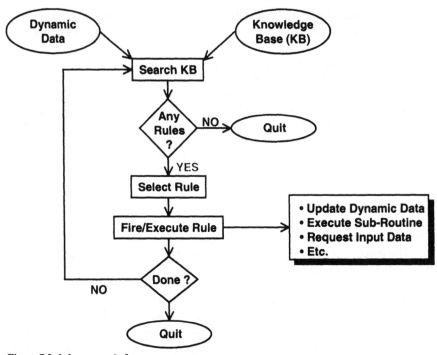

Figure 7.2 Inference cycle for an expert system.

The rule selection from among multiple applicable rules (termed conflict resolution) may utilize one or more of several strategies. These strategies include refraction, actuation, rule ordering, recency, specificity, and random choice. Each of these is summarized as follows:

- Refraction: Multiple instantiations of the same rule are avoided by marking a rule that has been used. Hence in subsequent searches of the knowledge base, the marked rule is avoided.
- Actuation: The dynamic data (i.e., working memory) is weighted dynamically in accordance with perceived value. These weights are used in prioritizing competing rules (i.e., the rule that incorporates the most important data is selected).
- Rule ordering: The rules in a knowledge base may be weighted a priori. Competing rules are selected based on their relative a priori weights.
- Recency: Incoming data to an expert system may be ranked by inverse time order in which the data are received. Recent data are ascribed to have more value than older data. The choice among competing rules is based on recency of the required data.

- Specificity: Weighting of rules is based on the number and type of conditions in the rule antecedent (i.e., left side). Rules that require more conditions to be satisfied are presumed to be more specific, and hence more likely to lead to a specific inference.
- Random choice: If no other technique resolves the conflict among competing rules (eventually this will happen during a search process), a rule is selected at random.

In most commercial expert system development tools, a combination of these conflict resolution strategies is implemented.

In Figure 7.2 the instantiation or execution of a rule is illustrated conceptually in the third step (in the block denoted *fire rule*). The rule execution may entail one of a number of possible actions. These include the following:

- Updating of the dynamic data (e.g., execution of the rule provides additional information that was not previously *known* by the inference process);
- A rule may request data from a system human user or access a database to retrieve data;
- Numerical or other subroutines may be executed. Thus, algorithmic techniques or models may be invoked in order to compute required quantities

Numerous commercial tools exist that support the implementation of rule-based systems. Annual reviews of such tools are provided in the *AI/Expert* magazine.

7.2.3 Forward and Backward Chaining

Given a rule set there are two basic approaches to the inference process: (1) forward chaining and (2) backward chaining. The forward chaining approach begins with a set of facts or observations and seeks to process rules in the knowledge base to determine if a conclusion can be found (this is the approach shown conceptually in Figure 7.2). Neopolitan [14] provides an example of the challenge of deducing the identity of a tree based on observations about the tree's stem, position, trunk, leaves, and leaf shape. In this example, the observations that a plant has a woody stem, an upright position, and a single main trunk would lead to the conclusion that the plant is a tree. Given that the leaves are not broad and flat, this leads to the further classification that the tree class is gymnosperm. Neopolitan provides a discussion of this example and pseudocode for the implementation of an inference engine.

By contrast, backward chaining involves beginning with a tentative hypothesis or conclusion (e.g., the observed plant is a tree of the gymnosperm class). The inference process seeks to determine if available evidence supports such a conclusion. In this case the necessary criteria are identified (namely that the plant must have a woody stem, upright position, single main trunk, and leaves that are not broad and flat), and evidence is sought to determine whether or not these conditions are met. Again, Neopolitan provides a further discussion of this example and pseudocode to implement an inference engine.

Neither one of these approaches is necessarily "better" or "worse" than the other. Instead the appropriate approach must be selected based on the type of evidence available and the operational concept concerning whether the reasoning system collects data and seeks to make inferences, or whether the system is used in an advisory role for a human who postulates potential hypotheses and seeks to determine their truth. Many of the commercial rule-based system tools support both forward and backward chaining.

More details about implementation of expert systems are provided in Section 7.3. We turn now to the issues of knowledge representation and search techniques.

7.2.4   Knowledge Representation

We have previously introduced several techniques for representing knowledge (e.g., rules, frames, scripts, and semantic nets, and direct methods). This section provides a summary of these techniques and a comparison of their utilization.

A summary of knowledge representation techniques is shown in Table 7.4. For each technique, the figure provides a brief definition or description of the technique, summarizes the formulation, and indicates the domain of applicability. Anton, Hemphill et al. [15] provide an example of knowledge representation for military problems.

Table 7.4

Summary of Knowledge Representation Techniques

Representation Technique	Description	Formulation	Applicability
Production rules	An antecedent, consequence pair	Syntactical representation: IF (evidence for x), THEN (do y)	Loosely coupled collection of facts
Networks	A description of hierarchical relationships between general entities and more specific instances of that object	Generally graphical formulation as semantic networks, belief networks, transition networks, or procedural networks	Hierarchies of stereotypes; patterns of causality; sequences of activities or events
Frames	A representation method designed to encapsulate multiple attributes of an object	Combined graphical and complex data record formulation; allows specification of entity hierarchies as well as attributes	Hierarchies of stereotypes; patterns; sequences of activities or events
Scripts/plans	Patterned after the technique used to direct ac-	Formulation specifies; active agents/objects,	Domains influenced by roles, contexts,

Table 7.4   (continued)

Summary of Knowledge Representation Techniques

Representation Technique	Description	Formulation	Applicability
	tors in a play; specification of active entities, general setting, sequence of actions	contextual setting, sequence of actions, interaction among entities	sequences of actions
Direct (analogical)	A class of representation schemes to represent knowledge about certain aspects of the domain	Various special representations such as maps, models, diagrams, and symbols	Specialized knowledge for a particular domain

Perhaps the most frequently used representation technique is production rules. Nearly every expert system skill tool provides for representation (and inference processing) using rules. Rules utilize one of two forms: production rules asserting (antecedent-consequence) relations or logic rules asserting (consequence-antecedent) relations. An example of a production rule for a medical diagnosis system is the following:

IF: The stain of the organism is gramneg, *and* the morphology of the organism is rod, and the aerobicity of the organism is anaerobic

THEN: There is evidence (with suggestive measure 0.6) that the identify of the organism is baceriodes.

Rules may incorporate uncertainty in the antecedent as well as the consequence. Rules may be developed to describe a variety of applications; however, a rule-based representation is especially applicable to domains characterized by a loosely coupled collection of facts. Large-scale expert systems may require hundreds or thousands of rules to adequately represent the required expertise. The program XCON (later named Rl) developed by the Digital Electronic Corporation with Carnegie Mellon University comprised more than 5,000 rules to describe the proper configuration of DEC's VAX computer systems.

A second method for representing knowledge is networks, including semantic networks, transition networks, and procedural networks. Semantic networks in particular are especially applicable for representing hierarchies of stereotypes. An example of a semantic network is a family tree. General objects (e.g., father, mother, grandfather, siblings, aunts, and uncles) can be identified. Each of these objects has characteristics that can be specified along with interrelationships with other objects. For example, the general object *father* has the characteristics of being a male person having one or more children. A semantic network can thus describe the general characteristics of familial relationships and associated hierarchy. In addition, specific instances of these objects may be identified (i.e., John Jones is the son of William Jones; John Jones has two siblings, Mary and Jim). Semantic networks are formulated to describe a network of objects and their relationships. A formal graphical scheme may be used in which circles or nodes identify objects; labeled arrows or links between nodes denote relations; and slots identify characteristics of a node. Again, for networks uncertainty can be represented. A specific type of networks called Bayesian belief nets is described in Section 7.5.

An important feature of networks is the concept of inheritance. The concept is that the identity of an object as a member of a general class allows us to assume information based on class membership. For example, if an aircraft is established as a member of the *fixed-wing, jet-propelled class of fighter aircraft*, then a great deal of information can be surmised based on this membership. In an inference process, inheritance procedures are used to seek all cases or examples of an object and fill in unknown properties of an object based on class membership.

In practice, knowledge representation via semantic networks is typically implemented in the form of *frames*. The notion of frames was developed and popularized by Minsky [4]. A frame is a collection of semantic net nodes and slots that together describe a stereotyped object, act, or event. Frames provide structured representations of objects or classes of objects. Frames are most applicable to representing hierarchies of stereotypes or patterns of data. Many of the large-scale commercial expert system shells provide the capability to represent knowledge via frames. Here again uncertainty can be represented in a frame.

The concept and use of scripts for knowledge representation was introduced by Roger Schank [16]. The concept is based on the technique used to direct actors in a play. A script proscribes scenes involving active agents, a general setting, and a sequence of actions and interaction between agents. This type of representation is especially suited to describing roles, contexts, and a group of actions habitually used to achieve some goal. Anton et al. [15] illustrate the use of a script to describe a military sortie to destroy an enemy bridge by friendly tactical aircraft. Scripts are not commonly available in commercial expert system development tools; nevertheless, they provide a useful concept for describing complex events and contexts. A special feature of scripts is that they allow a means of reducing the amount of *real-world* information that must be described. For example, by specifying distinct scenes and settings we can constrain the amount of information that must be provided.

A final *catch-all* category for knowledge representation is the direct use of analogical methods. This simply refers to any of a specialized set of techniques used to represent information about a particular domain. Examples include maps, diagrams, and mathematical or chemical symbols. Direct methods are tailored to a particular type of knowledge and as such may be a very efficient and effective way of describing knowledge. Uncertainty may also be incorporated into direct knowledge representation techniques.

How does the developer of a knowledge-based or expert system choose from among the techniques for representing knowledge? Unfortunately, the selection must be based on trial and error (i.e., simply attempting to represent knowledge for a specific domain to see which technique allows the easiest and most effective representation). While we have provided guidelines in Table 7.4, there is no ideal or universal knowledge representation technique. Fortunately, many of the commercial expert system building tools provide a variety of techniques to support knowledge representation.

The actual process of populating a knowledge base (i.e., developing rules, frames, or scripts) is termed knowledge engineering. A process for knowledge engineering is introduced in Section 7.3.

7.2.5 Representing Uncertainty

Each of the methods described for representing knowledge (i.e., rules, frames, networks, scripts, or analogical methods) allows for the possibility of incorporating uncertainty. Uncertainty may be introduced in two senses: first as a means of attributing uncertainty to observed data and second by specifying uncertainty of logical relationships. Thus, a rule describing the biological category of *mammal* may have the following form:

    IF   the animal has body hair

    AND   the animal is warm-blooded

    THEN the animal is a mammal

Uncertainty can be attributed to the observations (e.g., based on infrared data we might ascribe a probability or confidence that an observed animal is warm-blooded), and uncertainty can be assigned to the conclusion (i.e., even if we know with 100% certainty that an animal has body hair and is warm-blooded, then there may still be only a certain probability that the animal is a mammal). Similarly, observational and logical uncertainty may be represented in frames, scripts, networks, or other knowledge representation methods.

A variety of techniques may be used to represent uncertainty. In previous chapters, we have introduced the concept of probability and evidential intervals. Five techniques are summarized in Table 7.5 including probability, Dempster-Shafer's evidential intervals, generalized evidence processing theory, certainty factors, and fuzzy set theory. A good description of these methods for representing uncertainty is provided by [17-19]. Each of these uncertainty representation techniques is summarized as follows.

Table 7.5

Summary of Techniques for Representing Uncertainty

Representation Technique	Description	Uncertainty Propagation
Probability P(H)	Classical, empirical, subjective probability of a hypothesis, and conditional probability of evidence, given a hypothesis	Bayes rule and the classical laws of probability
Evidential intervals [Spt[A], Pls[B]]	Support (evidence supporting) a proposition, A, and the plausibility of a proposition (failure of evidence to refute A)	Dempster's rules of combination [20]
Generalized evidential processing theory	Similar to Dempster-Shafer evidential interval, except for a hypothesis rather than a proposition	Thomopoulos rules of combination [21]
Confidence factors	Numerical estimates of a fact, association, or rule; describes a degree of belief for validity	Combinatorial calculus for Boolean operations [22]
Fuzzy sets	Generalization of Boolean set theory; defines sets by specifying membership functions	Fuzzy logic defined as a generalization of Boolean (T,F) logic to incorporate fuzzy membership functions [23]

- *Probability:* Concepts of probability and its use in reasoning were introduced in the seventeenth century and used to predict the outcome of games of chance. In Chapter 6, a discussion of classical inference introduced the concepts of classical, empirical, and subjective probabilities. Probability may be ascribed to observations as well as to logical relations. The classical laws of probability (i.e., conditional probability rules) as well as Bayes inequality provide a basis for reasoning with probability. Probability seems intuitively appealing because of its formal grounding and familiarity in everyday life. Unfortunately, our intuition often provides fallacious results. Examples of these problems of reasoning with probability are provided by [24, 25]. Probability has a very strong theoretical background and correspondingly low computational complexity. Nevertheless, as discussed in Chapter 6, the use of classical and Bayesian inference, using probability concepts, demands careful attention to the underlying constraints (e.g., the need to define mutually exclusive hypotheses).

- *Evidential intervals:* The Dempster-Shafer theory introduced in Chapter 6 represents uncertainty by an evidential interval, the support for a proposition (based on the summed mass functions), and the plausibility of proposition (the lack of evidence that directly refutes the proposition). Dempster's rules of combination and an associated logical calculus allow evidential intervals to be used to represent and propagate uncertainty in an inference process. The Dempster-Shafer evidential intervals have a well-defined and strong theoretical background that relaxes some of the conditions required for Bayesian inference. As indicated in Chapter 6, however, the computational requirements for reasoning with evidential intervals is larger than that required for Bayesian inference. Moreover, the use of these intervals requires training for human analysts and system users.

- *Generalized evidence processing theory:* Thomopoulos [21] introduced a generalization to Bayes' inference as a variant to the Dempster-Shafer approach. Evidential intervals, analogous to Dempster-Shafer intervals are developed with an optimized decision approach. Chapter 6 provides an overview of this approach. While the theoretical basis for the generalized evidence processing theory is well-defined, the utilization of this technique requires training for users as well as increased computational complexity.

- *Certainty factors:* The use of certainty factors (CFs) were introduced by the development of the expert system MYCIN [26]. Certainty factors are numerical estimates (made by human analysts) of the certainty of a fact, association, or rule. The CF for a rule or logical relationship describes the degree of belief or validity placed in the relationships defined by the rule. Certainty factors range from -1 to $+1$. Positive values of the CF indicate that the verity of the premise (e.g., of a rule or relationship) should increase our belief in the conclusion, while negative values indicate that the verity of the premises should decrease our belief in the conclusion [14]. When assigned by a user (or sensor) to a fact or association, a certainty factor indicates the degree to which the user feels that fact or association is known. Often a confidence threshold is assigned to establish the level of confidence that must be achieved in the known facts and associations before the conclusion of a rule can be achieved. A set of rules has been established for propagation of the CFs in Boolean relationships. For example if a rule has an antecedent with uncertain data [i.e., CF(data)] and a conclusion whose confidence is CF(relationship) (given 100% confident data), then the confidence of the conclusion is simply

$$CF_{CONCLUSION} = CF(data) \times CF(relationship)$$

Similarly, combination rules define the propagation of confidence factors for complex rules involving Boolean expressions. The use of confidence factors is intuitively appealing and requires little computational resources. However, the theoretical basis for confidence factors is not well-defined and the interpretation of results may be ambiguous.

- *Fuzzy sets:* The concept of fuzzy sets was introduced in 1968 by Zadeh [27]. The basic concept seeks to address problems in which imprecision is an inherent aspect of a reasoning process. For example, some attributes and concepts commonly used by humans such as *tall, short, attractive,* or *personable* are inherently imprecise. That is, we may quantify a value for height (e.g., Bill Smith has a height of 6 feet, 1 inch), but still not be able to specify a value of Bill Smith's tallness (namely is Bill Smith tall relative to NBA basketball players, or tall compared to pygmies?).

Unlike the concept of probability we can actually measure height with great accuracy but still not be able to precisely specify the attribute of *tallness*. Zadeh introduced the concept of fuzzy sets and later the formulation of fuzzy logic [23] to address these issues. Zadeh argued that such imprecision was inherent in many ordinary concepts used in human reasoning. Therefore rather than ignore such imprecision, fuzzy sets and fuzzy logic provided a means to formalize the way in which the concepts are addressed. The formulation of fuzzy set theory using membership functions to represent uncertainty in reasoning is sometimes termed possibility theory. A generalization of the Dempster-Shafer theory to fuzzy sets has been developed by Yen [28].

Fuzzy sets are sets of ordered pairs [x, $\mu(x)$] where x represents an element defined to be in a set, A, and $\mu(x)$ is a membership function whose value ranges from zero to unity. The membership function,

$$0 \leq \mu(x) \leq 1 \qquad (7.1)$$

specifies the extent to which element x is a member of set A. In ordinary (Boolean set theory), sets are defined by specifying the elements. Thus for example, two sets, A and B, may be defined by

$$A = (x_1, x_2, x_3)$$

and

$$B = (x_2, x_3, x_4)$$

In this example, set A has members x_1, x_2, and x_3, while set B has elements x_2, x_3, and x_4. In Boolean set theory, an element x is either a member of a defined set or it is not. Boolean set theory defines classical set operations such as the union of sets, intersection of sets, and difference of sets.

By contrast, fuzzy set theory defines sets by ordered pairs:

$$A = [(x_1, \mu_A(x_1)), (x_2, \mu_A(x_2)), (x_3, \mu_A(x_3))]$$

and

$$B = [(x_2, \mu_B(x)), (x_3, \mu_B(x_3)), (x_4, \mu_B(x_4))]$$

In this case for each element of a set, an associated membership function prescribes the degree to which an element is a member of the designated set. For example, $\mu_A(x_1)$ defines the extent to which element x_1 is a member of set A. In fuzzy set theory, the same operations (e.g., inclusion, equality, complementation, union, intersection, and difference) are defined as in Boolean set theory. However, in each instance, rules are defined to describe the corresponding membership functions. In particular the following rules are defined for set operations:

$$\text{Inclusion } A \subset B \rightarrow \mu_A(x) = \mu_B(x) \tag{7.2}$$

$$\text{Inclusion } A \subset B \rightarrow \mu_A(x) = \mu_B(x) \tag{7.3}$$

$$\text{Equality } A = B \rightarrow \mu_A(x) = \mu_B(x) \tag{7.4}$$

$$\text{Complementation } \mu_{\overline{A}}(x) = 1 - \mu_A(x) \tag{7.5}$$

$$\text{Set Union } \mu(x)_{A \cup B} = \text{Maximum}[\mu_A(x), \mu_B(x)] \tag{7.6}$$

$$\text{Set Intersection } \mu(x)_{A \cap B} = \text{Minimum}[\mu_A(x), \mu_B(x)] \tag{7.7}$$

These rules to define membership functions for joint set operations are readily interpreted. For example, if the intersection of two sets, A and B, contain common elements, x_i, then the elements, x_i, are partly in set A and partly in set B. An element cannot have a greater degree of membership in the intersection set, $A \cap B$, than in either component set, A or B. This argues for the rule that,

$$\mu_{A \cap B}(x) = \text{Minimum } [\mu_A(x), \mu_B(x)]$$

(i.e., that the membership of element x in the union of sets A and B, is the minimum of its membership in set A, or in set B). Similar interpretations may be developed for the rules specified by (7.2)-(7.6).

In [29], Zadeh summarizes how fuzzy sets may be used to extend these concepts in a fuzzy logic. In particular, Zadeh defines the basic rules for fuzzy logic, incorporating such standard logic relationships as: conjunctive rules, cartesian products, projection, compositional rules, generalized modus ponens, and a dispositioned modus ponens. These allow incorporation of fuzzy set formulism in defining in precise concepts as well as reasoning with these concepts. Details of these formalisms are also provided in [30, 31].

Fuzzy set theory and membership functions allow quantitative representation of concepts that are inherently imprecise. In a medical example, a physician might supply the following rule [14]: IF (the growth is quite large), THEN there is a good chance that the tumor is cancerous. In this example, the physician replaces a statement about an actual physical dimension of the size of a tumor, and replaces it by a fuzzy variable "large." Similarly, for a threat recognition system, a tactical analyst might describe a target in terms of characteristics such as "target-like size," "target-like speed," and related fuzzy concepts. Sensor data and observations of size and speed are translated by a membership functional relationship. These relationships are developed a priori by user/analysts. The term fuzzification is used to mean the translation of a physical measurement, such as length, to a fuzzy variable such as target-like size. The concept of using fuzzy logic to fuse data is shown in Figure 7.3.

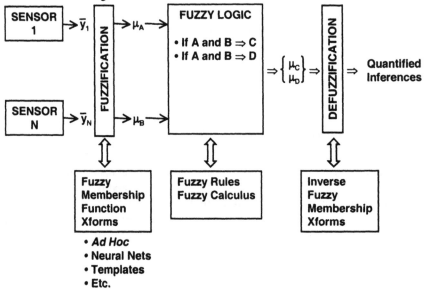

Figure 7.3 Concept of using fuzzy logic for fusion.

As the case with knowledge representation techniques, the choice of technique for representing uncertainty must be based on experimentation, to determine which method is most useful for a particular problem domain. Some work has been performed to compare the performance of these methods. An example of such a comparison is the experiment performed by [32]. In this experiment, several expert systems were constructed for the same simple problem; that of predicting the color of a wooden block based on observations of its shape (i.e., a number of colored blocks were postulated with a correlation between the block's shape and its color). Each expert system used MYCIN-type certainty factors, Bayesian probabilistic theory, or Dempster-Shafer theory to represent and propagate uncertainty. Using these techniques, numerous trials attempted to correctly predict block color. A comparison of the results is illustrated in Figure 7.4, which shows the number of guesses required to correctly identify color versus the number of training trials. After 81 training trials the results were:

MYCIN (CF)	85% correct
Dempster-Shafer	75% correct
Bayes	70% correct

After 324 trials, the results were

Dempster/Shafer	95% correct
MYCIN (CF)	93% correct
Bayes	85% correct

In general, the MYCIN certainty factors methods started well (i.e., with a high level of accuracy) but leveled off after relatively few trials. Bayesian inference was relatively inaccurate at first, but improved rapidly and was best for large trial sets. Finally the Dempster-Shafer evidential process performed reasonably well under all trial set sizes.

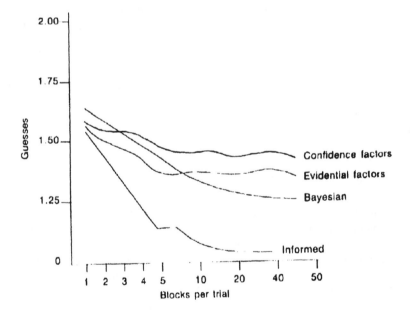

Figure 7.4 Comparison of uncertainty representation techniques [32].

With all of these methods, a critical factor is the ability to establish a priori estimates of confidence factors, probability, probability mass, or membership function. The determination of these data remains a difficult problem for real (as opposed to experimental or *toy*) applications. Specification of such data may out-weigh the considerations of underlying mathematical theory.

7.2.6   Search Techniques

In addition to establishing techniques for representing knowledge and reasoning with uncertainty, key elements of the expert system inference process are tech-niques for search. Figure 7.2 illustrated the expert system inference process that entailed a cyclical process of pattern matching, search through a knowledge base, and execution of rules (or their equivalent). Whatever method is used to represent knowledge, some technique must be used to search through a knowledge base to find applicable rules, frames, nodes within a network, or data.

The search process is often envisioned as analogous to finding a particular leaf on a large tree, beginning at the tree trunk and attempting to find a particular branch, twig, and leaf, by making informed choices at each juncture. A common example of a search problem is the traveling salesman problem, by which an agent begins at a starting point and seeks an end goal through a network, making choices at each node. Figure 7.5(a) illustrates such a problem (adapted from [33]). A traveling salesman begins at the node designated, *START,* and seeks to reach tile *GOAL* node by traveling through intermediate nodes (or cities). Allowable paths (i.e., roads) between nodes are shown in Figure 7.5 by the interconnecting lines. Associated distances are shown as numbers on the interconnecting lines. The problem can be formulated in a number of ways including the following:

1. Find a path that begins at *START,* ends at *GOAL,* and includes at least one stop at every intermediate node;
2. Find the shortest (i.e., *best*) path from *START* to *GOAL;*
3. Find any acceptable path from *START* to *GOAL.*

The later two formulations are simplifications of the general traveling sales-man problem and may be solved in significantly less computational time. The decision tree in Figure 7.5(b) shows the acyclic paths from START to the GOAL node. That is, beginning at *START,* one can choose a path (i.e., either go to node A or go to node D). At each subsequent node a decision can be made (namely at node A travel to node B or travel to node D), continuing until one either reaches a dead end or achieves the goal (arriving at node *GOAL*).

In Figure 7.5(b) the dashed line indicates three possible routes from *START* to *GOAL:*

1. START - A - B - E - F – GOAL;

2. START - A - D - E - F – GOAL;
3. START - D - E - F – GOAL.

In this example the shortest path is that defined by the third route, requiring a total distance of 13 units compared to a distance of 19 units for path 1 and 17 units for path 2. While it is easy to visualize possible successful paths, the actual search on a node-by-node basis requires search strategy.

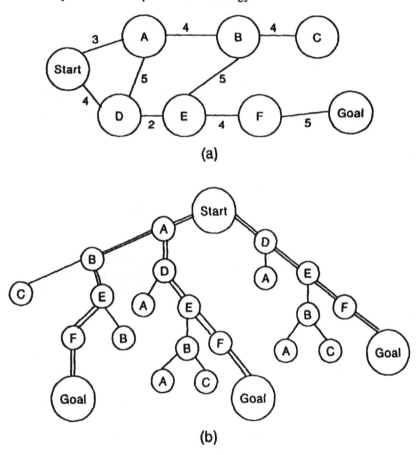

(a)

(b)

Figure 7.5 Examples of a traveling salesman problem (reprinted with permission from [33]).

A number of search strategies have been devised and used for these types of problems. Figure 7.6 shows an overview of some strategies. A complete discussion of these is provided in [33]. Three basic categories of strategies are shown. *Some path* methods such as Depth-first, Hill Climbing, Breadth-first, Beam, and Best-first methods provide strategies to find any acceptable path to the goal. For example, the depth-first algorithm arbitrarily chooses an alternative node, and continues to choose a path at each successive node until either a dead end is reached or until the goal is achieved. If a dead end is reached (where further downward motion is impossible) the search is restarted at the nearest ancestor node with unexplored paths. Hence in Figure 7.5(b) a depth-first search would proceed as follows (assuming the left-hand branch was arbitrarily the first choice at each node):

$$\text{START} \rightarrow A \rightarrow B \rightarrow C \rightarrow B \rightarrow E \rightarrow F \rightarrow \text{GOAL}$$

Depth-first searches are efficient when numerous alternatives do not exist at each node. Conversely, a breadth-first algorithm explores all alternative nodes, level-by-level, until the goal is reached. Alternatively, the hill-climbing technique uses a basic depth-first strategy, but tries to improve the process by ordering the choices at each node and searching the most promising branches first.

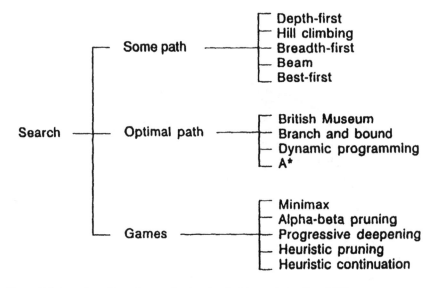

Figure 7.6 Overview of search strategies (reprinted with permission from [33]).

A second category of search techniques shown in Figure 7.6 is optimal path methods. These techniques attempt to find not simply an acceptable path from *START* to *GOAL*, but the best (e.g., shortest) path. Examples of these techniques include the British museum method, branch and bound techniques, dynamic programming algorithms, and the A* technique. Optimal path techniques may require much longer than *some* path methods. An exceptionally inefficient method (although one guaranteed to find the optimal path) is the British museum approach.

This method finds all possible paths, via either depth-first or breadth-first search, and selects the best path from among all possible paths. The A* procedure is relatively efficient and is commonly used in actual implementation of knowledge-based systems.

The third category of techniques shown in Figure 7.6 is adversarial techniques such as the mini-max, alpha-beta pruning, progressive deepening, heuristic pruning, and heuristic continuation techniques. These methods are used for searching in situations such as games in which an opponent or adversary may react to a chosen alternative by selecting an opposing move. Thus, in a game such as chess, when a player selects a move towards the ultimate goal of checkmate, the opposing player may make a countermove towards his or her own counter goal. The adversarial techniques attempt to perform searches (or determine moves) with the knowledge that an active adversary will react to those choices.

The selection of search techniques depends upon the basic domain being addressed, the nature of the search problem, available computing resources, and the goals of the system designers. Ultimate choice of a specific technique must be made in concert with related choices such as knowledge representation techniques, method(s) for representing uncertainty, and other basic issues.

7.2.7 Architectures for Knowledge-Based Systems

Architectures for an expert system vary from a simple implementation of an inference engine to process rules, to complex systems utilizing multiple methods for representing knowledge, hybrid methods for representing and reasoning with uncertainty, and the use of sophisticated search techniques. In Section 7.3 we describe the development process for expert systems and various available tools. Dennis Buede [34] has provided an analysis of techniques for automated reasoning systems and has developed a taxonomy of techniques for knowledge representation, inference/evaluation processing, and truth maintenance (e.g., search and control processes). A summary of this taxonomy is illustrated in Table 7.6. Many of the techniques described in this chapter, and those identified by Buede, have been used to develop knowledge-based systems. Which techniques (or combinations of techniques) are selected depend upon the specific problem domain.

Table 7.6

Summary of Techniques for Knowledge-Based Systems
(Adapted with permission from [34])

Knowledge Representation Scheme	Inference/Evaluation Process	Control Structure
Rules	Deduction	Search Techniques • Depth-first
Frames	Induction	• Breadth-first

Table 7.6   (continued)

Summary of Techniques for Knowledge-Based Systems
(Adapted with permission from [34])

Knowledge Representation Scheme	Inference/Evaluation Process	Control Structure
	Abduction	• A*
Hierarchical classification		
• Visual model	Analogy	Reason (truth) maintenance
• Object categories		
	• Plausible or default inference	Assumption-based
Semantic Net	• Classical statistics	Hierarchical decomposition
	• Bayesian probability	
Nodal graph	• Evidence theory	Control theory
• Bayes belief net	• Polya's plausible inference	
• Perceptual nets	• Fuzzy sets and fuzzy logic	Opportunistic reasoning
• Influence diagram	• Confidence factors	
	• Decision theory (expected utility theory and multi-attribute theory)	Blackboard architecture
Options, goals, criteria constraints		
		Intelligent agents
Scripts	Circumspection	
Time map		
• History		
• Interval		
• Chronicle		
Spatial relationships		
• Multifeatured maps		
• Coordinate systems		
• Templates		
Analytical models		

One particular architecture worth noting is the blackboard architecture originally developed by researchers at Carnegie Mellon University for the HEARSAY-II speech understanding system [35]. This architecture is based conceptually on the way a group of experts might use a classroom to solve a complex problem. Each expert contributes his knowledge, in his particular area of expertise, to an evolving solution to the problem at hand. The experts use a blackboard (a chalkboard) to communicate how their expertise is used to solve components of the problem. Another analogy is the method used by a group of people to solve a puzzle; each person solves a piece of the puzzle (e.g., one focusing on edge pieces, one focusing on pieces of a particular color), while a supervisor scans the entire puzzle, hoping to match components of the puzzle together.

The blackboard architecture simultaneously uses multiple knowledge sources and control structures to address a problem as illustrated in Figure 7.7. The blackboard stores evolving information about a problem domain. The data are structured via levels and nodes, with relational links. For each level and node, specialized vocabulary (data, logical relations, etc.) may be utilized. Independent knowledge sources focus on subproblems using special knowledge. Each knowledge source acts as an independent expert or problem solver. An overall control structure coordinates the processing of the knowledge sources and proceeds in an event-driven manner.

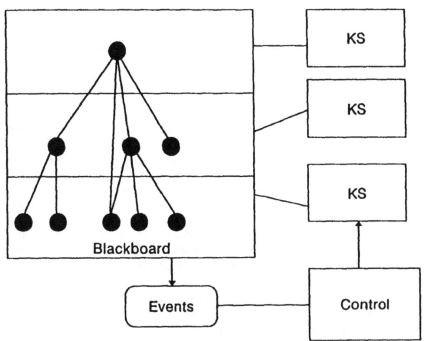

Figure 7.7 Concept of the blackboard architecture.

Blackboard models are particularly useful in the solution of the data fusion problem. A dynamic problem can often be partitioned into multiple aspects (e.g., definition of situation states, prediction of future activities or events, focus of attention on restricted geographical areas, analysis of types of data by hierarchy, time sequence, and logical function). Llinas and Antony [36] provide a summary of a number of blackboard systems developed for military data fusion systems.

7.3 IMPLEMENTATION OF EXPERT SYSTEMS

The implementation of expert systems provides a challenge compared to the implementation of ordinary software systems. Because there are significant differences between the development of conventional and expert systems, this section provides an overview of expert system development. Issues addressed include a life-cycle development model, knowledge engineering, test and evaluation, and the use of expert system development tools.

7.3.1   Life-Cycle Development Model for Expert Systems

Development of conventional software systems, especially for large-scale system applications, proceeds in accordance with highly structured standards and procedures. Typically, an established process model is utilized, which prescribes how a system is to be developed. A general discussion of large-scale software development and process models is provided by [37].

The methodology provides for a top-down development of a system that may include both hardware and software. Basically, the development of a system progresses from the abstract toward the tangible implementation in either hardware and/or software. This process is often partitioned into seven sequential steps or phases, which force the decomposition process to accomplish specific time-related goals with concomitant products. The above described process is known as *baseline* development, where a baseline is merely a point in time when a document or product has (1) been reviewed, (2) had all resulting action items closed, and (3) been approved. The official establishment of a baseline does not imply that the design is frozen at that stage; rather, it demands that future changes and updates shall be made via a set of established change procedures that are defined in a configuration management plan.

This seven-phase process for conventional systems and associated intermediate products is shown in the top portion of Figure 7.8. Key steps in such a structured development process include the following:

1. Requirements analysis: Aimed at establishing a system requirements specification, basic system test philosophy, and an initial allocation of requirements to subsystems;
2. Specification: Development of software or hardware requirements specification and associated development plan;

3. Preliminary design: Beginning actual subsystem and software design, resulting in a preliminary design specification document and a preliminary software integration and test I&T plan;
4. Detailed design: Completion of design to a component and unit level documented via a detailed design specification, final software I&T plan, and software acceptance test plan;
5. Code and unit test: Actual coding of software units and test in accordance with unit development folders;
6. Integration: Integration of software units into subsystems via progressive software builds;
7. Demonstration test and evaluation (DT&E): Functional testing to demonstrate that system components and complete systems perform in accordance with functional requirements.

Conventional Development Phases	Requirements analysis	Specification	Preliminary Design	Detailed Design	Code & Unit Test	Integration	DT&E Functional testing
	-System specification -Test philosophy -Functional allocation	-S/W requirements specifications - S/W plan	-Preliminary design specification - S/W I&T plan	-Final S/W I&T plan - Detailed design specification - S/W acceptance test plan	Integration ready reviews I&T plan Final S/W acceptance test plan	-S/W system	- Accepted system
Expert system development phases	Phase I: Modeling the requirements via rapid prototyping				Phase II: Modeling the solution via prototyping - Multiple iterations are performed		
	-Develop preliminary knowledge base - Rapid prototyping via expert system shell				-Multiple iterations of the knowledge base - Test and evaluate to refine knowledge base - Transition to final system		

Figure 7.8 Contrasting phases of development for expert and conventional systems.

The success of this approach lies in the ability of designers to define system and software requirements in advance of development. For the development of expert systems, however, baselines cannot be as firmly fixed (if they can really be defined at all), and phases must overlap.

By contrast, the development of expert systems requires an iterative multiphase process. Such a process is described by [38]. A three-phase process is recommended.

- *Phase 1:* Modeling the requirements recognizes that a number of issues need to be resolved before an expert system can be substantially developed. These issues include specification of techniques for knowledge representation, methods to represent uncertainty, identification of an appropriate architecture, definition of test concepts, and many other issues. The first phase of a life-cycle development model begins with experiments to address these issues. Such experiments often make use of commercial development tools aimed at developing a prototypical system. Implementation of the prototype eschews issues such as throughput, response time, ultimate implementation computer equipment, and other difficulties. Instead, a subset of the problem domain is selected. The aim of a prototype is to develop a representative knowledge base, and to validate the system by expert analysts or system users. In this way the requirements for an expert system are actually defined in the process of creating and demonstrating a prototype model.

- *Phase II:* A second development phase for expert systems proceeds by an iterative rapid prototyping and refinement of the initial prototype developed in phase I. During this phase the knowledge representation schemes and architecture recommended by phase I are used. The complete problem domain is addressed and the full knowledge base is built. The implementation proceeds iteratively as knowledge is added to the knowledge base, tested, and iteratively refined. During this phase, timing and throughput analyses are performed to determine the appropriate target architecture for phase III.

At this point in phase II of Figure 7.8, we have developed a prototype of the final expert system in a target form. The prototype is then *validated* by the expert and modified to reflect something closer to the final product. This is an iterative process with the prototype being refined in conjunction with experts' evaluation. It contrasts with the incremental builds associated with a traditional S/W effort where the system to be built may be viewed as independently developed units that must be integrated into a working system, usually in a top-down fashion with the skeleton of the system being integrated first. In a expert system this will often be the phase where some of the most sophisticated design will occur because the more obvious problems will have been addressed in earlier prototypes.

- *Phase III:* A third phase of development for an expert systems addresses the transitions from an experimental environment (e.g., expert system tool) to a final architecture and environment. The third phase may actually discard the prototype software (retaining the knowledge base) and reimplement the expert system software in a conventional language such as C++ or JAVA. This phase completes the development with a final test and evaluation, and integration into the final system.

The development model for expert systems, and associated standards and procedures, is still evolving. There is no equivalent to the standards used, for example, for DoD system development. It is expected, however, that such standards will emerge in the near future and become incorporated into conventional standards and procedures. The American Institute of Aeronautics and Astronautics has an artificial intelligence technical committee. This is one such group that is addressing these types of issues.

7.3.2   Knowledge Engineering

Knowledge engineering is concerned with extracting expertise from a human specialist (expert) and creating a knowledge base of rules, data, logic, or data relationships to form a knowledge base used by an inference engine in an expert system. Feignebaum, who correctly predicted that the most difficult aspect of creating nontrivial expert systems is the knowledge engineering process, coined the term knowledge engineering. Formulating a structured methodology for knowledge engineering continues to be challenging because much of human knowledge is procedural ("knowing how"), which is difficult to describe in words.

Part of the difficulty lies in the fact that knowledge engineering is not simply a process *of putting down on paper what an expert or specialist thinks.* Indeed, many experts are not consciously aware of the thought processes that they use to make inferences. Even if they were aware of these thought processes, the heuristic inferential process would still not be amenable to implementation via production rules, or other knowledge representation schemes. Knowledge engineering involves making an expert consciously aware of his or her inference process, creating a consistent and logical inference process, and transforming the knowledge to the selected representation form. In a sense the knowledge engineer works with an expert to create a new logical structure using a given representation scheme that can produce the same inferences as those made by the expert. This new logical structure generally does not even remotely emulate the expert's thought processes but it does produce the same inferences.

Because the success of an expert system critically depends on the underlying knowledge base, it is important to develop techniques for systematically performing knowledge engineering. This section describes one approach that has been synthesized to systematically populate a knowledge base with facts, rules, frames, networks, or scripts. A structured process useful to perform knowledge engineering consists of a five-step approach illustrated in Figure 7.9. These steps include: (1) identify and bound the problem, (2) acquire the basic knowledge, (3) select a representation scheme, (4) Interpret and refine the knowledge, and (5) encode the knowledge.

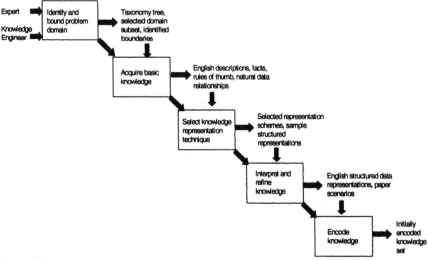

Figure 7.9 Knowledge engineering process.

Each of these steps is described in the following. It must be emphasized however, that this process merely develops a preliminary knowledge base, which must be subsequently validated and augmented via the iterative process described in the previous section. These steps provide a systematic approach to obtain the initial knowledge base:

1. Identify and bound the problem: The first step in creating a knowledge base is to identify and bound the problem domain. In the rush to *implement something* this important activity is frequently rushed. The specific problem domain must be carefully identified and bounded. The following issues should be addressed and the answers documented.

   - Who is (are) the specific experts to be emulated?
   - What aspect of the expertise will be emulated by the expert system?
   - What are the boundaries of the application domain (e.g., what are allowable and unallowable questions that the expert system will address)?
   - What simplifications are to be assumed?
   - How will the expert system be validated, (i.e., is there a set of test cases, standard inferences, or panel of experts to be used as judges of validity)?

   It is helpful to create written scenarios that document the typical environment of the human expert.

2. Acquire basic knowledge: Acquiring basic knowledge is the extraction and documentation of facts, rules of thumb, and inferential logic used by the expert. This requires many interviews of the expert by the knowledge engineer. Test cases (e.g., sample diagnoses and analyses) may be useful so that the expert can *step through* his or her process, while the knowledge engineer attempts to access what knowledge is useful. The knowledge engineer must be prepared to be very observant and aware of *rabbit's feet*. (A rabbit's foot is something that an expert believes is useful in making an inference, but that is actually independent of the result.) What the expert says he or she does, and what he or she actually does may be two different things. A rule of thumb here is that the knowledge engineer requires at least eight hours of preparation and analysis time for every one-hour of interview time with the expert. The knowledge engineer must become sufficiently familiar with the application domain to be able to speak the language of the expert. Conflicts may arise if the knowledge engineer continually uses AI jargon in his or her interaction with the expert. In addition, the knowledge engineer should exert sensitivity to implied professional threats that an expert system will readily replace the expert. Outputs from this step are natural language descriptions of the logic, rules of thumb, facts and data relationships (e.g., hierarchies of facts). As this basic knowledge acquisition progresses, the problem domain definition and boundary specification may be revisited.

3. Select knowledge representation technique: Having acquired a basic set of knowledge about the problem domain, the appropriate knowledge representation format can be selected. Choices included production rules, scripts, networks, frames, blackboards, and direct techniques. These representation techniques were summarized in Table 7.4. Each of these formats has special applicability to restricted classes of problems. None of the representation formats, however, is generally applicable to all realistic applications. The optimal choice of representation format is the format to which the basic knowledge can be most readily adapted. For example, production rules are readily adapted to diagnostic applications in which if-then logic is used. Once the knowledge representation format(s) has (have) been selected, a tool should be chosen to support the required format. Outputs from this step are a selected representation technique, and examples of the application domain knowledge transformed to the selected representation format.

4. Interpret and refine knowledge: Given a representation scheme and a set of basic knowledge it is necessary to interpret and refine the knowledge, to create an English version of the knowledge base. The idea here is to actually create the logical structure of rules or data relationships that emulate the expert's inference process. This step involves developing categories of data or inferences and a logical structure, which *on paper* will perform correct inferences in the specific application domain. A helpful technique is to use a single example or test case and develop a set of rules (or equivalent representation), which produces the correct inference. Even when an interactive

tool is available, it is helpful to develop a paper diagram or table that walks through the inference process for one complete case. This approach can result in insight in the formulation of the knowledge in the selected representation format. Output from this step is a preliminary knowledge base written in English but formatted in the form of the selected representation technique.

5. Encode knowledge: The final step of the knowledge engineering process is the actual encoding of the natural language knowledge base. This entails transforming the natural language knowledge representation into the syntax used by the specific implementation tool. For example, the natural language version of production rules might be transformed into OPS-5 production rules. This step is not merely a coding type of process, since insight will be gained about the characteristics of the tools syntax

7.3.3   Test and Evaluation

The test and evaluation (T&E) of any large-scale software system is difficult. In general it is not feasible to test every path, branch, and line of code in even a conventional system. Attempts to systematically test and evaluate conventional systems involve creation of a formal test plan and procedures, identification of processing threads with associated tests, and hierarchical testing of software routines, units, components, subsystems, and systems. These systematic methods prescribe strategies and techniques for T&E of conventional systems.

Unfortunately, the T&E of expert or knowledge-based systems presents even more difficulty than that of conventional systems. Because T&E for expert systems presents a special challenge, this section presents a discussion of the issues and strategies for expert systems. More details can be found in [39].

The difficulties associated with testing expert systems stem from the nature of the problems addressed, the structure of the systems, and the test environment. Since an expert system addresses a problem or domain, which is usually performed by a human specialist or expert, the very criteria for measuring success tend to be ill-defined. Can a computer program match the human specialist, under what conditions, and for how many trials? In some cases the nature of the expertise itself is uncertain with the result that human experts disagree on test cases. How, then, can a computer program be validated against an uncertain baseline? Typical problems addressed by expert systems involve large search spaces that are not amenable to modeling. Indeed, if the problem was amenable to modeling or exhaustive search strategies, the expert system paradigm would not be required or utilized.

Although the current development environment (tools, languages, and development procedures) for expert systems assists the development process, it complicates testing. Languages such as OPS-5 and PROLOG, which are especially suited for heuristic problems, tend to be difficult to apply to the installation of structured concepts. Further, the representation of knowledge for an expert system is a snythetic construct (e.g., rules, scripts, and frames) that attempts to encode the heuristic knowledge that would be applied by an expert. Thus, evaluation of the consistency, accuracy, and completeness of the knowledge base is rarely amenable to formal proofs.

Finally, the iterative development cycle for expert systems tends to obscure the separation between testing and development and the extension of the knowledge base. Each new test case may be considered to be a special case for which adjustments and additions are made to the knowledge base to treat that special case.

The test environment is also a complicating factor. In the absence of simulated data, test cases must be obtained on an opportunistic basis. Hence, for example, a medical diagnosis expert system such as MYCIN [26] or INTERNIST [40] must be validated with actual case studies. It may not be feasible to simulate viable test cases. The lack of a complete list of possible situations keeps the experimenter from estimating the accuracy or correctness of the test system. Here again, each test becomes a new test case in an endless sequence.

The actual testing of an expert system may be divided into two main phases: (1) the initial test and validation of the conventional software and (2) the test and validation of the knowledge base. Testing of the conventional software would verify that the software components such as the man-machine interface, the knowledge-base editor, database software algorithm/models, and the inference engine were all logically correct, coded correctly, and free from faults or bugs. Techniques for testing and validating conventional software, while not simple, are well known. Standard tests can be performed for each component with test cases defined and the results monitored. Although testing of this software is complicated by the use of unconventional languages and machines, the basic principles are straightforward.

The second phase of test and validation is the verification of the knowledge base. Issues of concern include: (1) the consistency of the knowledge base; (2) knowledge base completeness; and (3) system effectiveness. The consistency of the knowledge base concerns the extent to which the rules (or other knowledge representation mechanisms) in the knowledge base are internally consistent. Do they form a logically consistent set without one rule contradicting another? Do the rules form a minimum and sufficient set to perform logically correct inferences? Such questions are not easy to address for large and complex sets of rules.

The second issue is the question of completeness. Have we developed all the essential knowledge to cover a specified domain? Certainly a sufficient number of representative test cases need to be utilized. A special difficulty with determining the completeness of the knowledge base concerns what the precise boundaries of the domain are? Does a failure of a test case represent a peripheral lack of knowledge or is the test case outside of the area of expertise?

Finally, of course, having determined (to some sufficient degree of certainty) that a knowledge base is both consistent and complete, the question of system utility arises. Is this a worthwhile system?

Figure 7.10 illustrates a basic operational concept for testing the completeness of a knowledge base. It is generally necessary to test for completeness using two sets of test data (or cases). The first set is a training test set that has been selected by the knowledge engineer, domain expert, and system developer. These cases are given to the expert system one at a time. For each test case, the results are evaluated for correctness of the inference. If the test case fails, then the knowledge and domain experts analyze the cause of failure and potentially modify the knowledge base by modifying one or more rules or adding additional rules. Care must be taken in this process not to tailor the knowledge base simply to handle a finite set of cases. It is tempting to *tune* the knowledge base to successfully treat each test case on an individual basis, but this technique tends to create a knowledge base that is fragmented and unduly specialized. The test case should be viewed as a representative set and the knowledge base created to handle general cases in addition to the specific instances represented in the test set.

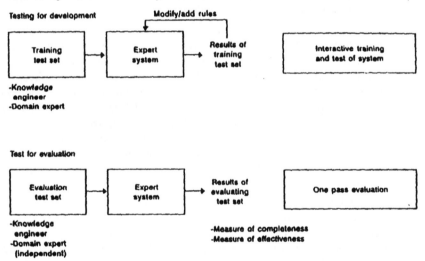

Figure 7.10 Concept for testing a knowledge-based system.

Upon completion of the tests using the training test set, a second set of tests is performed. The knowledge engineer and an independent domain expert should choose this evaluation set. The use of an independent domain expert ensures that the evaluation test set is representative of the types of sets that will be encountered in the real world.

The evaluation test set is used to evaluate the expert system. Each test case is treated by the expert system and the results are evaluated by the independent domain expert. During these tests, no modifications of the knowledge base are allowed, thus obtaining an independent evaluation of the system. The domain expert may also evaluate the effectiveness and utility of the system.

Strategies for testing the validity, consistency, and completeness of an expert system's knowledge base vary from formal methods to informal or ad hoc methods. Hall et al. [39] have developed an extensive taxonomy of evaluation paradigms and has summarized the advantages and disadvantages of the major strategies. Four basic categories identify formal, informal, semiformal, and empirical methods. Formal evaluation paradigms attempt to use mathematically based logical proofs to provide an evaluation of the knowledge base. Logical proofs may make use of the mathematics of predicate calculus, multi-valued logic, or fuzzy logic to demonstrate the correctness of inferences, made by an expert system. These proofs are generally at the level of individual rules or groups of rules. Statistical decision theories such as Bayesian techniques or maximum entropy formulations may also be used to establish formal proofs. Alternatively, linear programming models may be used to establish validity of the knowledge base. Formal proofs are pedagogically attractive, when available, because the results are quantitative and rigorous. It is easy to be misled by the formalism, however. Logical proofs are no better than the a priori assumptions and input data used. Often simplistic assumptions are required to force a logical model to fit the problem at hand. Validation of the assumed model may actually be more difficult than a direct validation of the expert system results. When logical proofs are applicable, they are an objective method for establishing the validity of the knowledge base.

Informal test strategies are more frequently used to evaluate expert systems. Techniques include: (1) ad hoc comparison of test results against an individual evaluator's subjective expectations; (2) comparison of the expert system results against preselected cases; (3) use of interactive simulation in which test cases are interactively selected and modified by humans; and (4) tests in which the expert system is compared against human expert performance (i.e., human experts and the expert system take the same tests). In all situations that utilize test cases, the test case selection and evaluation of results may be performed by a variety of people including the system developer, the sponsor of the system, or human experts.

The taxonomies of test strategies described here are useful in indicating the broad range of options that exist. Ultimately, however, the selection of a specific strategy must account for the realities of the situation at hand; the availability of experts, the extent to which formal models can be developed, and the time and funds available to support independent tests.

7.3.4 Expert System Development Tools

Numerous commercial tools have become widely available to support the development of expert systems. These tools range from simple, inexpensive, PC-based software to support the development of rule-based systems, to expensive and more complex tools designed to operate on high-end graphics workstations. The latter tools provide multiple forms of knowledge representation, sophisticated graphics interfaces featuring graphical display of data and symbols, and debugging support. Reviews of the features and utility of expert system development tools are provided annually in the *AI/Expert* magazine. Commercial competition for such tools, and rapidly evolving PCs and workstations, have encouraged the availability of powerful tools.

In essence, expert system tools provide a software shell containing the basic functions of an expert system, excluding a populated knowledge base. Analogous to commercial database management systems (DBMSs), which provide the tools to populate and access a database, expert system shells provide the mechanisms to populate and use (i.e., perform reasoning with) a knowledge base. The functional elements of an expert system toolset are illustrated in Figure 7.11, following [41]. Basic elements include: an interface to the system developer, a mechanism for populating a knowledge base, an inference engine, an end-user interface, and the ability to interface to other users. Each of these aspects is discussed below.

Expert system shells provide an interface for developers to populate, test, and evaluate a knowledge base. The most basic tools simply provide the ability to edit (create and modify) rules. More sophisticated tools provide a graphical interface to populate a knowledge base comprising rules, frames, networks, or other forms of knowledge representation. Features may also include the ability to display intermediate results of the inference process (e.g., a representation of a search tree), a debugging feature to allow a user to step through an inference, and graphical representations such as diagrams, icons, map displays, or other symbols. Finally the interface may provide tools to check the consistency of knowledge added to the knowledge base.

The knowledge base feature of an expert system shell provides the mechanism for representing knowledge. As previously described, these mechanisms may include rules, frames, networks, scripts, or analogical methods. Basic tools only provide a representation utilizing rules. More sophisticated tools incorporate networks or frames. Additionally, the knowledge base mechanism supplies the ability to represent uncertainty via probability, confidence factors, evidential intervals, or other forms.

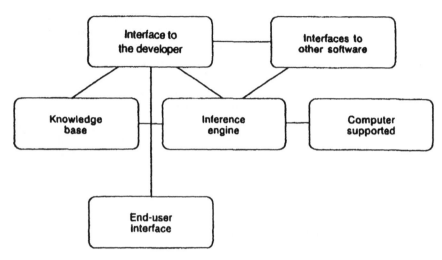

Figure 7.11 Functional elements of an expert system tool [41].

An inference engine in an expert system shell performs the pattern matching and search process. Typically these tools allow either a goal-driven or data-driven process with an optimized search strategy. The inference engine also utilizes one or more strategies for conflict resolution. Often, rule-based systems resolve conflicts by selecting the maximally specific rule (i.e., the one having the most conditions or antecedents). Some shells provide an ability to perform time-constrained searches in which an inference process is self-monitored to predict the time required to reach a conclusion.

An end-user interface for an expert system shell provides the human computer interface (HCI) for the user of the developed system. Like the development interface, such interaction may be obtained via menus, window displays, graphics, ICONs, and other diagrams. A special feature sometimes provided is an explanation feature. This feature allows a user to see a trace of the inference that resulted in a conclusion. That is, a trace or listing of the rules may be obtained to provide a user with insight into how the expert system reached a conclusion. Both the developer interface and user interfaces are a product of both the computer hardware environment (e.g., display equipment) as well as the software. Here again the evaluation of PCs and workstations has facilitated the availability of sophisticated graphical displays.

In some instances an expert system shell may provide support to interface to other software such as database management systems, numerical libraries, or user-built software. More typically, however, commercial tools are relatively self-contained and only interfaced to ancillary software through some development effort.

Expert system shells are often used to implement prototype or demonstration systems. The tools are generally easy to use and allow immediate experimentation with the development of a knowledge base. Because the tools provide an inference engine and user interface, a developer may concentrate on issues of knowledge representation, uncertainty representation, and knowledge engineering. There are some drawbacks, however, to such tools. Specifically the tools tend to be optimized to provide development flexibility vice efficiency. Hence the tools tend to utilize a significant amount of computer resources such as memory and processing time. Therefore, development of an expert system in a resource constrained environment, with throughput requirements (e.g., response time limitations) may require development of a *home-built* expert system, especially in an embedded environment.

7.4 LOGICAL TEMPLATING TECHNIQUES

Templating is the name given to logic-based pattern recognition techniques used in multisensor data fusion processing for event detection or situation assessment. These classes of techniques have been successfully used in data fusion systems since the mid 1970s. The term *templating* comes from the concept of matching a predetermined pattern (or template) against observed data to determine whether conditions have been satisfied, thereby allowing an inference to be made. A colloquial example of templating is the use of a grading template by a grade-school teacher. In this simple example, a piece of paper with a pattern of holes is laid over the student's completed answer sheet. For a multiple choice examination, a teacher can quickly scan the template and determine which questions have been correctly answered. The student's grade is determined by summing the number of correct answers against predetermined thresholds representing grade levels (i.e., A, B, or C,). The pattern is very simple and the matching is binary in that a student provides either a correct or incorrect answer.

This pattern-matching concept can be generalized for complex patterns involving logical conditions (e.g., Boolean relationships), fuzzy concepts, and uncertainty in both the observational data and the logical relationships used to define a pattern. The pattern defined by a template may represent a tactical event, such as a missile launch, an impending event or activity (e.g., movement of a battalion command post), or the recognition of a battlefield entity such as a weapon system or force. Hierarchies of templates may be defined to represent increasingly complex entities, events, or activities. Early templating systems, such as the IP system developed for the Rome Air Development Center (RADC) were implemented using conventional software approaches. More recent systems have made use of expert system techniques such as frame-based knowledge representation architectures.

The basic templating algorithm matches observational data against prespecified conditions to determine if the observations provide evidence to identify an entity, event, or activity. The input to the templating process is one or more observations, which may include parametric and nonparametric data, over a period of time. The output is a declaration of whether or not the observations match a predetermined profile. The output may also include an associated confidence level or probability of the event specified by the template.

Figure 7.12 shows a generic template. A generic template is illustrated since every application requires information specific to the application domain. However, some types of information are similar for all templates. The domain-independent information includes: template identification, threat type, acceptance threshold, rejection threshold, necessary conditions, sufficient conditions, and the components that describe or make up the threat/activity.

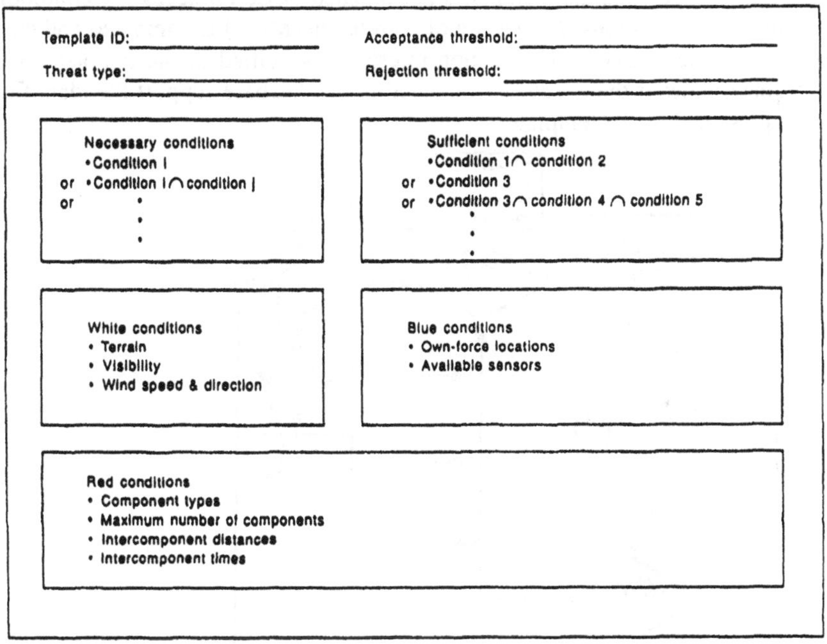

Figure 7.12 A generic template.

Template identification provides a means of retrieving the template from a database or of naming the template for system users. Threat type describes the category of tactical threat addressed by the template. Acceptance and rejection thresholds are user-specified numerical criteria that are utilized in the automatic template process. Necessary and sufficient conditions prescribe the required observations, logical relationships, and data patterns that cause the templating to be accepted as a candidate in the automatic templating process.

In Figure 7.12, *white conditions* refer to those environmental factors that influence both enemy activities and friendly or neutral force activities. These may include terrain, weather, or other effects. *Blue conditions* refer to predetermined conditions concerning friendly forces, such as locations and available sensors. *Red conditions* refer to observations made about enemy units or entities.

Figure 7.13 shows a basic template processing flow. This is a data-driven process in which new data causes a processor to determine which (if any) templates may be satisfied by the new data. The process may be formulated equally well in a goal-driven manner in which an existing database is searched against individual templates to seek data matches. The process is initiated when triggering information (new observational data) is received and used to retrieve candidate templates from the template database. Once the list of applicable templates is retrieved, the list may be culled based on filtering parameters such as distance or time. Each of the candidate templates is evaluated based on necessary conditions and sufficient conditions. A measure of correlation (MOC) is computed and compared to the acceptance and rejection thresholds specified in the template. This MOC determines if the events accumulated in the database support the identification represented by the template.

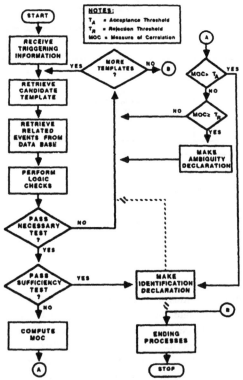

Figure 7.13 A basic template processing.

If necessary component sets are specified in the template, then the data are checked for these components. If they are not present, the template is rejected. When the necessary conditions are satisfied, then the template is activated (i.e., the situation or event specified by the template is identified as a likely inference) and no further processing is performed. If the sufficient conditions are not satisfied, then the MOC is calculated and compared against the accept and reject thresholds to perform a more flexible test. If the MOC is above the acceptance threshold, the identification declaration represented by the template is made. If the MOC falls between the accept and reject thresholds, an ambiguity declaration is made. If the MOC is below the rejection threshold, the identification is rejected. These comparisons are performed for each candidate template.

The template algorithm thus performs an iterative pattern-matching process that includes parametric and nonparametric data, as well as logical relationships. The process concludes when either one or more templates are identified as being supported by the data, or when none can be supported by the data. New data arriving into the template algorithm reinitializes the iterative process. In the event that multiple templates are identified as viable (i.e., for more than one template, the sufficient conditions are met, or the MOC exceeds acceptance thresholds), an ambiguity is declared that may be resolved by human intervention or a predetermined weighting of the templates. Such ambiguity resolution is analogous to the conflict resolution process in the expert system inference cycle.

Templates or various aspects of templates have been used with varying degrees of success in several data fusion systems. Examples of some of these systems include TCAC, ASAS, IP, and that reported by [42]. Other research is described by [43] and by [44].

Templates used in data fusion systems can be placed into four different categories: doctrinal, situation, event, and decision support. A doctrinal template describes enemy doctrinal deployment for various types of operations without weather and terrain constraints being imposed. The types of information provided in a doctrinal template are composition, formations, frontages, depths, and equipment numbers and ratios. This kind of template provides the basis by which enemy doctrine is integrated with terrain and weather data. A situation template describes how the enemy may deploy and operate within the contraints that weather and terrain impose and identifies critical events and activities that are expected to occur. These templates are used to predict time-related events within these areas. Event templates identify the participants, sequence of activities, and interrelationships to declare that an event has or is occurring. Finally, decision support templates depict decision points and target areas of interest that are keyed to significant events and activities. A decision support template provides a guide as to when tactical decisions are required relative to events on the battlefield.

Figures of Merit (FOM)

During the template inference process, observations are associated with each template, and observations are associated with other observations. Because the observations generally contain both parametric data (such as location, frequency, time, etc.) as well as nonparametric data (e.g., identity declarations from sensors), a measure of correlation (MOC) is used to qualify the degree of association. If only parametric data are involved, then statistics-based measures such as a Mahalanobis metric could be used to determine association.

The MOC is a numerical measure of the confidence that the two observations represent the same entity. An entity is described by definite attributes, and the MOC for two entities depends on the degrees of coincidence between the individual attributes of the entities. The association problem has two facets: that of finding measures to compare the individual attributes and that of devising a method to combine these measures to form the MOC. Following [45], we refer to the comparison measures as figures of merit (FOMs).

The FOMs must be tailored to the attributes of the individual entities. It is possible to use statistical measures as FOMs for numerical attributes for which the probability density descriptions are known. For example, for locations described by Gaussian vector processes, the normalized distribution for the displacement vector is a suitable FOM. For scalar numerical attributes with Gaussian distributions, the normalized chi-square distribution, which is a measure of the difference, may be useful as an FOM.

There is a class of numerical attributes in which the uncertainty is described in terms of a range of values for the attribute. An appropriate FOM for this class is the cross-correlogram of the two intervals describing the two attributes to be compared. This process yields a FOM that is unity in value when the intervals intersect completely. As the intervals become disjoint, the FOM value decreases linearly to zero when the intersection is null.

Finally, there are alphanumerical attributes such as equipment designators that require a different approach. On one extreme, the comparison can be performed as a simple Boolean test in which case the FOM is either one or zero. On the other extreme, the comparison can be performed at the individual character level yielding a continuous FOM function. The specific FOM function is often selected based on heuristic arguments rather than rigorous mathematics.

The method of combining FOMs to form an MOC is generally developed empirically. One approach is to simply compute the average of the weighted FOMs. The weights are chosen to reflect the uniqueness of a particular attribute in identifying the entity. The reliabilities (occurrences, uncertainties, etc.) of the individual attribute estimations/descriptions are reflected in the FOM functionals and not in the weights. Another possibility is to form the MOC by multiplying the weighted FOMs. The essential problem with this method is that the incremental sensitivity of the MOC value to variations of a particular FOM is a function of the values and weights of the other FOMs. Multiplicative FOMs have not been found to be useful in decision theoretic studies. The general form of the MOC equation is

$$MOC = \sum_{i=1}^{n} w_i FOM_i \qquad (7.8)$$

The computed MOC value is compared to a predetermined threshold value. If the MOC value exceeds the threshold value, it is judged that the two individual observations represent the same entity. If the threshold is not exceeded, the individual entities represent different entities.

It is possible to complicate this process with several thresholds, giving rise to more than two outcomes. For example, the judgements for two thresholds could be identical, different, and ambiguous. Such ambiguities require resolution by similar techniques utilized for expert system conflict resolutions.

7.5 BAYES BELIEF SYSTEMS

In the past 10 years, a special form of reasoning using Bayesian inference has become popular for data fusion. The method, called Bayesian belief nets, involves representing causal links or relationships using directed acyclic graphs (DAGs). A directed acyclic graph is simply a set of nodes and interconnections, with an indication of the relationship between nodes. An example of a directed acyclic graph is shown in Figure 7.14.

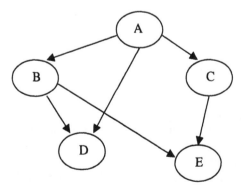

Figure 7.14 Example of a directed acyclic graph [14].

In this example, the node A is a parent of nodes B, C, and D. Node B is a parent of nodes D and E, and node C is a parent of node E (hence, E has two parent nodes, C and B). The arrows represent the direction of the graphical links and the graph is acyclic, since no node is linked "backwards" or "cyclically linked" to other nodes. The Baysian belief net representation uses directed graphs to show either causality, or simply links between variables. Jensen [46] points out that a Bayesian network is a compact (graphical) representation of the joint probability over its universe. Associated with each node is a conditional probability relationship (e.g., table). Each node represents a variable with a finite set of mutually exclusive states. The variables together with the directed edges form the directed acyclic graph. For each variable A with parents B_1, ..., B_n there is attached a conditional probability table $P(A/B_1, ..., B_n)$. If node (variable) A has no parents, then the table reduces to unconditional probabilities $P(A)$.

Jensen [46] [46] indicates that the key propagation relationship for Bayesian networks is the chain rule. Following Jensen's formulation, let BN be a Bayesian network over $U=[A_1, A_2, ..., A_n]$. The joint probability distribution $P(U)$ is the product of all conditional probabilities specified in BN:

$$P(U) = \prod_i P(A_i / pa(A_i)) \tag{7.9}$$

where $pa(A_i)$ are the parents of A_i. Let $e_1, ..., e_m$, be findings (namely evidence or observations), and then,

$$P(U,e) = \prod_i P(A_i / pa(A_i)) \prod_j e_j \tag{7.10}$$

and

$$P(A/e) = \frac{\sum_{U \ni A \in U} P(U,e)}{P(e)} \tag{7.11}$$

The knowledge contained in a Bayesian belief net includes the specification of the nodes (namely variables), the interconnections in the graph, the conditional probabilities computed at each node, and the specification of the a priori probabilities at the top of the graph. An inference engine begins with the prior probabilities and the evidence, and computes the resulting propagated probabilities using the graph structure and the conditional probability computations. As evidence is accrued (observations are obtained), the inference engine uses the chain rule and updates the values of the probabilities of the affected nodes. It must be noted that the solution of a Bayesian net is in general an NP hard problem. Neopolitan provides a discussion of various algorithms for addressing this problem. General discussions of Bayesian belief nets are provided by [14, 46-48]. There are a number of commercial packages for implementing Bayesian belief nets including the freeware, JAVABAYES. A list of Bayesian networks software tools is available on the Web site: http://excalibur.brc.uconn.edu/~baynet/tools.html. The Web site lists the tools, acceptable computer platforms, restrictions, references and contact information.

7.6 INTELLIGENT AGENT SYSTEMS

The most recent type of system/architecture for automated reasoning is software agents or intelligent agents. The concept is analogous to the development of robots in hardware—namely development of software entities that have a limited "awareness" of their situation, perform reasoning, and make decisions related to their own behavior and perceptions. A standard definition of agents or intelligent agents does not exist. However, [7] defines an agent as "anything that can be viewed as perceiving its own environment through sensors and acting upon that environment through effectors." Such a definition is related to intelligent control systems (i.e., in which a control system observes a system or environment and uses effectors to control the environment). In this sense, a simple thermostat or engine governor could be considered as an "agent." However, it is generally recognized that the term software agent or intelligent agent refers to a system that has more "awareness" than observing a single variable.

Bradshaw [49] lists the following as possible agent attributes:

- Reactivity: The ability to selectively sense and act;
- Situatedness: Being in continuous interaction with a dynamic environment, able to perceive features of the environment important to them, and effect changes to the environment;
- Autonomy: Goal-directedness, proactivity, and self-starting behavior;
- Temporal continuity: Persistence of identity and state over long periods of time;
- Inferential capability: Ability to act on abstract task specifications using prior knowledge of general goals and preferred methods to achieve flexibility, goes beyond the information given, and may have explicit models of self, user situation, and/or other agents;
- Adaptability: Being able to learn and improve with experience;
- Mobility: Being able to migrate in a self-directed way from one host platform to another across a network;
- Social ability: The ability to interact with other agents (and possibly humans) via some kind of agent-communication language, and perhaps cooperate with others;

- Knowledge-level communication ability: The ability to communicate with persons and other agents with language more resembling human-like "speech acts" than typical symbol-level, program-to-program protocols;
- Collaborative behavior: Ability to work in concert with other agents to achieve a common goal.

Woolridge [50] adds the following as possible agent attributes:

- Veracity: An agent will not knowingly communicate false information;
- Benevolence: Agents do not have conflicting goals, and hence every agent will try to do as asked;
- Rationality: An agent will act in order to achieve its goals, and will not act in such a way as to prevent its goals from being achieved (at least insofar as its beliefs permit).

Under this set of characteristics, an agent is a software entity with significant "intelligence" and ability to make decisions about its environment.

Russell and Norvig [7] provide an extensive discussion of agents and their utilization for automated reasoning. A summary of different classes or types of agents is provided in Table 7.7. These are listed in order of increasing complexity. A conceptual view of a utility-based cognitive agent is shown in Figure 7.15.

Table 7.7

Classes of Agents [7]

Type of Agent	Characteristics
Reactive agent	• Maintains an ongoing interaction with its environment
	• Responds (in a timely fashion) to changes that occur in it
	• Uses stimulus-response rules or condition-action rules
Agent with internal state	• Enable internal representation of the state of the world (internal model of the world)
	• Remembers the past as provided in earlier percepts
	• Sensors do not usually give the entire state of the world at each input, so perception of environment is captured over time
	• Allows ability to represent and determine changes in the world
Goal-based agent	• Proactive agent (takes the initiative and generates and attempts to achieve goals versus react to environment)
	• Agent maintains a list of one or more goals
	• A goal is a description of a desirable state or situation
	• Actions are chosen to achieve the goals
	• Deliberative—agent needs to reason about the actions to take to achieve goals
	• Goals achievement may involve long sequences of actions; may involve extensive search and planning

Table 7.7   (continued)

Classes of Agents [7]

Type of Agent	Characteristics
Utility-based agent	• If multiple sequences of actions are possible to achieve goals(s), how to decide which one is best or optimal
	• Uses a utility measure (e.g., cost, timeliness, likelihood of success, risk) to quantify the value or utility of a sequence of possible actions
	• Facilitates decisions between conflicting goals, likelihood of success and importance of goals

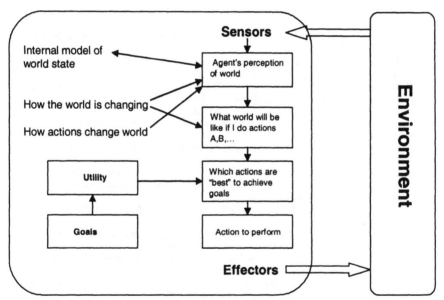

Figure 7.15 Concept of a utility-based cognitive agent.

Agent architectures and reasoning approaches have been used in a variety of applications. Yen and his students [51, 52] have developed a special representation language and formalism for modeling how humans collaborate as members of a team. They model the way human teams share a mental model and communicate in a proactive manner (to provide information to team members in anticipation of their needs). Currently, Yen's team is using such agent systems to model how humans perform situation assessment and threat assessment in a tactical command center. Other researchers have used agents for distributed sensor networks [53-56] systems for information retrieval, supply chain analysis, cognitive aides, and others. An excellent web site that provides links to tools, organizations and groups, theory and links, and projects is: http://www.ict.tuwien.ac.at/palensky/agents/.

REFERENCES

[1] Kurzwil, R., *The Age of Intelligent Machines*, Cambridge, MA: The MIT Press, 1990.

[2] Klix, F., ed. *Human and Artificial Intelligence*, Amsterdam: North-Holland Publishing Company, 1979.

[3] Pratt, V., *Thinking Machines: The Evolution of Artificial Intelligence*, Oxford, UK: Basil Blackwell, 1987.

[4] Minsky, M., *Society of Mind*, New York, NY: Simon and Schuster, 1985.

[5] Hofstadter, D., *Godel, Escher, Bach: An External Golden Braid*, New York, NY: Basic Books, 1979.

[6] Shapiro, S. C., ed. *Encyclopedia of Artificial Intelligence*, Vol. 1 and 2, New York, NY: John Wiley and Sons, 1987.

[7] Russell, S. J., and P. Norvig, *Artificial Intelligence: A Modern Approach*, 2nd ed., New York, NY: Prentice Hall, 2002.

[8] Hall, D., R. J. Linn, and J. Llinas, "Survey of Data Fusion Systems," *SPIE Conference on Data Structure and Target Classification*, Orlando, FL, 1991, pp. 13-36.

[9] Nichols, M., "A Survey of Multisensor Data Fusion Systems," in *Handbook of Multisensor Data Fusion*, ed. by J. Llinas, Boca Raton, FL: CRC Press, 2001.

[10] Jackson, P., *Introduction to Expert Systems*, 3rd. ed., Addison-Wesley Pub. Co., 1999.

[11] Benfer, R. A., E. E. Brent, and L. Furbee, *Expert Systems*, vol. 77, London, England: Sage Publications, 1991.

[12] Giarratano, J. C., *Expert Systems: Principles and Programming*, Pacific Grove, CA: Brooks/Cole Pub. Co., 1998.

[13] Stefik, J. J., *Introduction to Knowledge Systems*, New York, NY: Morgan Kaufmann, 1995.

[14] Neopolitan, R. E., *Probabilistic Reasoning in Expert Systems: Theory and Algorithms*, New York, NY: John Wiley and Sons, 1990.

[15] Anton, J., et al., "Artificial Intelligence for Problem Solving," *IEEE*, vol. CH1742-6/81/000/9232, 1981.

[16] Schank, R. C., and R. G. Childers, *On Language, Learning, and the Cognitive Computer*, Reading, MA: Addison-Wesley Publishing Company, 1984.

[17] Krause, P., and D. Clark, *Representing Uncertain Knowledge: An Artificial Intelligence Approach*, New York, NY: Kluwer Academic Publishers, 1993.

[18] Post, S., and A. P. Sage, "An Overview of Automated Reasoning," *IEEE Transactions on Systems, Man, and Cybernetics*, vol. 20, 1990, pp. 202-224.

[19] Ng, K.-C., and B. Abramson, "Uncertainty Management in Expert Systems," *IEEE Expert*, 1990, pp. 29-48.

[20] Shenoy, P. P., and G. Shafer, "Propagating Belief Functions with Local Computations," *IEEE Expert*, 1986, pp. 43-52.

[21] Thomopoulos, S. C. A., "Sensor Integration and Data Fusion," *Journal of Robotics Systems*, vol. 7, 1990, pp. 337-372.

[22] Adams, J. B., "A Probability Model of Medical Reasoning and the MYCIN Model," *Mathematical Biosciences*, vol. 32, 1976, pp. 177-186.

[23] Zadeh, L. A., "Fuzzy Logic and Approximate Reasoning," *Synthese*, vol. 30, 1975, pp. 407-428.

[24] Casti, J. L., *Searching for Certainty: What Scientists Can Know about the Future*, New York, NY: William Morrow and Company, 1990.

[25] Kahneman, D., P. Slovac, and A. Tversky, eds. *Judgement under Uncertainty: Heuristics and Biases*, Cambridge, UK: Cambridge University Press, 1982.

[26] Buchanon, B. G., and E. H. Shortliffe, *Rule-Based Expert Systems: The MYCIN Experts of the Heuristic Programming Project*, Reading, MA: Addison-Wesley Publishing Company, 1983.

[27] Zadeh, L. A., "Fuzzy Algorithms," *Information & Control*, vol. 12, 1968, pp. 94-102.

[28] Yen, J., "Generalizing the Dempster-Shafer Theory to Fuzzy Sets," *IEEE Tranactions on Systems, Man and Cybernetics*, vol. 20, 1990, pp. 559-570.

[29] Zadeh, L. A., "Fuzzy Logic," *Computer*, 1988, pp. 83-92.

[30] Zadeh, L. A., "The Role of Fuzzy Logic in the Management of Uncertainty in Expert Systems," *Fuzzy Sets and Systems*, vol. 11, 1983, pp. 199-227.

[31] Zadeh, L. A., "Syllogistic Reasoning in Fuzzy Logic and Its Application to Usuality and Reasoning with Dispositions," *IEEE Tranactions on Systems, Man and Cybernetics*, vol. SMC-15, 1985, pp. 754-763.

[32] Mitchell, D. H., S. A. Harp, and D. K. Simkin, "A Knowledge Engineer's Comparison of Three Evidence Aggregation Methods," *Uncertainty in Artificial Intelligence Workshop*, University of Washington, Seattle, WA, July, 1987, pp. 297-304.

[33] Winston, P. H., *Artificial Intelligence*, 2nd ed., New York, NY: Addison-Wesley Publishing Company, 1984.

[34] Buede, D. M., *Extension of the Data Fusion Process Model Based on Concepts of Automated Reasoning*, Rome, NY: Arvin/Calspan Corporation, 1991.

[35] Lessor, V. R., and L. D. Erman, "A Retrospective View of the Hearsay-II Architecture," *Fifth International Joint Conference on Artificial Intelligence*, Massachussetts Institute of Technology, Cambridge, MA, 1977.

[36] Llinas, J., and R. Antony, "Blackboard Concepts for Data Fusion and Command and Control Applications," *International Journal on Pattern Recognition and Arificial Intelligence*, 1992.

[37] Anderson, C., and M. Dorfman, eds. *Aerospace Software Engineering*, Progress in Astronautics and Aeronautics, Vol. 136, Washington, D.C.: American Institute of Aeronautics and Astronautics, 1991.

[38] Hall, D., and M. Crone, "Comments on the Procurement and Development of Expert Systems," *Expert Systems in Government Conference*, October, 1985.

[39] Hall, D., D. Heinze, and J. Llinas, "Test and Evaluation of Expert Systems," in *AI in Manufacturing: Theory and Practice*, ed. by A.L. Soyster: Institute of Industrial Engineering, 1989, p. 121-150.

[40] Miller, R. A., H. E. Pople, and J. D. Myers, "INTERNIST: An Experimental Computer-based Diagnostic Consultant for General Internal Medicine," *The New England Journal of Medicine*, vol. 307, 1982, pp. 468-476.

[41] Gevarter, W. G., "The Nature and Evaluation of Commercial Expert System Building Tools," *Computer*, vol. 20, 1987, pp. 24-41.

[42] Noble, D. F., "Template-based Data Fusion for Situation Assessment," *1987 Tri-Service Data Fusion Symposium*, Johns Hopkins University, Laurel, MD, 1987, pp. 226-236.

[43] Hall, D., and R. J. Linn, "Comments on the Use of Templating for Multisensor Data Fusion," *Tri-Service Data Fusion Symposium*, Johns Hopkins University, Laurel, MD, 1989, pp. 152-162.

[44] Gabriel, J. R., and M. H. Gabriel, "Data Representation and Matching for Events and Templates," *1989 Tri-Service Data Fusion Symposium*, Johns Hopkins University, Baltimore, MD, 1988, pp. 327-332.

[45] Wright, F. L., "The Fusion of Multisensor Data," *Signal*, 1980, pp. 39-43.

[46] Jensen, F. V., *Bayesian Networks and Decision Graphs*, New York, NY: Springer, 2001.

[47] Jensen, F. V., *An Introduction to Bayesian Networks*, London, England: Springer, 1996.

[48] Pearl, J., *Probabilistic Reasoning in Intelligent Systems*, San Mateo, CA: Morgan Kaufmann, 1988.

[49] Bradshaw, J., *Software Agents*, AAAI Press, 1977.

[50] Woolridge, M., *Introduction to MultiAgent Systems*, New York, NY: John Wiley & Sons, 2002.

[51] Yen, J., et al., "CAST: Collaborative Agents for Simulating Teamwork," *17th International Joint Conference on Artificial Intelligence (IJCAI, 2001)*, Seattle, WA, 2001.

[52] Yen, J., X. Fan, and R. A. Voltz, "On Proactive Delivery of Needed Information to Teammates," *International Joint Conference on Autonomous Agents and Multi Agent Systems: 2002 Workshop on Agent-based Teamwork and Coalition Forming*, Italy, August, 2002.

[53] Jayasimha, D. N., S. S. Iyengar, and R. L. Kashyap, "Information Integration and Synchronization in Distributed Sensor Networks," *IEEE Tranactions on Systems, Man and Cybernetics*, vol. SMC-21, 1991, pp. 1032-1043.

[54] Knoll, A., and J. Meinkoehn, "Data Fusion Using Large Multi-agent Networks: An Analysis of Network Structures and Performance," *International Conference on Multisensor Fusion and Integration for Intelligent Systems (MFI)*, Las Vegas, NV, 1994.

[55] Ross, K. N., et al., "Mobile Agents in Adaptive Hierarchical Bayesian Networks for Global Awareness," *IEEE International Conference on Systems, Man and Cybernetics*, 1998, pp. 2207-2212.

[56] Kerr, K. R., et al., "TEIA: Tactical Environmental Information Agent," *Intelligent Ships Symposium II*, Philadelphia, PA, 1996, pp. 173-186.

Chapter 8

Level 4 Processing: Process Monitoring and Optimization

Level 4 in the Joint Directors of Laboratories (JDL) data fusion process model is a *metaprocess,* a process that seeks to monitor the ongoing data fusion process and optimize the sensor utilization and algorithmic processing to achieve improved data fusion results. This chapter summarizes the concept and basic approaches for Level 4 processing.

8.1 INTRODUCTION

As indicated in Chapter 2, the level 4 process in the JDL data fusion process model is a metaprocess. That is, it's a process that monitors the ongoing data fusion process and seeks to optimize that process to achieve improved fusion products (e.g., improved estimates of the position, velocity, attributes, and identity of entities and improved situation refinement). Level 4 processing algorithms control the sensors and the algorithms in the fusion process. In addition Level 4 processing could be used to control network communications, algorithms, and all other system resources. Historically, level 4 processing was focused on the control of single sensors having limited parametric agility. For example, level 4 processing may have involved computing look angles for a single radar and predicting where to point the radar to achieve optimal performance (e.g., optimal surveillance to reduce possible unseen targets or optimize the tracking capability for a single target). Under these circumstances, the level 4 process entailed the computation of so-called look angles, used to determine where to point a sensor in order to track objects, and a straight-forward simple control process. Such a problem is amenable to optimization techniques, such as those described by [1-4].

In recent years, the level 4 process has become increasingly complex. First, data fusion systems frequently involve multiplatform, multisensor systems. The sensors and processing systems may be distributed across a wide area with multiple sensors, multiple processors, and a wideband communications system to interconnect the sensors and processors. Second, intelligent sensors exhibit wide variability in parameter space control (e.g., different selectable waveforms) and processing parameters. Smart sensors allow variations in waveforms, internal processing, and utilization of sensing resources such as transmission power. Third, evolving concepts of human-in-the-loop computer assisted cognition suggest that the data fusion system should adapt to the dynamic preferences and needs of the human decision-maker. An excellent review of sensor management and scheduling is provided by McIntyre [5].

Figure 8.1 provides an overview of basic functions performed in level 4 processing. These categories of functions are described as follows.

- *Mission management:* In the top level of the JDL data fusion process model, the level 4 process is usually shown as being partially inside, and partially outside of, the data fusion process. This is a deliberate representation. This representation explicitly recognizes the fact that level 4 processing cannot be performed solely to satisfy the needs of the data fusion process. Instead, the level 4 process must also account for constraints and needs of the overall mission supported by the fusion system. For example, in order to obtain an improved estimate of a target's position, the level 4 process may recommend that active radar be utilized. However, if the mission involves stealth, then this recommendation would be counter to the needs of the mission. Similarly, when monitoring a complex mechanical system it may be desirable to collect information such as lubrication samples; however, collection of these data would require stopping a machine (not particularly feasible, for example, when flying a helicopter).
- *Target/ entity prediction:* In one sense, a data fusion system is unable to perceive targets or entities that it does not expect to observe. In order to compute "look angles" (namely to tell the sensor system or information source where to look for data) models are required of the entities to be observed. For mobile entities this includes models of target motion. In addition, models are needed to predict the emissions or observable characteristics or attributes of the entities to be observed. Note that this is relatively straightforward for discrete physical objects such as vehicles (aircraft or ground vehicles). However, for complex or abstract targets (e.g., a distributed military unit or potential event), the link between target characteristics and observable phenomena may be complex.

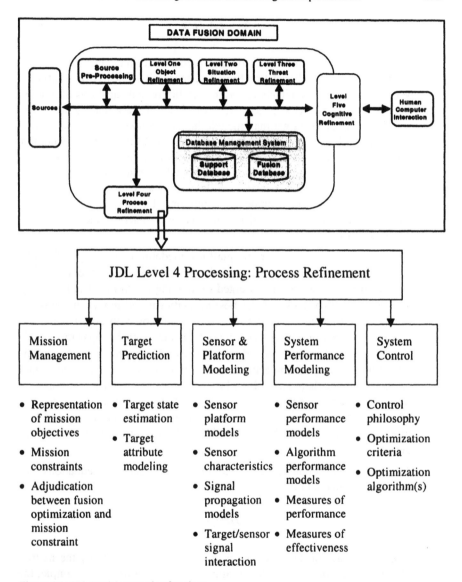

Figure 8.1 JDL level 4 processing functions.

In particular, for any given target and specified observation time, t_i, we need to be able to predict what a sensor would be able to "observe." As indicated in chapter 4, this entails modeling the evolution or motion of a target as a function of time (to transform the target's state vector from an epoch time, t_0, to the time of the observation, t_i). This is accomplished via equations of motion,

$$x(t_i) = \Phi(t_i, t_0)x(t_0) \tag{8.1}$$

The term, Φ, represents a transition matrix to transform the target state vector at time, t_0, to the observation time, t_i. The transition matrix may actually be computed using simultaneous, nonlinear differential equations of motion that account for factors such as terrain and roads (for a ground-based vehicle). The state of the target may include factors such as position and velocity at an epoch time as well as factors to represent the target's orientation (e.g., attitude vector) with respect to the observing sensor. For entities such as emitters, it is necessary to model the signal emissions anticipated from the entity. This can be especially challenging if an emitter is hosted on a platform in realistic environmental conditions. Factors such as the platform-antenna interaction, local terrain, and atmospheric conditions affect the emission and collection process.

- *Source Requirements* (sensor and platform modeling): Given specific entities to be observed, models are needed to characterize the sensors or information sources. For sensors mounted on moving platforms, it is necessary to model the flight paths of the hosting platform and model the sensor's performance (namely to predict what the sensor would observe given a hypothesized target or entity). Such models include signal propagation models, sensor/target interaction and sensor noise. Algorithms for such modeling span the range from detailed physical models of entities, to signal propagation models (e.g., to describe the path of electromagnetic radiation from the sensor to the target and its return), statistical noise models, models of internal sensor processing (to characterize signal conditioning and feature extraction processing), and even hierarchical syntactical models of targets.

The predicted observation may be computed via observation equations represented by the relation,

$$z(t_i) = Hx(t_i) + \eta \tag{8.2}$$

The actual form of the observation predictions, represented by the matrix, H, typically involves multiple coordinate transformations. For example, the equations may provide components to transform from the coordinate system in which the target state vector, x, is represented to the frame of reference of the sensor platform, to the coordinate frame of the observation. There are no general expressions for these transformations. Instead, they must be developed on a case-by-case basis. However, texts such as [6] provide equations for ground-based tracking of space satellites, and are useful as examples.

The relationship between targets and sensors can be very complex. In general, it is necessary to understand what can be observed and how well the sensors perform. An example of such analysis for sensors onboard a tactical aircraft is provided by Steinberg [7], as illustrated in Figure 8.2.

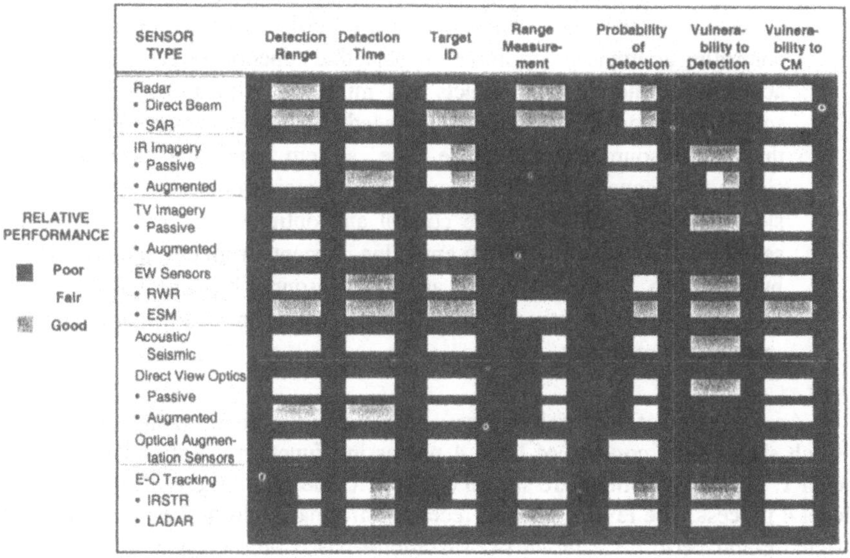

RELATIVE PERFORMANCE	SENSOR TYPE	Detection Range	Detection Time	Target ID	Range Measurement	Probability of Detection	Vulnerability to Detection	Vulnerability to CM
	Radar							
	• Direct Beam							
	• SAR							
	IR Imagery							
	• Passive							
	• Augmented							
	TV Imagery							
	• Passive							
	• Augmented							
	EW Sensors							
	• RWR							
	• ESM							
	Acoustic/ Seismic							
	Direct View Optics							
	• Passive							
	• Augmented							
	Optical Augmentation Sensors							
	E-O Tracking							
	• IRSTR							
	• LADAR							

Relative Performance: ■ Poor, Fair, ▦ Good

* Steinberg, DFS-87.

Figure 8.2 Example of the sensor performance against specified target [7].

- *System performance modeling:* In addition to the specific models for sensor performance, the overall data fusion system needs to be modeled. How does the system perform including sensors, communications components between the sensors and the processing components, computer computational performance, algorithm performance, and (most challenging) the human-in-the-loop processing? We seek to develop measures of performance (MOP) and measures of effectiveness (MOE) to characterize how the data fusion system is performing (given the observed targets, utilization of sensors and information resources, specified fusion algorithms, and the human-in-the-loop). Given this characterization of the system, control strategies can be developed to optimize the performance. This is analogous to characterizing an information system via quality of service (QoS) parameters. There have been numerous MOP and MOE developed for target tracking systems. Waltz and Llinas [8] provide an excellent discussion of the issues associated with MOP and MOE, and they present a hierarchical concept for linking sensor performance ultimately to system effectiveness.

- *System planning:* A key aspect of level 4 processing is system planning. This involves development of plans for sensor utilization and processing to meet anticipated mission requirements, namely transformation of mission plans and objectives into anticipated data collection needs.
- *System control:* Finally, a basic component of the level 4 processing is the actual control function. This entails classic feedback and control in which a controller attempts to optimize an objective function (e.g., maximize the accuracy of the fusion products and minimize utilization of system resources). Fusion control requires knowledge of the entities being observed, the sensor/source performance, the system performance, and mission/system constraints. Techniques for fusion control are drawn from classic methods in multiobjective control and optimization. More recent research has focused on utilizing emerging concepts from e-commerce and e-business (e.g., using intelligent agents as resource bidding agents). These are particularly attractive for distributed systems involving heterogeneous information resources.

Another view of the level 4 process is shown in Figure 8.3 (adapted from [8]) which shows their concept of level 4 processing functions. Notice that the JDL level 1, 2, and 3 functions are shown on the left side of the figure as inputs to the level 4 process. The level 4 process receives information from the evolving target database and performs a target prioritization—based on the types of factors described in Section 8.3. High-priority targets (or observable events, entities, or activities) are nominated for further observation and a sensor-target computation determines which sensor(s) could observe the priority target or event. A target/sensor request is input to a function that performs that allocation, scheduling, and control of sensor resources. Specific spatial and temporal control parameters are computed and used as a basis for commands to the sensors via a sensor interface module. This information is linked to sensor performance models and to event prediction models. External information may be provided to support the level four process including an electromagnetic concept, cues from external sources, and manual priorities provided by a human-in-the-loop decision maker. Data and status from the sensor suite are inputs to the level 4 process.

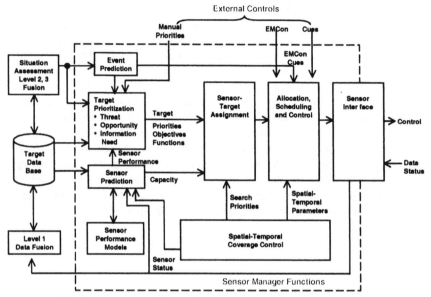

Walt, E. and Llinas, J., *Multisensor Data Fusion"*, Artech House, Boston, MA, 1990.

Figure 8.3 Example of level four functions [8].

8.2 EXTENDING THE CONCEPT OF LEVEL 4 PROCESSING

The traditional level 4 processing monitors the evolving results of the level 0-3 processing and provides directions to the sensors, and to the Level 1 functions to optimize the fusion performance. In order to quantify the level 4 process, measures of performance (MOP) or measures of effectiveness (MOE) are computed using information from the level 1 process (e.g., state vectors, identity information, and covariance error data). Generally, there are a number of MOP or MOE parameters that could be computed such as size of the error ellipse for tracked targets, number of false alarms, probability of correct identification, number of dropped targets or number of unassociated observations. Sometimes these parameters are combined into a single, scalar MOP or MOE using an ad hoc figure of merit function [9] or utility functions. Rothman and Bier [10] discuss the evaluation of sensor management systems and list 10 categories of MOP for tactical military operations, including detection and leakage, target tracking parameters, identification of entities, raid assessment, kill assessment, emission, sensor utilization, system response, and computational performance. Within these categories, 45 measures of performance are listed that can be used to assess how well a sensing system is performing.

The traditional level 4 process uses information from the level 1 process and data from the sensors to compute a MOE or MOP. The MOE or MOP is optimized with respect to sensor control parameters, subject to one or more constraints. Constraints could include limits on individual sensors, limitations on sensing resources (e.g., due to power or communication bandwidth for dispersed sensors), or mission limitations (e.g., to minimize the use of active sensors for stealth purposes). This constitutes a classic multivariable constrained optimization problem. There are numerous techniques for addressing these problems [2-4]. The optimization is performed sequentially in a time frame that is commensurate with the sensor observation data rate (or a multiple of that data rate). As shown in Figure 8.4, a connection is also available to a human operator. This allows the operator/analyst to provide directions to the optimization process or to override the process to provide directions to the sensors or algorithms. In the traditional approach to level 4 processing, there is only a loose connection with the human operator. The process is also not generally connected with the level 2 or level 3 processes, because of the difficulty in quantifying the MOE/MOP for level 2 and level 3.

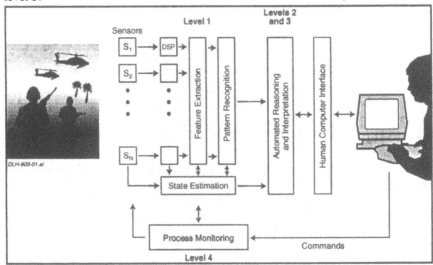

Figure 8.4 Traditional view of level 4 processing.

While the traditional approach to level 4 processing is well posed (yet still remains challenging), there are emerging technologies that change the dimensions of the level 4 problem. The first technology involves the advent of *smart* sensors. These sensors are enabled by MEMS [11] and increasing computational power available from microprocessors. These technologies allow implementation of low-cost, compact, low-power sensors that may perform a number of functions including self-location via organic GPS receivers, communication with other sensors using spread-spectrum communications, self-calibration, observation and characterization of local environmental conditions, and local signal processing and pattern recognition. The availability of advanced microprocessors allows extensive processing to be performed at each sensor (or sensor node). These computations could include the level 0 signal processing and feature extraction, and level 1 state estimation and pattern recognition functions Accurate predictions of sensor performance can also be made using sophisticated models of the local environment (e.g., utilizing signal propagation models). These predictions have the potential to significantly improve the data fusion process by providing accurate information on sensor performance. The use of simplistic models of sensor performance (e.g., modeling sensors as unbiased, *Gaussian* sensors) can result in fused results that are actually less accurate than the performance of the best individual sensor. Finally, local wideband communications and accurate internodal timing (via GPS time synchronization) allows the possibility of coherent multisensor processing. All of these advances in sensors have implications for Level 4 processing.

A second area that has the potential to affect level 4 processing is the concept of using an active *human-in-the-loop*. Traditional level 4 processing allows a human operator/user to provide directions to the level 4 process, but does not allow significant interaction. A study by Llinas, Drury et al. [12] has indicated that decision support systems are more effective if they allow more human control and direction. Phoha, Eberbach et al. in [13, 14] have investigated the concept of an active *human-in-the-loop* for control systems involving a hierarchical set of semi-autonomous robotic vehicles. They have developed formal mathematical techniques to model the role of the human. It is anticipated that an intelligent control system could adapt to the information needs and preferences of a human user, and *learn* from the human. A truly intelligent controller/level 4 process would adapt to the human user (e.g., to adapt to information needs, preferences, decision styles, response to stress) in a manner similar to the current way that level 4 processing adapts to the sensors. This changes the level 4 process from simply a process controller to an intelligent companion to the system user. Such an intelligent process would make the fusion process much more effective for the human user and address the battlefield information overload issues described by Hall and Llinas [15].

This concept of the enhanced role of level 4 processing is shown in Figure 8.5. level 4 processing is expanded to include interaction with the sensors, the level 1 process, the level 2 and level 3 process, and with the human operator. The intelligent level 4 process accepts commands and preferences from the human operator, and also evolves to meet the information needs of the operator. It is also made to be context-sensitive to changes in the mission and the condition of the human. The expanded level 4 process also monitors the evolving level 2 situation assessment, and the level 3 threat assessment to determine the information needs for improved effectiveness beyond that of the level 1 process. Accurate information from the *smart* sensors, concerning dynamic performance, also allows a major improvement in the level 1 fusion by an improvement in sensor utilization. Thus, the improved level 4 process can significantly improve the fusion process and the effectiveness of the fusion system for the human operator/user.

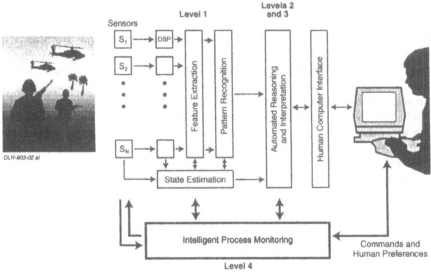

Figure 8.5 Enhanced view of level 4 processing.

8.3   TECHNIQUES FOR LEVEL 4 PROCESSING

This section provides a discussion of specific techniques for level 4 processing including sensor management functions, optimization of computing resources, auction-based methods, and measures of performance (MOP), and measures of effectiveness (MOE).

8.3.1   Sensor Management Functions

The design and evaluation of sensor management systems is becoming increasingly important. Modern sensors have embedded processors and extensive agility in control. On a modern tactical aircraft, for example, the aircraft platform has an integrated avionics system and sophisticated sensors that can be dynamically programmed to change their emission and receiving patterns. However, the environments faced in tactical engagements are very fast moving and characterized by electronic warfare (i.e., jamming and deception) and low-observable situations (e.g., stealth aircraft and camouflaged targets). Thus, on one hand, the flight space of such an aircraft may include an increasing number of opponents, large combinations of enemy weapons and sensors, and difficult observing environment. On the other hand, the increased computing capability onboard the aircraft, coupled with the agility of the onboard sensor systems, provides an opportunity to develop an effective level 4 processing subsystem.

A challenge of sensor management is that the sensor management functions are (or at least should be) coupled with the functions and capability of the sensor suite and the data fusion processing. This concept is illustrated in Figure 8.6.

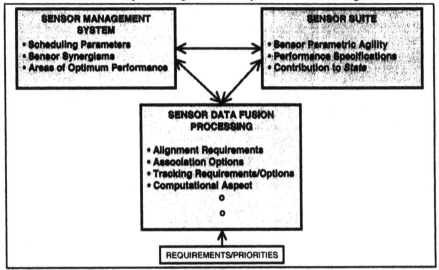

Figure 8.6 Coupling of sensor, processing, and sensor management functions.

The design of a sensor management system must be integrated with the design of the sensor suite and the data fusion processing. All too often, these systems are treated as independent entities. For example, the sensor suite is generally selected first (or is treated as given inputs to the data fusion system), followed by the design of the sensor management system, and the data fusion processing is treated independently. These three systems are integrally related and should be designed together. These three subsystems are summarized as follows.

- *Sensor suite:* As discussed in Chapter 1, the sensor suite should be selected after a careful analysis of a variety of factors. These include the physics of the targets and signal environment, an understanding of how the sensors interact with the host platform (e.g., how an antenna is affected by the aircraft and other sensors), and an understanding of what decisions need to be made to support the tactical mission. Key factors include what the sensors can physically observe (e.g., radar cross-section, radio frequency, and infrared spectra) and how these observable quantities contribute to the knowledge of the target's state vector. An understanding is needed of the sensor performance specifications as well as an understanding of how the sensors perform in a realistic observing environment. The parametric agility (i.e., the ability to control factors such as transmission waveform, and transmit power) affects the design of the sensor management system and the data fusion processing.

- *Sensor management system:* The sensor management system acts as a facilitator between the sensing system and the sensor data fusion system. The sensor management system seeks to optimize the performance and utilization of the sensor suite to obtain an optimal set of data fusion products. Ideally, the sensor management system will exploit synergy among the sensors and capitalize on their coordinated strengths. These strengths may change as a function of the targets being observed and the observing environment. At a minimum the sensor management system computes the appropriate look angles (to point the sensors either mechanically or electronically to acquire and track targets) and performs a scheduling function. This scheduling function determines what sensors should be used at what times, and how to effectively use system resources. In some cases the rate and type of data collected will be determined via context-based reasoning. For example when monitoring the condition of a complex system, the rate of data collection may be changed from a nominal monitoring rate to intense data collection (e.g., as a system transitions from normal operation to a critical or near-failure mode).

- *Sensor data fusion processing:* Previous chapters of this book have described data fusion processing functions. These include data alignment, data association/correlation, and state-vector estimation.

8.3.2   General Sensor Controls

Modern sensors may be very complex, containing multiple sensors and processors within a single system. Moreover, control may involve factors such as control of observing mode, spatial pointing and control, temporal control, and reporting controls. A summary of general sensor controls and control functions is shown in Table 8.1. A brief description of each control category is provided in the following.

Table 8.1

General Sensor Controls

Category of Control	Control Functions input to Sensor

Mode control functions	• On/off control
	• Sensor mode selections
	    ○ Power level (active sensors)
	    ○ Waveform or processing mode
	    ○ Scan, track, or track-while-scan
	• Sensor processing parameters

Table 8.1 (continued)

General Sensor Controls

Category of Control	Control Functions input to Sensor
	o Decision thresholds
	o Detection, track, ID criteria
Spatial control functions	• Pointing coordinates (center of field of view)
	• Field of view selection
	• Scan/search rate
	• Scan/search pattern selection
	• Parameters to control individual sectors
	o Sector coordinates
	o Modes within sectors
	• Parameters for designated targets
	o Target or track index number
	o Coordinates or search volume
	o Mode to be used
	o Predicted time of appearance
	o Dwell time on target
Temporal control functions	• Start/stop times for modes and sector control
	• Specified sensor look time
	• Specified dwell time on target and search
	• Maximum permissible emission duty cycle
Reporting control functions	• Report filters based on target attributes
	o Friend, foe, or both
	o Filter by class or type
	o Filter by lethality
	• Report filters based on spatial attributes
	o Min/max range limits
	o Altitude layer filters
	o Spatial region filters
	• Priority of designated targets (by index)

• *Mode control functions:* Modern smart sensors generally have multiple modes of operation. These modes can be controlled by either an external command to the sensor or by allowing the sensor to make automated mode changes to achieve optimal sensing. For example, the automatic gain control may adjust the relative difference between a target's signal and the background noise to some a priori level for improved detection and tracking (e.g., much like a copy machine that adjusts the contrast between light and dark on a copied image). For active sensors, the mode control may include transmit power levels and choice of waveform used. Other mode controls may include whether a sensor is used in a scanning or staring manner, tracking mode, or track-while-scan mode. The mode control may also change decision thresholds (for reporting a detection) or identification criteria. One control not often mentioned by the sensor manufacturer is the ability to turn a sensor on or off. In some environments, a sensor may be so prone to false alarms that turning the sensor off is preferable to corrupting the data fusion process by including erroneous observations. This concept is explored mathematically in critic-based voting identification processes (in which an automated critic judges the potential value of a sensor's input and may deny the use of the sensor's input to improve the fused result [16]).

- *Spatial control functions:* Most sensors do not have an omni-directional field of view (analogous to a fish-eye camera). Instead, many sensors have a limited field of view and must be pointed, either physically steered or electronically steered (e.g., a phased array radar), to track particular targets or survey specified areas of interest. Spatial functions include providing pointing coordinates or so-called bore-site angles for a sensor, selection of a field of view, control of search or scan patterns, and parameters for tracking an individual target. As sensors increase their directivity (e.g., laser radar having a very narrow angular beam), there is need for improved accuracy in estimating the future location of a target and corrections for atmospheric and other effects.

- *Temporal control functions:* The timing of sensor observations can be controlled including start and stop time for observing, look or dwell time on target, and maximum duty cycle for energy emission (for active sensors). Such controls may be provided externally or based on internal context-based reasoning. For example, a sensor may detect that a target is maneuvering and increase the rate of data collection to improve the ability to track the target.

- *Reporting control:* Finally, a number of parameters may be useful to control how and under what conditions a sensor will provide reports to various data users, including data fusion processes. The sensor may have internal filters to allow reports to be suppressed unless specified conditions are met. This is especially useful when there is limited communications bandwidth available, or when one seeks to limit communications to avoid detection or reduce power consumption requirements (e.g., for a ground-based autonomous sensor). Filters and templates can be used to suppress only certain types of targets (e.g., report only identified enemy targets or certain types of threats).

8.3.3   Optimization of System Resources

The overall level 4 process seeks to optimize the utilization of system resources while simultaneously maximizing the measures of effectiveness (MOE) and measures of performance (MOP) described below. In general a MOP is an objective characterization of system performance (e.g. sensitivity, bandwidth, probability of correct classification) while MOE is a measure of utility in some task or mission. Hence, we seek to maximize MOP or MOE, subject to constraints on the following:

- Computer performance;
- Communications bandwidth and load (between data fusion nodes and sensors);
- Sensor locations and performance;
- Targets and target characteristics;
- Observing conditions (environment, terrain);
- Data fusion algorithm performance;
- Perceived need for information.

In general, this is a multiobjective optimization problem with multiple constraints to maximize (or minimize) as appropriate,

$$\text{MOE} = \sum_{i=1}^{N} w_i \text{mop}_i \tag{8.3}$$

Subject to constraints,

$$\text{mop}_i \leq C_i$$

For example we might want to maximize the probability of detection and range of target detection subject to constraints on sensor tasking, communications bandwidth, and processing capability. This is a classic optimization problem. Unfortunately, optimal performance may request system services that are limited. For example, to maximize the probability of detecting a target, it may be desirable to maximize the observation and reporting rate of all system sensors, which in turn overloads the communications links between the sensors and the processing nodes, and in turn overloads the computational resources available.

Techniques for solving this problem are drawn from operations research. Methods include linear programming, quadratic programming, and goal programming. Descriptions of these techniques can be found in multiple texts on optimization, such as [2, 3]. Alternatively, one might use methods such as the covariance error analysis technique described in Chapter 4 to assess system performance. This allows exploration of factors such as data fusion accuracy as a function of sensor placement, target characteristics, observing conditions, and other factors. However, the covariance error analysis method is primarily useful for a priori analysis and design rather than on-the-fly dynamic optimization of the system operation. Finally, one might turn to Monte Carlo simulation methods to evaluate the system performance [17, 18]. In this approach, the MOE are modeled as a multidimensional integral whose value is estimated by evaluation of random samples.

8.3.4   Measures of Effectiveness and Performance

The optimization of the data fusion system performance requires the specification of quantitative measures of performance (MOP) or measures of effectiveness (MOE). These MOP and MOE are computed and used as an objective function or functions for the optimization process. Examples of possible MOP and MOE are shown in Table 8.2. These are only a few examples. Numerous MOE/MOP have been developed and used to support the optimization process.

The specific form or representation of the MOP and MOE can be selected from a variety of choices. For example, one might represent certain MOP or MOE using probability functions (e.g., track variance), linear distances (e.g., target range at detection), scalar measures (number of undetected target per surveillance volume), and physical measures such as power consumption or other measures. For general concepts such as information value one might select Shannon's information metric.

Regardless of the specific metric, it might be necessary to convert the quantity into a measure of utility for the system users. For example, we might desire "high accuracy" from a tracking system. However, the value of a target's positional accuracy is not linear. Knowing where a target is within 1 millimeter is not 1,000 times better than knowing its position within 1m. Utility theory [19, 20] was originally developed in the 1800s to assist in understanding how humans bet on games of chance. Humans do not appear to bet in a way that is linearly related to a monetary payoff. Utility theory develops functions to map parameters to their perceived value or use. For data fusion systems, this concept can be used to account for such nonlinearity in utility of MOP such as track accuracy or timeliness.

Another approach to translate physical parameters such as tracking accuracy into useful MOP or MOE involves the use of fuzzy membership functions. Fuzzy set theory and logic are extensions of Boolean set theory and logic to allow reasoning with imprecise quantities. Fuzzy logic was originally developed by Zadeh [21]. In fuzzy set theory, sets are comprised of ordered pairs [x, $\mu(x)$]. A quantity, x, is specified as a member of a set by the value, $\mu(x)$, whose value ranges from 0 to 1. The quantity, $\mu(x)$, is termed a fuzzy membership function. For example, we might declare that a person of height 6 feet is a member of the set of tall people by the value ½. Conversely, a person of height 6 feet, 4 inches, might be assigned as a member of the set of tall people by the value of ¾. Hence, in this example we are converting a measurement of a person's height to the attribute "tallness" using a membership function. Similarly, we may use such transformations to transform values of position covariance error to a "low," "medium," or "high" value of "accuracy." In this way we can transform quantities such as covariance error, track drop rates or percent load of a communications link to values that are meaningful to a human user. It must be cautioned, however, that the definition of the fuzzy membership functions are strictly ad hoc and defined a priori by system designers or users. An excellent discussion of fuzzy logic is provided by [22].

Table 8.2

Examples of MOP/MOE for Sensor Management

MOP/MOE Category	Examples of MOP/MOE
Detection and leakage	• Range at target detection
	• Percent of target detection at specified distance
	• Percent of target detection at minimum specified distance
	• Number of undetected targets in the field of regard
	• Percentage of detected targets in the field of regard
Target tracking	• Track variance
	• Number of tracks divided by number of targets
	• Number of track miscorrelations
	• Number of dropped tracks
	• Revisit frequency for all tracks
Identification	• Percent targets correctly identified
	• Range of identification for targets
	• Time in sensor view prior to identification
Emissions	• Total power emitted
	• Power emitted per unit time
Sensor utilization	• Percent sensor idle time
	• Sensor time spent reacquiring targets
	• Number of impossible sensor attempts
System response	• Task turn-around time
	• Critical request reporting time
	• Number of starved (unserved) tasks
Computational performance	• MIPS utilized
	• Memory utilized
	• Bus bandwidth utilized

8.4 AUCTION-BASED METHODS

Finally we note that there are new classes of techniques for resource optimization available from economic theory using auction concepts. These methods are rapidly being adopted for applications on the Internet. A common example is the Internet auction, e-Bay. Similarly, the service Amazon.com offers a brokering service in which users (potential book buyers) are dynamically linked with book suppliers (including private individuals, book stores, and publishers). In this operation a user "logs on" to Amazon.com with a request for a book. The Amazon service searches a database of suppliers and provides lists of suppliers and associated prices and book conditions. Then, the user chooses a supplier and the "deal" is brokered by Amazon.com. The information science community has conducted extensive research to develop advanced intelligent agents to handle the negotiations and brokering.

This concept can be applied to resource optimization for data fusion systems. The following discussion is based on research being conducted by Mullen [23]. A market-based approach appears suitable for managing the key resources for a multisensor data fusion system including sensors, communications, fusion processing algorithms, and related limited resources. The concept uses innovations in market-based economic theory to improve the capability of a data fusion to achieve accurate inferences about the identity, location, and characteristics of multiple targets.

To transform a resource allocation problem into a computational economy, it is necessary to identify the scarce resources, namely sensor attention (e.g., sensor interface control) and estimator process focus, as well as the associated energy (e.g., battery power), computational resources, and network resources necessary to run the sensors and estimators, and to transport their information reports across different platforms. Within a computational economy, information fusion processes become the end consumer and are assigned a budget based on their relative importance and any other relevant criteria. These information consumers use their budget constraints to drive the collection and processing of target information through bids on various estimator services. These bids are based on an information fusion process's perceived utility, or expected value, of an estimator's future information. A second tier of market entities coordinates the information needs of the estimator processes with those of the sensor resources available. Through market activities, scarce sensor and estimator attention can be dynamically reallocated over time according to changing target priorities, different environmental conditions, and new sets of available sensors. Changes in sensor and estimator configurations and numbers can be made dynamically without requiring system reconfiguration.

The use of sensor or estimator processing could be modeled as a service that can be purchased through a contract, or service level agreement (SLA). Long-term information fusion tasks (e.g., scanning for new targets) can buy necessary processing services via long-term contracts in a multiattribute auction. Multiattribute auctions allow negotiation over multiple attributes simultaneously, where these attributes might be particular service elements (e.g., what volume to scan) or contract terms (e.g., duration or whether the service can be preempted for some more urgent task). High-priority interrupt tasks can purchase services through a spot market, where the market clearing occurs instantaneously. Thus, long-term information service contracts allow information to be pushed to the buyers; short-term spot market purchases mean that urgent new information can be obtained on-demand by buyers. The concept is illustrated in Figure 8.7.

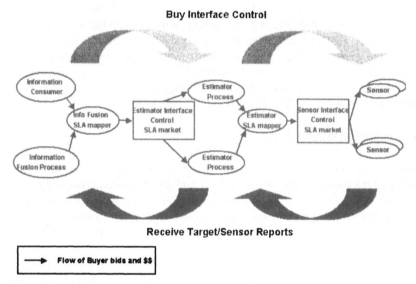

Figure 8.7 Multisensor data fusion economy overview [23].

8.4.1   Market Components

The raw resources that drive this economy are the sensors' interface control. Controlling a sensor's activities effectively bounds the reports it can generate. It also constrains the sensor's behavior to act in accordance with any task or mission constraints. Sensors act as producers in our market system. In economic terms, producers are defined by a technology function that describes their production constraints, in this case based on energy and computation. For example, given x units of battery power and y units of computational resources (e.g., CPU), a sensor can produce z units of scanning capabilities. Thus, before placing sell bids for various SLAs, the sensor agent must consider trade-offs in its technology function regarding the kinds of scanning it can do versus the battery drain.

The end consumers in this system are information fusion processes and information consumers. While information fusion processes are software modules, information consumers might be various software modules or they might be humans. In a market system, consumers are defined by their budgets and preferences over how to spend that budget on various goods. In the case of software modules, preferences can be represented explicitly as a utility function. An information fusion process's utility function would be based on the expected value of information acquired from an estimator's processing via the resultant target track reports.

Estimators act as intermediaries in the economy, transforming raw sensor resources (e.g., sensor activities and resultant reports), into value-added information goods for the consumer (e.g., target track reports). Estimators are generally software modules, but could also be human analysts. Like sensors, estimators define their production constraints via a technology function. In addition to computational resources, estimators also face production restrictions on the types of information they can get from sensor reports. An estimator process that needs an extremely accurate target state vector may need to get sensor reports from three commensurate sensors in that sector to achieve the desired accuracy. Or using an alternative production technology, the accuracy might be achieved using two non-commensurate sensors.

8.4.2   Multiattribute Auctions

Currently, most automated auctions used in e-commerce are auctions where the attributes of the item are fixed ahead of time and agents can only negotiate over price. Research is needed to investigate auctions that negotiate over attributes other than price including multi-item auctions, and combinatorial auctions, where bidders submit bids on packages or bundles of items rather than just for individual items.

Solving the auction allocation optimization problem generally relies on integer or linear programming algorithms. Alternative algorithmic approaches include dynamic programming and algorithms that take advantage of any special-case approximations specific to a given domain or problem. However, the potential explosion of possible item combinations means that unless carefully crafted, combinatorial auction implementations can raise issues about their computational complexity. Often these issues can be addressed by placing restrictions on allowable item combinations. This approach can also be used to reduce the information load on the bidders, which is of particular importance when the bidders are humans.

Modifications of the auction bidding mechanism to account for real-time constraints are also being developed. These approaches incorporate design trade-offs between local reasoning versus communication bandwidth. For example, instead of bidding at every new auction round, agents may keep standing bids either at the auction or with a bidding proxy. Such proxy agents can also engage in bid preprocessing, so that the auction need only be updated when a change in resource allocation is necessary. Most auction algorithms are built around exchanging rival resources, or resources that can only be used by one agent at a time. The output of information services, such as sensor or target reports, can be easily replicated and shared among many participants simply for the cost of transporting them across the network. Two or more consumers, each of whom wants compatible sensor reports from the same sensor, should be able to share this nonrival service. The combination of both rival and nonrival uses of sensor activities requires modification to standard auction mechanisms.

8.4.3   Multiattribute Auction Algorithms

Mullen's research has focused on the use of auction models involving multiple buyers and multiple sellers (i.e., a many-to-many negotiation). While combinatorial auction algorithms have been developed for many-to-many negotiations, existing work on procurement multiattribute auctions generally looks at a single buyer who is procuring goods from many sellers (i.e., a one-to-many negotiation). The buyer's utility function is used as the scoring rule for the auction, which determines how sellers' items compare to one another. One example of a many-to-many multiattribute auction is OptiMark, a new trading system that uses a multiattribute auction to match buyers and sellers based on price, quantity, and willingness to trade. Complicated multistep scoring rules are used to determine how to match buyers and sellers.

8.5   RESEARCH ISSUES IN LEVEL 4 PROCESSING

There are a number of research issues that should be addressed in order to effect the intelligent level 4 processing described in this chapter. Resolution of the issues which are summarized in the following would enhance the discipline of data fusion by enabling more intelligent level 4 processing, and hence implementation of more effective data fusion systems.

- *Measures of performance/measures of effectiveness:* In order to develop an optimization criterion for Level 4 processing, it will be necessary to establish quantitative measures of performance (MOP) and measures of effectiveness (MOE) for the fusion system. The focus of level 4 processing will be extended across the complete range from energy detection (at the sensor level), through target tracking and identification (level 1), information synthesis (at levels 2 and 3), and ultimately to the decision level (including the perception and decision-making of a human user). MOP and MOE are needed to quantify these factors. At this time, such MOE and MOP do not exist, with the exception of ad hoc expressions
- *Multitime scale control and optimization:* Extension of the level 4 processing beyond sensor control and Level 1 processing requires consideration of multiple simultaneous time scales. At one level, the process must be synchronized with the rate of sensor observations and updates to target tracks.

At a slower rate, the process must address the time scale related to situation assessment and threat assessment. The process must also be cognizant of the time scale of human decision making. Finally, the intelligent level 4 process must allow learning and evolution over multiple missions or engagements. Such multitime scale optimization processes do not currently exist.

- *Cognitive models for human decision-making:* Cognitive models are not available to allow modeling and adaptation to a human decision-making human-in-the-loop. Significant research is required to develop models to emulate the time frames, sensitivities, and effects of human perception of fusion data and the subsequent assimilation/decision process. This is a very challenging area. A need especially exists to model and account for issues related to human biases, selection effects, effects of stress, and differences in decision styles.

- *Adaptive learning for control systems:* Strategies need to be developed to allow a level 4 process to learn and evolve over time to improve performance. This is difficult because the Level 4 process must be sensitive to new information, different users, and new missions, without adapting so rapidly that historical information is ignored or forgotten. Another challenge involves how to learn from humans (including preferences, corrections, and adaptation to make the fusion system/human component more robust in the decision-making process).

- *Incorporation of dynamic sensor performance information:* As sensors become smarter, and exhibit the ability to provide accurate information on their own performance (namely accurate information on sensor accuracy), it will be necessary to effectively use this information in the level 4 process (as well as the level 1 process). New methods will be required to represent and process this data about sensor performance.

While this list of research issues is not exhaustive, it does indicate a number of challenges that need to be addressed in order to allow the implementation of intelligent level 4 processes.

REFERENCES

[1] Wagner, H. M., *Principles of Operations Research with Applications to Managerial Decisions*, Englewood Cliffs, NJ: Prentice-Hall, Inc., 1969.
[2] Press, W. H., et al., *Numerical Recipes: The Art of Scientific Computing*, New York, NY: Cambridge University Press, 1986.
[3] Nocedal, J., and J. Wright, *Numerical Optimization*, Berlin: Springer-Verlag, 1999.
[4] Fletcher, R., *Practical Methods of Optimization*, New York, NY: John Wiley & Sons, 2000.
[5] McIntyre, G. A., *A Comprehensive Approach to Sensor Management and Scheduling*, in *Department of Information Technology*: George Mason University, 1998.

[6]   Escobol, R. R., *Methods of Orbit Determination*, Melbourne, FL: Krieger Pub. Co., 1976.

[7]   Steinberg, A. N., "Threat Management System for Combat Aircraft," *1987 Tri-Service Data Fusion Symposium*, Johns Hopkins University, Baltimore, MD, June, 1987, pp. 532-554.

[8]   Waltz, E., and J. Llinas, *Multi-sensor Data Fusion*, Norwood, MA: Artech House, Inc., 1990.

[9]   Wright, F. L., "The Fusion of Multisensor Data," *Signal*, 1980, pp. 39-43.

[10] Rothman, P. L., and S. G. Bier, "Evaluation of Sensor Managment Systems," *1991 Fusion Systems Conference*, Johns Hopkins University Applied Physics Laboratory, Baltimore, MD, 1991, October, 1991.

[11] Jones, A., *Micro-electrical-mechanical Systems (MEMS): A DoD Dual Use Technology Assessment*, Washington, DC: Department of Defense, 1995.

[12] Llinas, J., et al., *Studies and Analyses of Vulnerabilites in Aided Adversarial Decision-making*, Buffalo, NY: State University of New York at Buffalo, Department of Industrial Engineering, 1997.

[13] Phoha, S., et al., "A Coordination of Engineering Design Agents for High Assurance in Intelligent Systems," *Third IEEE High Assurance Systems Engineering Symposium*, Washington, DC, November 13-14, 1998.

[14] Phoha, S., E. Eberbach, and E. Peluso, "Design Coordination Network for Evolutionary Integration of Structural Components of Complex Dynamic Systems," *Automatica*, 2000.

[15] Hall, D., and J. Llinas, "From GI Joe to Starship Trooper: The Evolution of Information Support for Individual Soldiers," *International Symposium on Circuits and Systems*, Monterey, CA, June, 1998.

[16] Miller, D., and D. Hall, "The Use of Automated Critics to Improve the Fusion of Marginal Sensors in ATR and IFFN Applications," *National Symposium on Sensor Data Fusion*, John Hopkins Applied Physics Laboratory, Baltimore, MD, May, 1999.

[17] Shreider, Y. A., ed. *The Monte Carlo Method*, Oxford: Pergamon, 1966.

[18] Robert, C. P., and G. Casella, *Monte-Carlo Statistical Methods*, Berlin: Springer-Verlag, 2000.

[19] Luce, R. D., *Utility of Gains and Losses: Measurement -Theoretical and Experimental Approaches*, New York, NY: Lawrence Erlbaum Associates, 2000.

[20] Eatwell, J., M. Milgate, and P. Newman, eds. *Utility and Probability*, New York, NY: W. W. Norton & Company, 1990.

[21] Zadeh, L. A., *Fuzzy Sets and Systems*, Amsterdam: North-Holland, 1978.

[22] Yen, J., and R. Langari, *Fuzzy Logic: Intelligence, Control, and Information*, Upper Saddle River, NJ: Prentice Hall, 1999.

[23] Mullen, T., *Concepts of Auction-based Methods for Resource Allocation in Multi-sensor Data Fusion Systems*, State College, PA: The Pennsylvania State University, School of Information Sciences and Technology, 2002.

Chapter 9

Level 5: Cognitive Refinement and Human-Computer Interaction

This chapter provides an introduction to the concepts of level 5 processing in the JDL data fusion process model: cognitive refinement and human-computer interaction. Level 5 processing involves supporting a human decision-maker in the loop. Functions include cognitive aids and human-computer interaction.

9.1 INTRODUCTION

Traditionally, data fusion systems have been developed to ingest information from multiple sensors and sources to provide information for automated situation assessment, or to assist a human in development of situation assessments. Extensive research has focused on the transformation from sensor data to target locations, target identification, and (in a limited cases) contextual interpretation of the target information. Numerous systems have been developed to perform this processing to achieve or support situation assessment. The JDL augmented model shown in Figure 9.1 shows the level 5 process recommended by Hall, Hall, and Tate [1] to explicitly account for functions associated with human-computer interactions (HCI). The level 5 process was added because of the importance that information representation and human-machine interaction have in most data fusion systems.

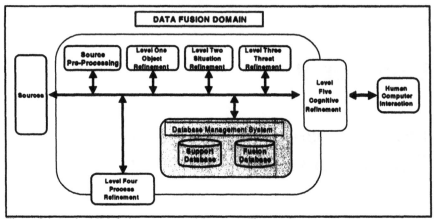

Figure 9.1 Revised JDL fusion process model (same as Figure 1.4).

The process of this transformation from sensor data to a graphics display is well known and illustrated in Figure 9.2, which shows that data from sensors (e.g., signals, images, or vector data) are successively transformed to high-level inferences using functions such as target detection, position and kinematic estimation, target identification and situation assessment or interpretation. This is illustrated conceptually on the left-hand side of Figure 9.2. The techniques to perform this hierarchical transformation include signal or image processing, statistical estimation, feature extraction and pattern recognition, and interpretation via automated reasoning techniques (as illustrated on the right-hand side of Figure 9.2).

For Department of Defense (DoD) applications, the hierarchical transformation usually begins with images, signals, and other sensor data and results in overlays on a geographical display. Extensive research has focused on the presentation of this visual data [2-10]. There has also been rapid evolution of technology to support three-dimensional visualization environments (CAVE) and haptic interfaces [11]. The focus on image displays and map overlays is pervasive as evidenced by DoD programs focused on development of a *common operational picture* (http://www.dtic.mil/jv2010/briefings.html). However, there is more to HCI for data fusion than simply visual displays. The concept of level 5 processing involves developing functions to support a human user in a collaborative human-computer environment. The next section discusses issues of cognition for situation assessment.

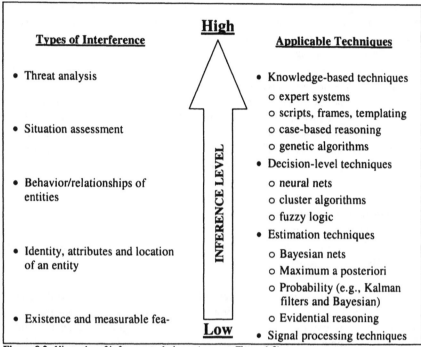

Figure 9.2 Hierarchy of inference techniques (same as Figure 1.3).

9.2 COGNITIVE ASPECTS OF SITUATION ASSESSMENT

What do human analysts do when they perform situation assessment and how does the human-computer interaction affect the analysis process and the resulting products? Are there individual differences that affect this process and the resulting efficacy of the analysis? What is the mental process for analyzing multisensor data to achieve an inference? For example, how does a maintenance technician assess multiple types of data (e.g., smell, touch and sight) to diagnose a mechanical problem? The short answer to these questions is that we simply do not know. This problem has been under investigation for a number of years in both the intelligence and DoD communities and more recently in the information science community. The essence of the level 5 processing is to assist the human in performing the inference process. A brief review of progress is provided here.

In the DoD and intelligence community, Wohl [12] and his colleagues developed the Stimulus-hypothesis-option-response (SHOR) cognitive model in the mid-1980s to describe the iterative process performed by tactical commanders in understanding the situation and developing responses. This model is analogous to the observe-orient-decide-act (OODA) loop model currently used to describe the tactical analysis and decision process (described in Chapter 2). Wohl performed some interesting experiments involving antisubmarine warfare (ASW) analysts to show how relatively simple tools could be used to improve the decision process. These improvements exceeded the improvements obtained from improved sensors and data fusion processing.

More recently, Heuer [13] has described cognitive issues related to intelligence analysis. He provides insight into the analysis process and discusses cognitive challenges and aides. Hall et al. [14] have described the process of situation assessment using a problem-centered composition/de-composition model. They performed an assessment of the state-of-the-art in data fusion and automated reasoning and analysis. They conclude that challenges for problem-centered composition and decomposition still include the following,

- Automated reasoning for data interpretation;
- Fusion of noncommensurate data;
- Human-computer interaction;
- Transformation of general concepts/queries into data collection requirements (and the converse).

Effective cognitive models have yet to be developed to understand the analysis process (and to serve as a basis for improving the process via supporting tools). However, there is extensive ongoing research in the cognitive sciences. Pinker [15] described some of the attempts by cognitive scientists to understand basic human cognition and sensing. In particular, Pinker discusses the concepts of visual intelligence and human ability to understand the physical world based on a preprogrammed *naïve physics* that provides a basis for interpretation of the data received from our visual senses. Pinker identifies basic strengths and weaknesses of our visual intelligence. The left-hand side of Figure 9.3 shows an example of a normal figure (on the top of the left-hand side) and an optical illusion on the bottom of the left-hand side of Figure 9.3. It is clear that we have certain visual logic or *built-in* expectations and we recognize that the bottom left-hand side of Figure 9.3 is not physically possible (namely it is an optical illusion). Interestingly, developers of computer games make use of this visual intelligence to improve the sense of reality of a computer game while reducing the computational load [16]. For example, when an object obscures (or blocks the view of) another object, our visual intelligence tells us that the obscured object still exists in the world. In a computer game, an obscured object is removed from the virtual world to reduce storage requirements.

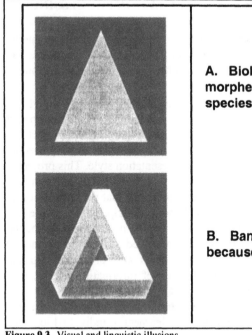

A. Biologists find the A-spinelli morphenium an interesting species to study.

B. Bananas automobile because mathematics specify.

Figure 9.3 Visual and linguistic illusions.

Pinker [15, 17] and Devlin [18] have described the role of language as part of our intellectual tools for understanding the external world. Pinker describes our innate language ability and discusses the components of our language intelligence that allow manipulation of words to create complex statements to communicate and to describe the external world. Devlin discusses the history of logic and mathematicians and philosophers search for a theory of thought. Devlin presents an example of our ability to distinguish a correct sentence from an incorrect sentence. The example is shown on the right-hand side of Figure 9.3. Sentence A in Figure 9.3 is judged to be a correct sentence while sentence B is incorrect (it is the syntactic equivalent of an optical illusion). Interestingly, sentence A contains words (i.e., *spinelli* and *morphenium*) that have been *made up* by Devin. Thus, sentence A is complete nonsense but we judge it as a feasible sentence. Pinker and Devlin marvel at our ability to manipulate language symbols and to perform logic (even though there are well-known cognitive problems with human reasoning [18-20]). A basic question addressed later in this chapter is whether our language capabilities can be deliberately exploited to improve situation assessment.

9.3 INDIVIDUAL DIFFERENCES IN INFORMATION PROCESSING

In addition to research on the differences between language and visual intelligence, extensive work on individual differences in human use of computers has been motivated by the need to improve computer-based training, computer-assisted group decision making, and computer-assisted information retrieval.

M. J. Hall [21, 22] and S. A. Hall [23] investigated the effects of individual differences on the efficacy of human-computer interfaces for computer-based training (CBT). In a research project conducted under a Phase II SBIR project, M. J. Hall developed a CBT module that allowed training material to be presented emphasizing either a visual, aural, or kinesthetic presentation style. This project is discussed in Section 9.7. Experiments conducted with Penn State University ROTC students and USN trainees at Dam Neck indicated that different learning information presentation styles affected the ability of participants to acquire and retain information. Psychometric instruments were used to determine participants' learning preference modes (namely whether they preferred to receive information visually, aurally, or kinesthetically). When participants were presented information in a mode that matched their preferred mode of information access, they learned better and retained information better than if the information style was mismatched. Another factor that overshadowed the information access style was whether or not participants were oriented as group learners or individual learners. If participants were determined to be group learners (as measured by a psychometric instrument) then they had a very difficult time accessing information via a computer, regardless of how the information was presented. These results were confirmed and analyzed in more detail by S. A. Hall [23].

Aalbersberg [24] and Spink and Saracevic [25] have investigated the effects of individual differences (including gender differences) and preferences in information retrieval (IR), especially from the Internet. They are developing cognitive models to assist the understanding of IR. McNeese [26] has focused on investigations of HCI for group decision-making. Allen et al. [27] present an overview of conversational human-computer interactions. Finally, Wang, Li and Wiederhold [28] have developed some advanced techniques for linking graphics/pictures and textual representations of data.

It is anticipated that rapid advances in HCI will continue to be made in the commercial world. These advances are motivated by the competition for improved access to the Internet and for data mining. These advances should be closely monitored and leveraged for improved data fusion systems.

9.4 ENABLING HCI TECHNOLOGIES

As shown on the right-hand side of Figure 9.4, the interface between a human operator/user and a complex system is implemented by a computer system that may include display devices, audio capability, and haptic interface devices. This section provides a summary of the state of the art of enabling HCI technologies. Table 9.1 provides a summary of this assessment. A detailed discussion is provided below. In this section the sense of smell and taste will be ignored. While there is ongoing research to develop sensor-processor systems to emulate the sense of smell, these are currently of limited value for human-computer interfaces for data fusion applications.

Human User   **Complex System**

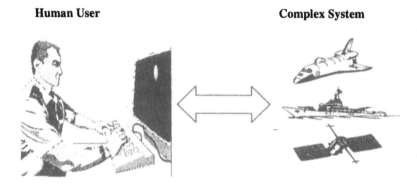

Cognitive Factors:

* Individual Characteristics
* Decision Style
* Demographics
* Training

Figure 9.4 Concept of human-in-the-loop control of a complex system.

9.4.1   Visual and Graphical Interfaces

Perhaps the most spectacular part of a human-computer interface is the visual interaction. Humans tend to be visually oriented and are attracted to computer interaction by seeing a visual image or graphic icon. Current computer interfaces are dominated by computer terminals with color screens that range in size up to 25 inches (the diagonal measurement across the face of the screen). Current visual displays often use special icons and window-based textual templates (e.g., pull-down or pop-up menus). These displays, which are two-dimensional, feature images such as geographical information system (GIS) displays showing maps, terrain data, political information, roads, and other information. Other examples of common displays include weather information and environmental information (e.g., crop yields, location of environmental resources, and information on the spread of contaminants). An excellent general discussion of how to create figures and charts to convey information is provided by Tufte [29].

Two broad forces drive display technology. First, the rapid evolution and competition for attention on the Internet provides a motivation to create eye-catching displays so that people browsing the Web are motivated to linger at a commercial site. Examples of these displays include moving icons, graphics, images, and cartoon figures. Another commercial driver involves the design of computer games [16] Laird and Lent describe the rapid evolution of games to provide increasing realism. Display effects include two- and (simulated) three-dimensional images, dynamic effects such as changing lighting, damage to walls, control of game characters, and recognition of visual physics to emulate a real world. Games include action games such as *Doom*^{TM}, *Soldier of Fortune*^{TM}, *Half-life*^{TM}, *Duke Nuke'em*^{TM}, and *Tomb Raider*^{TM}. Other types of games involve a simulation of sports such as football, soccer, and basketball. Finally, there are games that allow a user to create a miniature world (e.g., *Simcity*^{TM} and *The Sims*^{TM}).

Table 9.1

Summary of HCI Enabling Technologies

HCI Component	Description	Current Technology	Technology Trends
Visual displays	Devices for display of information to a human operator. Examples include alphanumeric displays, icons; and graphic devices; devices may include optical devices to monitor the eye motion of a user to determine *where the user is looking*	• Large, high-fidelity graphic screens • High density television • High-end 3-D devices such as the CAVE^{TM}, Immersadesk^{TM}, and Infinity Wall^{TM} • Personal displays associated with wearable computers	• Proliferation of 3-D full immersion environments (e.g., for collaborative data analysis) • Drive towards increased reality via gaming industry • Movement towards interactive versus passive displays (e.g., interactive TV, etc.)
Aural interaction	Software and hardware to provide either aural feedback (speech synthesis or special tones) to a user; also, mechanisms for voice recognition and natural language processing	• COTS voice recognition software • Web search engines using pseudo natural language (NLP) • Boolean keyword searches and limited thesaurus-based searches	• Increased interactivity (voice recognition and voice feedback) • Emulation of human facial expressions and voice inflections • Transition from keyword search to

Table 9.1 (continued)

Summary of HCI Enabling Technologies

HCI Component	Description	Current Technology	Technology Trends
		• Some syntactic analysis	semantic-based tools • 3-D sound
Haptic devices	These are devices that allow a user to provide commands to a computer system (e.g., via mouse, joystick, touch screen or other devices to touch or move); in addition it includes devices that provide physical feedback to a user to simulate touch; at the extreme end this could include emulation/simulation of whole body motion	• Standard mechanical interfaces (joy stick, mouse, touch screen) • Experimental haptic devices with emulation of touch • Limited devices for computer access by physically impaired users	• Increased sensitivity and feedback • Trend towards full-body haptic interfaces • Link between visual interface, acoustic, and haptic • Wireless interfaces to computers

New full-immersion displays provide the illusion of three dimensions (literally being immersed in a display). An example is the CAVE™, Immersadesk™, and Infinity Wall™ commercial equipment. These products are expensive (the computing hardware and software can cost up to $1 million) and require extensive space. A typical full immersion facility may require a 30' X 30' room (to house the optics for a 10' x 10' visualization room along with the associated computing facilities). However, these devices are increasingly used for applications including analysis of seismic data by oil companies to search for oil deposits, scientific visualization for aerodynamic and hydrodynamic modeling, and visualization of environmental data. The CAVE™ system allows a user to walk into a 10' x 10' room with translucent walls. Graphical images are projected onto the walls, floor, and ceiling of the room. A user wears special glasses that alternate his or her vision from one eye to the other at a rate of 90 Hz. The user can see three-dimensional objects, walk through and around objects, and control the graphic images (e.g., he or she can scale the image from being inside a display to a *gods-eye* view). Information about the CAVE can be found at http://www.ncsa.uiuc.edu/VEG/ncsaCAVE.html.

An example of a full-immersion display is shown conceptually in Figure 9.5. Similarly, the Infinity Wall™ provides a one-wall display that gives a 3-D visual illusion, while the Immersadesk™ allows a user sitting in front of a terminal to see a 3-D illusion. Recently, the Elumen Company has developed desktop system that allows immersive visualization experiences. This system, depicted in Figure 9.6, provides a full-color, immersive display, with 360-degree projection and a 180-degree field of view. The hemispherical screen is positioned vertically to fill the field of view of the user, creating a sense of immersion. The observer loses the normal depth cues, such as edges, and perceives 3-D objects beyond the surface of the screen.

Figure 9.5 A full-immersion 3-D human computer interface environment at Penn State.

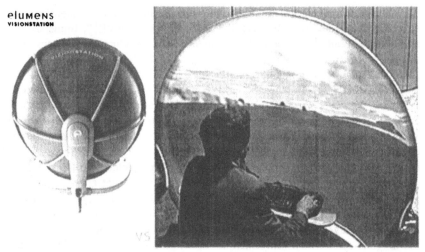

Figure 9.6 Elumens' VisionStation®.

9.4.2   Aural Interfaces and Natural Language Processing (NLP)

The second area of HCI involves providing capabilities for a computer to understand spoken commands and respond to a user's voice input. This involves two issues: (1) how to provide for voice recognition and (2) how to make a computer understand *natural language*. The first area is relatively straightforward. It is necessary to convert voice sounds into a digital signal, and then perform recognition of the digitized voice to recognize words and phrases. Commercial-off-the-shelf (COTS) software products such as the IBM product ViaVoice™ dominate current technology. This software is relatively robust, but must be trained for each individual speaker. A user speaks a number of words into a microphone, and the software attempts to convert these spoken words into text on the computer. The user can subsequently correct the converted spoken words-to-text as it appears on the screen. These speech recognition tools are becoming increasingly sophisticated and with training can recognize over 90% of spoken words. However, some dialects are particularly difficult for these types of programs to recognize accurately. In addition, the English language can be ambiguous. For example, many words, called homophones, sound similar but are spelled very differently (e.g., two, to, too; merry, marry, Mary; and Terry, tarry). These must be interpreted syntactically and semantically in the context of different phrases. A review of conversational computer interface technology is provided by Zue and Glass [30].

Researchers at the Oregon Graduate Research Institute (OGRI) have developed the concept of a *talking head* interface. A dynamic graphic image of a human face is used to augment an aural interface. When text input is converted into synthesized speech, the *talking head* graphic mimics the facial expressions and movements of a speaker's mouth. An issue here is that understanding spoken language involves more than just converting words to text. There are pragmatic aspects of language that are difficult to define. These include gestures, facial expressions and prosodic aspects of speech that convey much of the message in spoken language. Aspects such as these are not readily programmable for these complex systems. Having said this, experiments conducted by the Oregon Graduate Research Institute suggest that the use of a *talking head* for a computer interface may increase communication fidelity by as much as 25%. The OGRI web set at www://murray.newcastle.edu/au/users/staff/speech/home_pages/link.html provides links to Internet resources on commercial tools, on-line articles, speech databases, and links to research activities at other universities and companies.

True natural language processing involves trying to make a computer understand the meaning of spoken English. It is one thing for a computer to convert voice into text, it is quite another for a computer to understand the meaning of phrases and sentences. The evolution of NLP can be seen by the progress in Internet search engines. These are currently dominated by keyword and Boolean keyword capabilities. Some metasearch engines such as *AskJeeves* (www.askjeeves.com) provide a pseudo NLP. Perhaps the most famous NLP program was ELIZA [31]. This program emulated the interaction with a Rogerian therapist (i.e., using so-called reflective listening to interact with a user). While ELIZA was a very simple pattern matching program, it illustrated the promise of NLP interaction. A generic block diagram for a typical conversational interface (including speech recognition and language understanding) is shown in Figure 9.7.

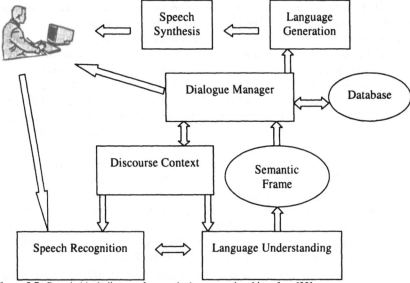

Figure 9.7 Generic block diagram for a typical conversational interface [30].

Current trends focus on development of semantic processors that understand the meaning of natural languages. This requires a significant amount of encoded real-world knowledge. Pinker [15] and Devlin [18] discuss how difficult it is to make computers understand English. Examples of semantically ambiguous English phrases include the following:

- Polish the Polish furniture.
- The dump was so full it had to refuse refuse.
- After a number of Novocain injections, my jaw got number.
- The buck does funny things when the does are present.

Humans have a very robust capability for resolving these apparent ambiguities. Current research in this area uses techniques such as an electronic thesaurus and semantic distance calculations to access general contextual information [32]. Butler [33] provides an excellent review of Web-based search engines and standards. A summary of search engines and their Internet addresses is provided in Table 9.2.

Table 9.2

Examples of Internet Search Engines

Search Engine	Internet Address
AltaVista	http://www.altavista.com
Ask Jeeves	http://www.askjeeves.com
Direct Hit	http://www.directhit.com
Excite	http://www.excite.com
FAST Search	http://www.alltheweb.com
Go (Infoseek)	http://www.go.com
Lycos	http://www.lycos.com
HotBot	http://www.hotbot.com
Inktomi	http://www.inktomi.com/products/portal/search/tryit.html
Northern Light	http://www.northernlight.com
Google	http://www.google.com
IBM Clever Project	http://www.almaden.ibm.com/cs/k53/clever.html
Research Index	http://www.researchindex.com
Sherlock Hound	http://www.sherlockhound.com
Autonomy	http://www.autonomy.com
Kenjin	http://www.kenjin.com

9.4.3   Haptic Interfaces

Haptic interfaces involve nonvisual and nonaural interfaces to computers. Examples include the common keyboard, joysticks, mouse, touch screens, wands, and other interfaces that involve the use of human motion or touch. We commonly think of haptic interfaces used to control a computer. This involves a device for the human to provide input to the computer (e.g., moving a cursor to a place on a screen). However, haptic interfaces can also provide feedback and information to the human user.

Haptic devices are being developed to allow a user to simulate a sense of touch. For example, research funded by the Defense Advanced Research Projects Agency (DARPA) has sought to develop realistic interfaces for remote surgery (i.e., to allow a surgeon to manipulate robotic hands and receive sensitive feedback necessary for surgery at a distance). Other examples of haptic interfaces include the early use of devices by the nuclear industry to provide a user with *remote hands* to manipulate radioactive material.

A key issue with haptic interfaces is how the feedback *feels*. In some cases, such as surgery, the haptic design seeks to provide the most realistic feedback possible to emulate touch. In other cases, such as *fly-by-wire* applications, the designer creates an artificial feedback that allows a user to control a complex device with a simple touch. Ellis, Ismaeil, and Lipsett [11] discuss the design of haptic interfaces. In addition, researchers at the Stanford Dexterous Manipulation Laboratory are performing extensive research in tactile sensing, haptic exploration, and dexterous telemanipulation (see www-cdr.stanford.edu/Touch/touchpage.html).

9.4.4   Gesture Recognition

Perhaps the most recent type of human computer interaction involves systems that recognize human gestures. Such systems are especially useful for team-based interaction and decision-making. Cai [34] and his colleagues have developed a combined speech and gesture recognition system for interaction with geographical information systems (see Figure 9.8). In this system, multiple users can speak and point at a computer screen to retrieve data, change displays, provide annotations, and issue commands for the computer system. Sharma et al. [35] have developed a speech-gesture-driven multimodal interface for crisis management.

The use of gestures is intended to augment human speech communication. In observing human-to-human communications, gestures are seen to play a significant role in improving the understanding of speech. Examples of gesture motion include movement of the body and limbs, facial expressions, positioning of the body (relative to the communication recipient), and related factors. Creation of a gesture recognition system is challenging. It typically requires processing of video data and sequences, tracking of individual humans and their limb motion, and development of a parameterized visual model of the gesturer. The use of haptic devices such as pen and touch systems or cybergloves simplifies the gesture recognition. However, these devices are restrictive and less "natural" than the ability of a system to visually track the user. Sharma et al. [35] provide a detailed discussion of speech-gesture-driven systems and develop an architecture and prototype system for use with geographical information systems.

Figure 9.8 An example of a gesture recognition interface for a geographical digital library [34].

9.4.5   Wearable Computers

The final enabling technology for advanced interfaces is wearable computers. The rapid advances in micro processing and HCI devices have enabled the emergence of wearable computers [36]. Wearable computers involve the use of very compact personal computers and portable peripheral devices to allow a user to actually wear the computer and use it as a portable aid during job functions. A near-term DoD application of wearable computers involves use for maintenance technicians. This allows the technician to "carry along" an interactive maintenance manual and to create "augmented reality" in which the technician can be shown a diagram or annotations projected as he is looking at a mechanical component.

Commercial wearable computers include Xybernaut's Mobile Assistant V, VIA II's wearable personal computer, and new systems being developed by IBM. These computers use processors such as Intel's low-power 600-Mhz Pentium III and Transmeta Corporation's 5400 Crusoe processor running up to 700 Mhz. These processors are coupled with new interfaces such as the Twiddler chorded keyboard that allows one-handed keyboard interface and video display using a MicroOptical Corporation device (shown in Figures 9.9 and 9.10, respectively).

Clearly, the technology for human-computer interfaces has made major strides. The continual evolution of lower power, faster, increased-memory micro-processors should enable remarkable changes in how we view computers. However, it remains to be seen how this technology will be used and adapted for use in everyday jobs and life.

Figure 9.9   The Twiddler chorded keyboard [36].

Figure 9.10   Georgia Institute of Technology researcher using a portable video display [36].

9.5   COMPUTER-AIDED SITUATION ASSESSMENT

9.5.1   Computer-Aided Cognition

The previous section provided a brief discussion of the technology for the development of human-computer interfaces. There has been significant progress in computer displays, natural language interfaces, and haptic devices. As previously noted, the commercial gaming industry and so-called e-commerce have driven rapid advances.

By contrast this progress is not as readily seen in understanding the human part of the human-computer interface (the left side of Figure 9.4). While much work has been performed in the area of ergonomics (see for example, [37]), less research has been performed in cognitive-based HCI. Research in cognitive science has focused on two broad areas:

1.   Understanding of the physiology and biochemistry of the brain;
2.   Research in understanding thought processes and perception.

While it is beyond the scope of this chapter to summarize progress in understanding the physiology and biochemistry of the brain, there are some excellent texts that provide summaries of recent findings. These include [38-41].

In the area of cognitive function, research has focused on understanding cognition and perception. This is a well-established field of study, yet one that remains very challenging. Davis [42] summarizes efforts of mathematicians and philosophers from Aristotle to Leibnitz, Boole, Frege, Cantor, Hilbert, Godel, and Turing to understand logic and automated reasoning. Evans et al. [19] discuss the process and limitations of reasoning. Pinker [15] provides a discussion of the evolution of human cognitive abilities such as language, reason, and visual perception/interpretation. Limitations in human cognition are described by Piattelli-Palmarini [20]. The impact of these limitations for military intelligence analysis is summarized by Heuer [13]. Finally, Claxton [43] discusses alternatives to logic in the cognition process. In addition, Gardner [44] has written several books on our evolving understanding of intelligence and cognition. General thoughts on the evolution of HCI are presented by [39, 45-49].

While much research has been performed on reasoning, logic, and perception, these concepts have not made significant changes or improvements in computer-based systems for learning or control of complex systems. In the multisensor data fusion community, Waltz and Llinas [50] noted that the overall effectiveness of a data fusion system (from sensing to decisions) is affected by the efficacy of the HCI. Llinas and his colleagues [51] investigated the effects of human trust in aided adversarial decision support systems, and [52] identified computer aided cognition as a key research need for data fusion.

9.5.2   Utilization of Language Constructs

As previously noted, the standard process for assimilation of data to perform situation assessment is to transform image or signal data into information overlaid on a geographical display. Because this approach is so common, it is easy to forget that the data are not directly displayed, but rather are preprocessed via a hierarchical series of transformations. For example, an infrared or visual image from a sensor is actually a two-dimensional matrix of pixel values that are converted into a display overlaid on a situation map. Similarly, multisensor directional information (e.g., lines of bearing or angular information) about a target may be converted into coordinates on a situation map. The uncertainty of the target location (represented in the original data stream by a state-vector covariance matrix) is often converted into an error ellipse for display on the map. Finally, target attribute information such as size, shape, and signal characteristics may be converted into a declaration of target identity and represented by an icon displayed on a map.

The use of graphic displays allows analysts to use their visual cognition to rapidly assess the portrayed situation. Our human vision and visual reasoning provide a powerful method of understanding the situation. Humans are capable of recognizing patterns, identifying geographic proximity of targets, and looking for relationships between target entities and man-made entities such as bridges, roads, buildings, and airstrips. Unfortunately, even well-intentioned displays such as the use of error ellipse to represent uncertainty can be deceiving. Perceptually, an analyst may view these as a *bullseye* that surrounds the *exact* position of a target or entity. It is difficult for a person *not* to see the centroid as the precise location of the target or entity, rather than as a representation of uncertainty. Similarly the use of icons to represent target identity may obscure the uncertainty of that assignment or declaration.

Interestingly, a similar type of approach is used for a wide variety of applications such as control of complex machinery (e.g., nuclear power plants or spacecraft ground control systems) and for medical applications such as monitoring the health of humans in an operating room. Woods [53, 54] has performed research on human-computer interactions for these types of applications. His research suggests that humans mentally transform data displays into verbal descriptions (e.g., transforming sensor changes into adjective descriptions such as *increasing temperature*, becoming hotter or excessive vibration). Based on these transformations, the human analyst can perform mental reasoning regarding how to react to the display. Woods suggests that these typical system data displays may be relatively ineffective because they require a continual mental transformation process.

In order to enhance the analysis process, we suggest that this hierarchical transformation should be augmented by a textual description that is developed as part of the hierarchical transformation process. Instead of having an analyst look at graphic displays and mentally transform the displayed data into semantic concepts for internal reasoning, why not assist by explicitly performing these transformations to assist the analyst. This concept is illustrated in Figure 9.11. This approach would allow an analyst to use both his or her visual reasoning and language reasoning skills. Analysts can employ sophisticated language reasoning skills to assist in the analysis and partially mitigate the cognitive problems with visual reasoning. In addition, this approach allows analysts to use multiple modes of accessing information to match their inherent information access preferences. The SBIR experiment described in Section 9.7 suggests that differences in visual versus aural presentation of information can make a statistically significant difference in the ability of users to learn and retain information provided via a computer.

The intent is not to replace graphic displays and the corresponding transformation between signal and image data to graphic displays and overlays. Instead, the graphical approach should be augmented by a parallel language transformation. Figure 9.11 shows a typical transformation from image and signal data to identity information via feature extraction and pattern recognition techniques. Figure 9.11 shows the parallel transformation using standard image/signal processing techniques and language constructs.

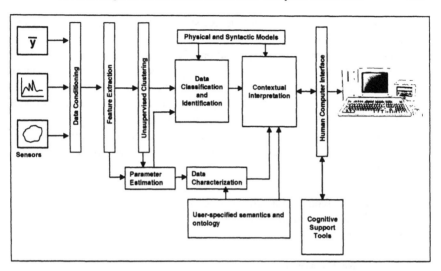

Figure 9.11   Parallel transformation of data.

The transformations between data and language constructs include the following.

- *Image/signal processing*: Application of image and signal processing algorithms to prepare the data for extraction of features and parameter estimation.
- *Feature extraction methods*: Representation of image and signal data using a vector format (e.g., using time-domain, frequency-domain, or time-frequency domain methods).
- *Estimation techniques:* Use of statistical methods for refinement of parameters (e.g., target characterization and attribute estimation or tracking).
- *Pattern recognition techniques*: use of methods such as cluster analysis or neural networks to translate feature vectors to semantic labels or identity declarations.
- *Characterization functions*: Translation of feature vector data, state vector data, or covariance matrix information to characterization labels (e.g., "exceeds threshold level," "increasing," "decreasing," and "low probability") These methods include fuzzification and de-fuzzification techniques, parametric and logical templating, and use of ontological transformations. Other techniques include the use of semantic distance calculations and electronic thesauruses.
- *Decision-level fusion:* Decision-level fusion algorithms such as Bayesian inference, voting, Dempster-Shafer's method, and associative memory

functions can be used to combine decision-level information (e.g., declarations of target identity).

- *Contextual interpretation*: Application of automated reasoning methods for textual data and signal/image data to interpret the information in the context of an evolving situation. Types of methods include hybrid reasoning methods (e.g., neural networks modeling explicit and implicit information), syntactic models, model-based reasoning, and intelligent agents.
- *Cognitive support tools:* Tools can support an analyst decision process including intelligent search agents, interpretive models, analogy tools, and explanation functions.

Together these types of functions can be applied to transform sensor data (and background information from models or databases) to assist data interpretation and understanding.

9.5.3   Areas for Research

Ultimately, the results of this dynamic process are displayed for a human user or analyst. We note that this description of the data fusion process has been greatly simplified for conceptual purposes. Actual data fusion processing is much more complicated and involves an interleaving of the level 1-level 3 (and level 4) processes. Nevertheless, this basic orientation is often used in developing data fusion systems—namely the sensors are viewed as the information source and the human is viewed as the information user or sink. In one sense, the rich information from the sensors (e.g., the radio-frequency time series, imagery) is compressed for display on a small two-dimensional computer screen.

Bram Ferran, the vice president of R & D Disney Imagineering Company, has pointed out that this approach is a problem for the intelligence community [55]. Ferran argues that the broadband sensor data are funneled through a very narrow channel (namely the computer screen on a typical workstation) to be processed by a broadband human analyst. In his view, the HCI becomes a bottleneck or very narrow filter that prohibits the analyst from using his or her extensive pattern recognition and analytical capability. Ferran suggests that the computer bottleneck effectively defeats one million years of evolution designed to make humans' excellent data gatherers and processors. Interestingly, Stoll [56, 57] makes a similar argument about personal computers and the multimedia misnomer.

We believe that there is extensive research required in this area. Instead of allowing the HCI for a data fusion system to be driven by the *latest and greatest* display technology, research should be performed to develop adaptive interfaces to encourage human-centered data fusion. This would break the HCI bottleneck (especially for nonvisual, group-oriented individuals) and leverage the human cognitive abilities for wideband data access and processing. That was the basis for recommending that this area be explicitly recognized by creation of a level 5 data fusion process in the data fusion process model.

Conceptually, HCI processing functions should be augmented by functions to provide a cognitive-based interface. What are the functions to be included in the new level 5 process? The following are examples of new types of algorithms and functions for level 5 processing.

- Deliberate synthesia: Synthesia is a neurological disorder in humans in which the senses are cross-wired [58]. For example, one might associate a particular taste with the color red. Typically, this disorder is associated with schizophrenia or drug abuse. However, such a concept might be deliberately exploited for normal humans to translate visual information into other types of representations such as sounds (including direction of the sound) or haptic cues. Algorithms could be implemented to perform sensory cross-translation to improve understanding.

- Time compression/expansion: Human senses are especially oriented towards change detection. Development of time compression and time expansion replay techniques could assist the understanding of an evolving tactical situation.

- Negative reasoning enhancement: For many types of diagnosis such as mechanical fault diagnosis or medical pathology, experts explicitly rely upon negative reasoning [59]. This approach explicitly considers what information is not present which would confirm or refute a hypothesis. Unfortunately, however, humans have a tendency to ignore negative information and only seek information that confirms a hypothesis (see Piattelli-Palmarini's [20] description of the three-card problem). Negative reasoning techniques could be developed to overcome the tendency to seek confirmatory evidence. Note that there are two senses of negative reasoning that should be developed: (1) negative reasoning in the sense of "absence of evidence," and (2) negative reasoning in the sense of "counter-evidence."

- Focus/de-focus of attention: Methods could be developed to systematically assist in directing the attention of an analyst to consider different aspects of data. In addition, methods might be developed to allow a user to defocus his or her attention in order to comprehend a broader picture. This is analogous to how experienced Aikido masters deliberately blur their vision in order to avoid distraction by an opponent's feints [60].

- Pattern morphing methods: Methods could be developed to translate patterns of data into forms that are more amenable for human interpretation (e.g., the use of Chernoff faces to represent varying conditions or use of Gabor-type transformations to leverage our natural vision process [61]).

- Cognitive aids: Numerous cognitive aids could be developed to assist human understanding and exploitation of data. Experiments should be conducted along the lines initiated by [12].
- Uncertainty representation: Finally, visual, auditory, and haptic techniques could be developed to improve the representation of uncertainty. An example would be the use of three-dimensional icons to represent the identity of a target. Blurring or transparency of the icon could represent the uncertainty in the identification. For location uncertainty blurred error ellipses may be useful. In addition, second-order information (i.e., uncertainty of the uncertainty estimate) could be represented by a tactile interface; representing second-order uncertainty by the softness of the error ellipsoid.

These areas only touch the surface of the human-computer interface improvements. By rethinking the HCI for data fusion, we may be able to re-engage the human in the data fusion process and leverage our evolutionary heritage.

9.6 AN SBIR MULTIMODE EXPERIMENT IN COMPUTER-BASED TRAINING

The final section of this chapter describes an experiment on multimode presentation of information for enhancing computer-based training. The experiments provide insight into how users may be positively or negatively affected by the mode (e.g., visual, aural, or kinesthetic) in which information is presented, and how this is affected by individual characteristics of the user. The research has implications for implementation of multisensor data fusion systems.

9.6.1 SBIR Objective

Under a phase II small business innovative research (SBIR) effort (Contract No N00024-97-C-4172), *Tech Reach Inc.* (a small company located in State College, Pennsylvania) was tasked with designing and conducting an experiment to determine if a multimode information access approach improves learning efficacy. The basic hypothesis involved the research hypothesis that computer-assisted training, which adapts to the information access needs of individual students, significantly improves training effectiveness while reducing training time and costs. Specific objectives of the phase II effort included the following.

- Design, implement, test, and evaluate a prototype computer-based training (CBT) system that presents material in three formats (emphasizing aural, visual, and kinesthetic presentations of subject material);
- Select and test an instrument to assess a student's most effective learning mode;

- Develop an experimental design to test the hypothesis;
- Conduct a statistical analysis to affirm or refute the research hypothesis.

9.6.2   Experimental Design and Test Approach

The basic testing concept for this project is shown in Figure 9.12 and described in detail by Hall [22]. The selected sample consisted of approximately 100 Penn State ROTC students, 22 selected adult learners (namely post high school adult education students), and 120 USN enlisted personnel at the USN Atlantic Fleet Training Center at DAM NECK. This sample was selected to be representative of the population of interest to the U. S. Navy sponsor.

As shown in Figure 9.12, the testing was conducted using the following steps.

1. Initial data collection: Data were collected to characterize the students in the sample (including demographic information, a pretest of the students' knowledge of the subject matter, and a learning style assessment using standard test instruments).
2. Test group assignment: The students were randomly assigned to one of three test groups. The first group used the CBT that provided training in a mode that matched their learning preference mode (as determined by the CAPSOL learning styles inventory instrument [62]). The second group was trained using the CBT that emphasized their learning preference mode (as determined by the student's self-selection). Finally, the third group was trained using the CBT that emphasized a learning preference mode that was deliberately mismatched to the student's preferred mode (e.g., utilization of aural emphasis for a student whose learning preference is known to be visual).
3. CBT training: Each student was trained on the subject matter using the interactive computer-based training module (utilizing one of the three information presentation modes: visual, aural, or kinesthetic).
4. Post testing: Post testing was conducted to determine how well the students mastered the training material. Three post tests were conducted: (1) an immediate post test after completion of the training material, (2) an identical comprehension test administered one hour after the training session, and (3) an identical comprehensive test administered one week after the initial training session.

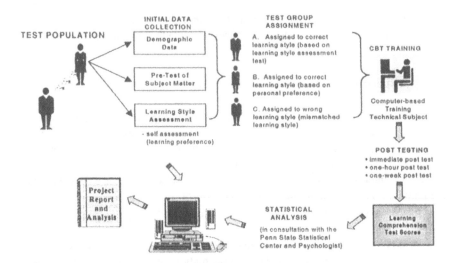

Figure 9.12 Overview of test concept.

The test subjects were provided with a written explanation of the object of the experiment and its value to the DoD. Test conditions were controlled to minimize extraneous variations such as the use of different rooms (e.g., for pre-tests and post tests), different time of day for learning versus the one-week post test, and different instructions provided to test subjects.

9.6.3   CBT Implementation

The computer-based training (CBT) module for this experiment was a training module describing the functions and use of an oscilloscope. The training module was implemented using interactive multimedia software for operation on personal computers. The commercial authoring shell, *Toolbook,* developed by Asymetrix Corporation was used for the implementation. DoD standards were followed for the design and implementation of the CBT module. An example of the CBT display screens is shown in Figure 9.13.

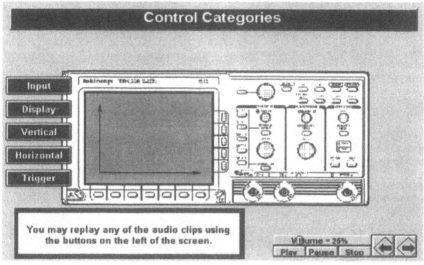

Figure 9.13(a) Example of an aural CBT display screen.

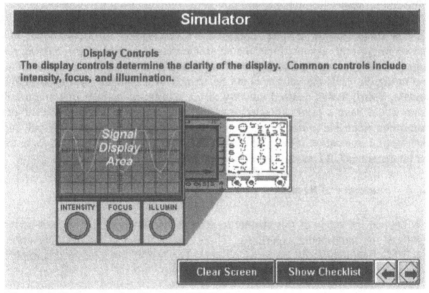

Figure 9.13(b) Example of an kinesthetic CBT display screen.

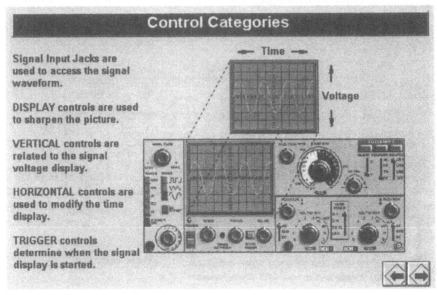

Figure 9.13(c) Example of a visual CBT display screen.

The subject matter selected, operation and functions of an oscilloscope, was chosen for several reasons. First, the subject matter is typical of the training requirements for military personnel involved in equipment operation, maintenance, and repair. Second, the subject could be trained in a coherent, yet small, CBT module. Third, it was deemed unlikely that a significant number of test participants would have a priori knowledge of the subject matter. Finally, the subject matter was amenable to implementation with varied emphasis on aural, visual, and kinesthetic presentation styles. All of the CBT screens, aural scripts, and logic for the implemented CBT modules are provided in [63].

9.6.4   Summary of Results

It is beyond the scope of this chapter to present the details of the analysis or the results of the multimedia experiment. These can be found in [64, 65]. Nevertheless a brief summary will be provided here. The results obtained in this experiment support several conclusions.

- The way (or mode) information is presented has an effect on how well that information is learned and retained.
- Learning and retention are enhanced when the information is presented in a mode that matches the learning style (or information access mode) preferred by the individual learner.

- Individuals who are group-oriented seem to have difficulty learning and retaining information in the computer-based training format regardless of the mode in which the information is presented.

While the results are statistically significant, there are a number of issues with the experiment and associated data. These are described below.

A surprising result from the original SBIR study was the factor of individual versus group learning. This factor was identified as a potential factor that might overshadow the effect of matching or mismatching information preference style. There is evidence to support the conclusion that there is a significant difference between individual and group learners. It is interesting to note that the group learners had more a priori knowledge than shown by individual learners. Despite this, the group learners responded poorly to the CBT training module. The CBT actually provided a very small increase in knowledge for group learners (namely the post-test average results were marginally better than the average pretest results). This was a very small sample. However, if verified by a more extensive sample, it could have serious implications for computer-based training and HCI for group-oriented people.

The concept of group versus individual learning style preference is intriguing. Why does a CBT training module challenge an individual with a group learning orientation? What factors affect this learning problem? If the CBT could be personalized (e.g., using aural information provided by a *friendly* voice) would this mitigate the group-learning effect? Do group-oriented people have other common characteristics?

9.6.5   Implications for Data Fusion Systems

We note that this was a very basic experiment using a homogeneous, highly motivated group of ROTC students. All of these students were highly computer literate. While preliminary, the results indicate that factors such as group versus individual learning style can significantly affect the ability of an individual to comprehend and obtain information from a computer. This suggests that efforts to create increasingly sophisticated computer displays may have little or no effect on the ability of some users to understand and use the presented data. Many other factors such as user stress, the user's trust in the decision-support system, and preferences for information access style, also affect the efficacy of the human-computer interface.

REFERENCES

[1]   Hall, M. J., S. A. Hall, and T. Tate, "Removing the HCI Bottleneck: How the Human Computer Interface (HCI) Affects the Performance of Data Fusion Systems," *2000 MSS National Symposium on Sensor and Data Fusion*, San Diego, CA, June, 2000, pp. 89-104.

[2] Neal, J. G., and S. C. Shapiro, "Intelligent Integrated Interface Technology," *1987 Tri-Service Data Fusion Symposium*, Johns Hopkins University, Laurel, MD, June 9-11, 1987, pp. 428-436.

[3] Nelson, J. B., "Rapid Prototyping for Intelligence Analyst Interfaces," *1989 Tri-Service Data Fusion Symposium*, Johns Hopkins University Applied Physics Laboratory, Laurel, MD, 1989, pp. 329-334.

[4] Marchak, F. M., and D. A. Whitney, "Rapid Prototyping in the Design of an Integrated Sonar Processing Workstation," *1991 Joint Service Data Fusion Symposium*, Johns Hopkins University Applied Research Laboratory, Laurel, MD, October, 1991, pp. 606-627.

[5] Pagel, K., "Lessons Learned from HYPERION, JNIDS Hypermedia Authoring Project," *1993 Sixth Joint Service Data Fusion Symposium*, Johns Hopkins University, Applied Physics Laboratory, Laurel, MD, 1993, June 14-18, 1993, pp. 555-564.

[6] Clifton, T. E., III, "ENVOY: An Analysts' Tool for Multiple Heterogeneous Data Source Access," *1993 Sixth Joint Service Data Fusion Symposium*, Johns Hopkins University Applied Physics Laboratory, Laurel, MD, June 14-18, 1993, pp. 565-570.

[7] Hall, D., and J. H. Wise, "The Use of Multimedia Technology for Multisensor Data Fusion Training," *1993 Sixth Joint Service Data Fusion Symposium*, Johns Hopkins University Applied Physics Laboratory, Laurel, MD, June 14-18, 1993, pp. 234-258.

[8] Kerr, R. K., et al., "TEIA: Tactical Environmental Information Agent," *Intelligent Ships Symposium II*, Philadelphia, PA, November 25-26, 1996, pp. 173-186.

[9] Brendle, B. E., Jr., "Crewman's Associate: Interfacing to the Digitized Battlefield," *SPIE: Digitization of the Battlefield II*, Orlando, FL, April 22-24, 1997, pp. 195-202.

[10] Steele, A., B. Marzen, and B. Corona, "Army Research Laboratory Advanced Displays and Interactive Displays - Fedlab Technology Transitions," *SPIE Digitization of the Battlespace IV*, Orlando, FL, 1999, April 7-8, 1999, pp. 205-212.

[11] Ellis, R. E., O. M. Ismaeil, and M. Lipsett, "Design and Evolution of a High-performance Haptic Interface," *Robotica*, vol. 14, 1996, pp. 321-327.

[12] Wohl, J., et al., "Human Cognitive Performance in ASW Operators," *1987 Tri-Service Data Fusion Symposium*, Johns Hopkins University Applied Physics Laboratory, Laurel, MD, 1987, June, 1987, pp. 465-479.

[13] Heuer, R. J., *Psychology of Intelligence Analysts*, Washington, DC: Central Intelligence Agency, 1999.

[14] Hall, D., "Beyond Level N Fusion: Semantic Based Fusion," *National Symposium on Sensor and Data Fusion*, San Diego, CA, 2002, June, 2002.

[15] Pinker, S., *How the Mind Works*, London: Penguin Books, 1997.

[16] Laird, and V. Lent, *Tutorial on Computer Gaming*, Washington, DC: Outrageous Entertainment, Inc., 2000.

[17] Pinker, S., *The Language Instinct: How the Mind Creates Language*, Harper Perennial, 2000.

[18] Devlin, K., *Goodby Descartes: The End of Logic and the Search for a New Cosmology of Mind*, New York, NY: John Wiley & Sons, 1997.

[19] Evans, J. S. B. T., S. E. Newstead, and R. M. Byrne, *Human Reasoning: The Psychology of Deduction*, Hillsdale: Lawrence Erlbaum Associates, 1993.

[20] Piattelli-Palmarini, M., *Inevitable Illusions: How Mistakes of Reason Rule over Minds*, New York, NY: John Wiley & Sons, 1994.

[21] Hall, M. J., "A Multi-mode-informational Approach to Improved Computer-based Training," *1996 Intelligent Ships Symposium*, Philadelphia, PA, 1996.

[22] Hall, M. J., *Adaptive Human Computer Interface (HCI) for Improved Learning in Electronic Classrooms: Investigation of the Effects of Adaptive Information Access Modes on Computer-Based Training*, State College, PA: Tech Reach, Inc., 1998.

[23] Hall, S. A., *An Investigation of Factors that Affect the Efficacy of Human-Computer Interaction for Military Applications*, in Aeronautical Science Department: Embry-Riddle University, 2000.

[24] Aalbersberg, I. J., "Incremental Relevance Feedback," *15th Annual International SIGIR Conference*, 1992.

[25] Spink, A., and T. Saracevic, "Interactive Information Retrieval: Sources and Effectiveness of Search Terms during Mediated On-Line Searching," *Journal of the Society for Information Science*, vol. 48, 1997, pp. 741-761.

[26] McNeese, M. D., "Socio-cognitive Factors in the Acquisition and Transfer of Knowledge," *Cognition, Technology and Work*, vol. 2, 2000, pp. 164-177.

[27] Allen, J., et al., "Towards Conversational Human-computer Interaction," *AI Magazine*, 2002.

[28] Wang, J. Z., J. Li, and G. Wiederhold, "SIMPLicity: Semantics-sensitive Integrated Matching for Picture Libraries," *IEEE Transactions on PAMI*, vol. 23, 2000, pp. 1-21.

[29] Tufte, E. R., *The Visual Display of Quantitative Information*, 2nd ed., Graphics Press, 2001.

[30] Zue, V. W., and J. R. Glass, "Conversational Interfaces: Advances and Challenges," *Proceedings of the IEEE*, vol. 88, 2000, pp. 1166-1180.

[31] Weizenbaum, J., *Computer Power and Human Reason II*, W. H. Freeman and Co., 1997.

[32] Byrne, C. C., and B. McCracken, "An Adaptive Thesaurus Employing Semantic Distance, Relational Inheritance and Nominal Compound for Linguistic Support of Information Retrieval," *Jounal of Information Science*, vol. 25, 1999, pp. 113-131.

[33] Butler, D., "Souped-up Search Engines," *Nature*, vol. 405, 2000, pp. 112-115.

[34] Cai, G., *GeoVIBE: A Visual and Gesture Recognition Interface for Geographical Digital Libraries*, D.L. Hall, Editor, State College, PA, 2002.

[35] Sharma, R., et al., "Speech-gesture Driven Multi-modal Interfaces for Crisis Management," *Proceedings of the IEEE*, vol. TBD, 2002.

[36] Ditlea, S., "The PC Goes Ready to Wear," *IEEE Spectrum*, 2000, pp. 35-39.

[37] Sanders, M. S., and E. J. McCormick, *Human Factors in Engineering Design*, 7th ed., New York, NY: McGraw-Hill, Inc., 1993.

[38] Austin, J. H., *Zen and the Brain: Toward an Understanding of Meditation and Consciousness*, Cambridge, MA: MIT Press, 1998.

[39] Reingold, H., *Tools for Thought: The History and Future of Mind-Expanding Technology*, 2nd ed., Cambridge, MA: MIT Press, 2000.

[40] Kotulak, R., *Inside the Brain: Revolutionary Discoveries of How the Mind Works*, Kansas City, KS: Andres McMeel Publishing, 1997.

[41] Marshall, L. H., and W. W. Magoun, *Discoveries in the Human Brain: Neuroscience Prehistory, Brain Structure, and Function*, Totowa, NJ: Humana Press, 1998.

[42] Davis, M., *The Universal Computer*, New York, NY: W. W. Norton and Co., 2000.

[43] Claxton, G., *Hare Brain, Tortoise Mind: Why Intelligence Increases when You Think Less*, Hopewell, NJ: The Ecco Press, 1997.

[44] Gardner, H., *Frames of Mind: The Theory of Multiple Intelligences*, New York, NY: Basic Books, 1983.

[45] Carroll, J. M., "Human Computer Interaction: Psychology as Science of Design," *Annual Review of Psychology*, vol. 48, 1997.

[46] Ford, K., C. Glymour, and P. Hayes, "Cognitive Prosthesis: Computers as Equalizers," *AI Magazine*, vol. 18, 1997.

[47] Lewis, M., "Designing for Human-agent Interaction," *AI Magazine*, vol. 19, 1998.

[48] Lieberman, H., "Intelligent Graphics: What's a New Paradigm," *Communications of the ACM*, vol. 39, 1996.

[49] Norman, D. A., *Things That Make Us Smart: Defending Human Attributes in the Age of the Machine*, New York, NY: Addison Wesley, 1993.

[50] Waltz, E., and J. Llinas, *Multi-sensor Data Fusion*, Norwood, MA: Artech House, Inc., 1990.

[51] Llinas, J., et al., *Studies and Analyses of Vulnerabilities in Aided Adversarial Decision-making*, Buffalo, NY: State University of New York at Buffalo, Department of Industrial Engineering, 1997.

[52] Hall, D., and J. Llinas, "An Introduction to Multisensor Data Fusion," *Proceedings of the IEEE*, vol. 85, pp. 6-23.

[53] Woods, D., *Studies Show that Some Medical Devices Are too Complex*: Ohio State University, 2001.

[54] Woods, D., *Space Mission Control as Natural Laboratory: Modeling and Supporting Collaboration in Anomaly Response and Replanning Tasks*, P.S.U.S.o.I.S.a. Technology, Editor, State College, PA, 2001.

[55] Ferran, B., U.S.G. National Reconnaissance Office, Editor, Chantilly, VA: Available on video tape from the NRO office in Chantilly, VA, 1999.

[56] Stoll, C., *High Tech Heretic: Reflections of a Computer Contrarian*, Anchor Books, 2000.

[57] Stoll, C., *Silicon Snake Oil: Second Thoughts on the Information Highway*, Anchor Books, 1996.

[58] Bailey, D., *Hideaway: Ongoing Experiments in Synthesia*, Baltimore, MD: University of Maryland, 1992.

[59] Hall, D., R. J. Hansen, and D. C. Lang, "The Negative Information Problem in Mechanical Diagnostics," *Transactions of the ASME*, vol. 119.

[60] Dobson, T., and V. Miller, *Aikido in Everyday Life: Giving in to Get Your Way*, North-Atlantic Books, 1993.

[61] Olshausen, B. A., and D. J. Field, "Vision and Coding of Natural Images," *American Scientist*, vol. 88, 2000.

[62] CAP-WARE, *CAP-WARE: Computerized Assessment Program*, Mansfield, OH: Process Associates, 1987.

[63] Hall, M. J., *Product Drawings and Associated Lists*, State College, PA: Tech Reach, Inc., 1998.

[64] Hall, M. J., *Adaptive Human Computer Interface (HCI) for Improved Learning in the Electronic Classroom*, State College, PA: Tech Reach, Inc., 1998.

[65] Hall, S. A., *An Investigation of Cognitive Factors that Impact the Efficacy of Human Interaction with Computer-based Complex Systems*, in *Aeronautical Sciences*: Embry-Riddle University 2000.

Chapter 10

Implementing Data Fusion Systems

This chapter provides insights and guidelines on the implementation of data fusion systems.

10.1 INTRODUCTION

Some issues related to the implementation of data fusion systems are addressed in this chapter. While each data fusion system constitutes a separate problem in systems engineering, there are some general remarks that can be made about requirement analysis, sensor selection, architecture selection, algorithm selection, software implementation, and test and evaluation. These are summarized below and in subsequent sections of the chapter. Additional information on implementation and systems engineering for data fusion systems can be found in [1-8].

- Requirement Analysis: The initial step for the implementation of any complex system is to perform a requirement analysis. In a typical structured system development approach, formal analyses are performed, culminating in a requirement review and associated documentation. This approach is also recommended for the development of data fusion systems. It should be remembered that data fusion is a derived requirement. Strictly speaking there is no such thing as a data fusion system. Instead there are systems designed to address specific applications or to achieve specified missions. These systems may utilize data fusion algorithms to achieve the application or mission goals. Under this perspective, the system designer must understand what the mission goals or application goals are. What targets or entities must be detected, characterized, or identified? Who are the intended users of the system and what decisions must these users make (to achieve a mission)? This approach focuses on the end user of the data fusion system rather than on the sensors. All too often, system designers treat a data fusion system as a *bucket* used to collect or process the sensor data, rather than a

345

system to support an end user. The requirement analysis must consider the effects of the observing environment, the end user, the platform (on which the sensors are located), communication constraints, and computing limitations. Key issues include the observing and decision time line and required level of specificity and accuracy. An extensive discussion of requirement analysis for data fusion system is available in [9].

- Sensor selection and analysis: The sensors associated with a data fusion system are often specified a priori. This is frequently the case for existing platforms such as aircraft or ships for which a data fusion system must be developed. However, even if the sensors are specified a priori, it is necessary to understand their performance. In general, there is no single sensor that can detect, locate, characterize, and identify the targets of interest under all circumstances (i.e., there is no perfect sensor). Indeed, the lack of a perfect sensor is one of the motivations for developing a data fusion system. A key component of the system development process is to understand how the sensors perform, both individually and in concert, to contribute to inferences sought by the data fusion system. Sensor selection and analysis must determine what can be observed, and how these observable quantities relate to the targets or inferences of interest. Steinberg [10] provides an example of this type of analysis for a tactical aircraft. The system designer should develop or utilize high-fidelity models that predict how sensors will perform in realistic environments. These models should include the effects of the target, the signal propagation environment, the location of the sensor antenna on an observing platform, and the internal sensor processing. In addition, real data should be collected and analyzed. It is important to understand and accurately model sensor performance. These estimates should be used to help weight the sensor data in the data fusion process. If the performance estimates are not accurate, then it can corrupt all of the downstream fusion processing.

- Functional allocation and decomposition: Data fusion systems often entail a distributed set of sensors and processing systems interconnected by a wideband communications system. How are the functions of data fusion to be allocated across multiple sensing and processing nodes? Bowman and Steinberg [8] describe a process for systematic definition of data fusion nodes and allocation of requirements and processes among these nodes. In many cases, each node may replicate functions performed at other nodes (e.g., to distribute the processing of data from different sources or of different types of data). Bowman and Steinberg provide guidelines for how to optimize the system design based on issues of processing capabilities, data storage, communications utilization, and other factors.

- Architecture trade-offs: In a distributed data fusion system, multiple nodes may perform a combination of sensing and processing. Examples include systems onboard a single platform (e.g., an aircraft) linked via a computer network such as Ethernet, and unattended ground sensor systems in which multiple sensors are linked using wireless communications. In these systems, processing is performed both at the sensors and at a centralized processing computer. A basic question for the system designer is to determine where in the processing flow the fusion should be performed. That is, within and across fusion nodes, where should the "fusion" occur? Conceptually, there are three basic types of architectures for data fusion: (1) centralized fusion, (2) distributed fusion, and (3) hybrid fusion. The centralized fusion architecture involves passing raw sensor data (e.g., time series, images, and vectors) or sensor-derived data from the sensors to a central fusion processing function. In this approach the raw sensor data must be associated, correlated, and fused. In the distributed fusion approach, individual sensors process their own data to result in state vector estimates for observed targets. These estimates may contain estimates of the target's position, velocity, attributes, and identities. These state estimates are fused to provide a combined estimate of target locations, identities, and characteristics. Finally, hybrid fusion architectures combine both a centralized approach and a distributed approach. This allows the system to limit communication resource utilization (using the distributed mode of operation), or to improve accuracy by processing the raw sensor data. Hybrid fusion may provide the most flexible type of architecture at the expense of added computational overhead to address the decision processes associated with operating the hybrid processing.

- Algorithm selection: Perhaps the most controversial issue in data fusion is the selection of algorithms. There are many algorithms that can be applied to the different processes within the data fusion process. For example, many different algorithms have been developed for target tracking and target identification. Even for a function as basic as data correlation, a wide variety of techniques can be applied. Llinas et al. [7] identified over 50 different techniques used for data correlation and association (although the > 50 techniques are not alternatives for performing the same function). The methods involve different assumptions concerning the nature of the input data, understanding of the uncertainty of the input data, available computing resources, and other factors. One challenge in the implementation of a data fusion system is how to select from these different techniques. Hall and Linn [4] provide general guidelines for the selection of data fusion algorithms. However, there is no prescription that provides a unique specification of which algorithms to use. Unfortunately, many proponents of specific techniques (e.g., Bayesian inference, Dempster-Shafer's method, and multiple-hypothesis tracking) argue that their method is the *only* method to be used for every application. The choice of a set of algorithms should be based on a system's engineering approach. The designer should have a clear understanding of the algorithms (including the underlying assumptions and required a priori data), the processing constraints of the fusion system, and the limitations in the observing environment.

- Database definition: A significant fraction of a data fusion system involves the database software. A survey conducted by [11, 12] indicated that over 25% of the software developed for data fusion systems involves database functions (despite the use of sophisticated commercial off-the-shelf DBMS). Database support is particularly complex for data fusion systems for two reasons. First, data required for such a system tends to be very diverse and large. Data may include images, signals, vectors, scalar data, and textual information. The information may also include representation of abstract knowledge via rules, frames or scripts. The second reason that database definition and design is challenging is the requirement to ingest data at a real-time rate to accept sensor inputs, and simultaneously support complex information queries from users who want rapid responses. These two requirements are in general conflicting. Special representation techniques may need to be developed to support their simultaneous satisfaction. Additional information on database design and management is provided by [13-16].

- HCI design: Chapter 9 provides a discussion of the role of humans in data fusion systems and the need for creative human-computer interfaces that support understanding of data and computer-aided cognition. New display devices, haptic interfaces, wearable and hand-held computers, and acoustic devices (which enable sound augmentation of displays and natural language interfaces) and even gesture recognition, provide a variety of options to create effective human-computer interfaces. The design of such interfaces must consider not only the specific applications (and corresponding appropriate types of data representations and displays) but also the individual needs of users. Interfaces are needed to act as a way of reducing the impedance between the user and the computer rather than simply treating HCI design as a way of constructing graphic displays for a computer screen. There are a several hundred books that address design of human-computer interfaces. Examples of these include: [17-19] and the classic text by Card, Moran and Newell [20].

- Software implementation: Data fusion systems are generally implemented in software. The rapid evolution of commercial off the shelf (COTS) software [5] provides the opportunity to assemble a data fusion system using primarily commercial components. However, care must be taken not to fall into the trap of using too much existing data fusion software [21]. In general these are "point solutions" and not particularly portable to other applications. The equivalent of a numerical methods library for data fusion does not exist. However, there are some tool kits that are beginning to emerge such as *Khoros* (http://www.khroral.com) and the data fusion tool kit developed at the Penn State Applied Research Laboratory [22]. It must be remembered that the bulk of the software for a data fusion system involves the infrastructure for database management, communications, human-computer interaction, sensor control, and other functions. These follow standard systems design and software development approaches.

- Test and evaluation: Finally, test and evaluation involves the traditional problems of testing and evaluating large-scale hardware and software systems. There are additional challenges for systems involving data fusion. These have been described by [6, 9, 23, 24]. Because of the complexity of data fusion applications, the community has experienced difficulty in obtaining data sets for both algorithm development and evaluation. For systems involving level 2 or level 3 processing, there are special challenges in testing because the systems are attempting to emulate human-level reasoning [24].

10.2 REQUIREMENTS ANALYSIS AND DEFINITION

The derivation or analysis of requirements for a multisensor data fusion system must begin with the recognition of a fundamental principle: *There is no such thing as a data fusion system*. Instead, there are applications to which data fusion techniques are applied. Specific applications may range from military systems such as automated target recognition and threat assessment systems, to nonmilitary applications such as medical diagnosis, monitoring of complex machinery, intelligent buildings, or systems to monitor the environment. The implication is that generating the requirements for a generic data fusion system is not particularly useful. Instead, the particular application or mission of the system must be considered as a basis for deriving requirements that drive the specific requirements for algorithms, database components, and human-computer interaction.

Figure 10.1 [25] shows that the requirements analysis process begins with an understanding of the overall mission or application requirements. What decisions or inferences are sought by the overall system? What decisions are to be made by human users? What would constitute a successful system (e.g., as measured by the ability of the system to support a human user for a specified application or mission)? It is necessary to understand the decision environment, the anticipated entities to be observed, the observation (and signal propagation environment), limitations on communications and computer resources, issues with observability, required performance goals and objectives. The process shown in Figure 10.1 provides a hierarchical requirements derivation approach that begins with the understanding of the application and mission to ultimately derive requirements for sensors, communications, processes, algorithms, display requirements, and test and evaluation.

Ideally, the system designer has the luxury of analyzing and selecting sensors and information sources. This is shown in the middle part of Figure 10.1. Realistically, however, the designer may face an environment in which the sensors are already "given" or prespecified. In any event, the designer should understand how the sensors perform and how the observables map to the entities (or activities or events). Often, this mapping will be complex. Indeed the reason that data fusion is required is due to the potentially weak link between observable parameters and entities or inferences of interest. The designer performs a survey of current sensor technology, analyzes the observational phenomenology, and information sources. The result of this process is a set of sensor performance measures that link sensors to functional requirements and an understanding of how the sensors could perform under anticipated observing conditions.

The flow-down process continues as shown in Figure 10.1. The subsystem design and analysis process is shown within the dashed frame. At this step, the designer explicitly begins the process of allocating requirements and functions to subsystems such as the sensor subsystem, the processing subsystem, and the communications subsystem. These must be considered together because the design of (and requirements for) each subsystem affect the others. In addition, the requirements levied to these subsystems have implications for the architecture. For example, if all of the sensors are required to perform target location and identification based on single sensor data, this reduces the communications requirements (i.e., it eliminates the need to pass raw data to the processing subsystem) and permits the use of a centralized architecture. The processing subsystem design entails the further selection of algorithms, the specific elements of the database and the overall fusion system architecture.

This requirements analysis process should result in well-defined and documented requirements for the sensors, communications, processing, algorithms, displays, and test and evaluation requirements. If performed in a systematic and careful manner, the analysis provides a basis for an implemented system that supports the application to meet mission needs.

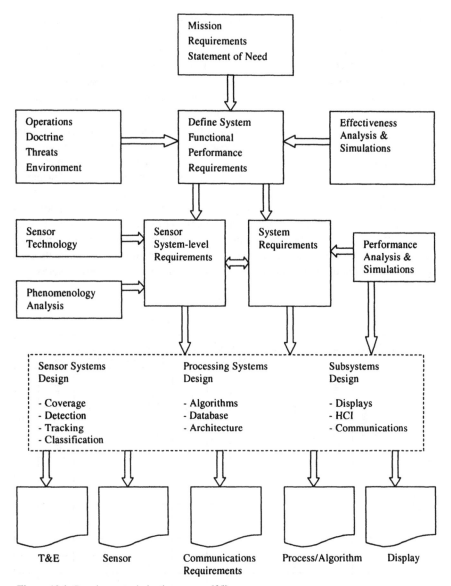

Figure 10.1 Requirements derivation process [25].

10.3 SENSOR SELECTION AND EVALUATION

Sensor data is often the primary source of input to data fusion systems. Additional inputs may include information from other data fusion nodes, prior models, human user inputs, and input from response systems such as sensor control, process control, offensive/defensive weapons (for military systems), and model refinements. Clearly, it is important both to select an optimal set of sensors or information sources to "feed" the data fusion system as well as to understand how those sensors and sources perform. In this section we will discuss the sensor selection and evaluation, but we admit that the system designer should perform analysis on all information sources for the fusion system.

Bowman and Steinberg [8] suggest that the analysis of a data fusion system should begin with an assessment of the system inputs and outputs. For sensor data, they suggest that all inputs be reviewed and analyzed to produce a matrix as shown in Table 10.1. This table provides an assessment of the sensor performance related to quality, availability, and timeliness of data (as it relates to the ability to detect, characterize, locate, and identify entities to be observed by the fusion system). The analysis needs to identify what the key entities are to be observed, what physical phenomena may be observable by given (or potential) sensors, and how these observations are impacted by real observing environments with effects such as weather, terrain, or confounding effects such as electronic countermeasures.

Table 10.1

Matrix of Data Fusion System Input Requirements

Sensors and Sources	*Quality*				*Availability*				*Timeliness*			
	Accuracy		Resolution		Coverage		Reporting		Responsiveness		Adequacy	
	Kinematic parameters	ID and Attributes	Entity separability	% reports	% area	% entities	% total reports	% sole coverage	Latency	Update rate	Situation warning	CM effect
Visible E-O												
Radar												
Infrared												
Acoustic												
...												
...												
...												

When completed, Table 10.1 assists in understanding the characteristics of the input data including the following:

- Reported continuous kinematic data types (e.g., location, altitude, course, speed, angle-only, and slant range);

- Reported parametric data types (e.g., frequency, radar cross-section, and infrared spectral characteristics);
- Reported discrete entity identification inputs (e.g., entity classification, type, and individual identification);
- Reported entity attributes (e.g., track number and scan type).

The elements of Table 10.1 are qualitative assessments that should be augmented by models of the sensor performance including error rates, predicted signal characteristics, effective range, sensitivity to the signal propagation environment and environmental effects, and effects of countermeasures.

In addition to this general analysis, the system designer should understand the "inner workings" of each individual sensor or sensor type associated with the data fusion system. In particular, extensive processing often occurs inside each sensor. In order to effectively treat the sensor input in the data fusion system, it is necessary to understand that internal processing. Figure 10.2 shows the concept of a "generic" sensor. In general, a sensor is either active or passive. An active sensor emits some form of energy (e.g., electromagnetic energy) to interact with the target (e.g., reflect off the target), and observes the response. A passive sensor simply observes the energy produced by an entity (e.g., infrared emissions associated with the engine of a tank). In either case the sensor must have collection or receiving apparatus such as an antenna. Figure 10.2 shows such a sensor concept. The left-hand side of the figure shows two "antenna," one for generating energy (for an active sensor) and the second for receiving energy from the target (for both active and passive sensors). In practice, active sensors may use the same antenna for both transmission and receipt of energy. Emission of the active transmission of energy is controlled by emission sensor elements shown in the upper portion of Figure 10.2. Processing within the sensor includes signal conditioning, signal processing, information processing, and decision-making and output processing. These functions are described in the following. (The output of the generic sensor is a signal, image, scalar or vector quantity illustrated on the right hand of Figure 10.2.)

- Signal conditioning: The signal received from the antenna or collection system must be conditioned for subsequent processing. Examples of signal conditioning functions include analog-to-digital (A/D) conversions, digital-to-analog (D/A) conversions, translation of the input energy from one form into another (e.g., translation of vibration or received visual light into an electrical signal), and detection. The latter usually involves measurement of a target signal above some background noise. Ideally, the signal conditioning process should not corrupt the content of the received energy.
- Signal processing: The output from the signal conditioning process may be a scalar, vector, time-series, or image. Signal processing can be performed to enhance the information associated with the (presumed) target or entity

versus the background. There are numerous signal processes that could be performed. These include filtering, Hilbert transformations, thresholding functions, data compression techniques, spectral algorithms, and a wide variety of special transformations or operations. The reader is referred to classical texts on digital signal processing and image processing to provide more in-depth descriptions of these functions (see for example, [26, 27]).

- Information processing/decision-making: After signal processing functions are applied to enhance the ability to distinguish the target data from the background data, information processing can be performed to represent the target information. Examples of information processing and decision-making algorithms include use of look-up tables, application of decision rules, and statistical estimation techniques. In this step the signal (or image) data is transformed into information about the observed entity (e.g., target location, identity, characteristics, or attributes).

- Output processing: Output processing could be performed prior to outputting the data and information from a sensor. This can include data compression, coordinate transformations, data conversions, filtering and smoothing operations, and sensor corrections.

- Sensor guidance and control: Within the sensor, guidance and control functions may be performed. These can include computation of sensor look angles (namely how to point the sensor to observed a specified location), and computation of sensor control parameters.

The output from a sensor may involve a scalar measurement (e.g., slant range), a vector (range, range-rate, azimuth and elevation), a time series (e.g., radar cross-sections versus time and aspect angle) or an image. These are shown on the right-hand side of Figure 10.2. In addition, note that input to the sensor may include tasking (when to turn on and off), tip-off information (identification of a potential target at a specified location), and environmental data. Figure 10.2 is strictly a generic/functional view of a sensor to provide an understanding of how a sensor performs prior to fusion of data in subsequent processes.

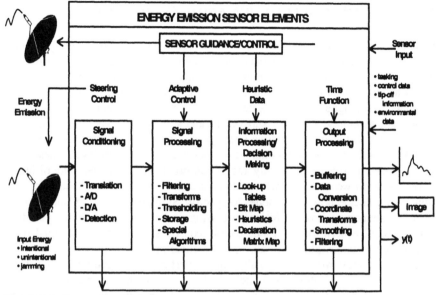

Figure 10.2 Concept of a generic sensor.

An overview of the types of algorithms associated with individual sensor processing is shown in Figure 10.3.

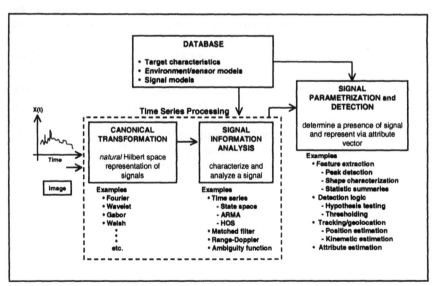

Figure 10.3 Single sensor target detection and parameter estimations.

10.4 FUNCTIONAL ALLOCATION AND DECOMPOSITION

How does one decide how to allocate data fusion functions across multiple processing nodes? For example, a data fusion system dedicated to monitoring the health of a complex machine such as a helicopter may contain smart multisensing nodes to monitor individual components of the machine (e.g., gears, shafts, and bearings), with higher level nodes dedicated to monitor subsystems (e.g., gearbox, drive-shaft assembly, and rotor subsystem) and still higher level nodes related to systems such as the engine or drive-train system. Hence, a data fusion system may be implemented at a sensor, component, subsystem, or system level. These data fusion nodes must act individually and in concert to provide a hierarchy of inferences. How do we decide which functions are allocated to which nodes? Indeed, how do we decide how many nodes to specify? Similarly, a military surveillance fusion system may involve fusion/processing nodes associated with individual sensors (e.g., a network of ground-based acoustic sensors) and individual platforms such as tanks, planes, or ships, and tactical operations centers. These data fusion nodes must work individually (e.g., to perform automatic target detection, location, and identification for an individual sensor system) and in concert to perform higher level situation assessment. As a result, the allocation of functions to fusion nodes requires a systematic design process in which system-level functions and requirements are decomposed to requirements and functions for individual fusion nodes.

A structured approach for this functional allocation and decomposition has been developed by Bowman and Steinberg [1, 8]. This section provides a very brief description of that methodology. For detailed discussions, the reader is referred to the previous references.

The approach developed by Bowman and Steinberg is built upon two basic principles: (1) that data fusion systems (or nodes) may coexist with a resource management component at every level (namely that each data fusion node may contain both data fusion functions and resource management functions), and (2) that the system engineering process for functional decomposition and allocation for data fusion can mirror successful techniques previously developed and successfully used for resource management/control systems. The high-level concept for the system engineering approach is shown in Figure 10.4.

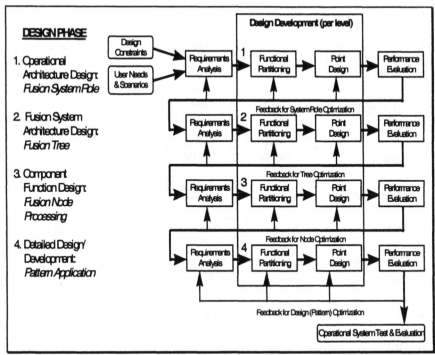

Figure 10.4 Data fusion system design process [8].

The design approach involves a sequence of phases including; (1) operational architecture analysis and design, (2) fusion system architecture design, (3) component function design, and (4) detailed design and development. As shown in Figure 10.4, the analysis proceeds from high-level architectural and operational concepts to detailed design and implementation. This process is amenable to iterative refinement as in spiral development methods. At the initial phase, a strawman data fusion tree is defined. The data fusion tree simply specifies potential fusion nodes based on an operational concept involving specification of sensors, platforms and subsystems. An initial allocation may simply specify data fusion nodes for each platform and sensor system, or each major function within a platform. There are a number of different strategies that could be used for identifying (and creating) nodes in hierarchical fusion architectures. Examples of possible strategies are shown in Figure 10.5. Thus, for example, we might allocate conceptual nodes based on the following.

- "One node for each source" (e.g., for a tactical aircraft, one node associated with the onboard radar system, one node associated with the onboard infrared system and one node for ingestion of tracks received from off-board sources);

- "One node for each data type" (e.g., for a multisource system, a node is created for each data type such as radar, infrared, acoustics);
- "One node for each level of aggregation" (e.g., a different node for each level of fusion).

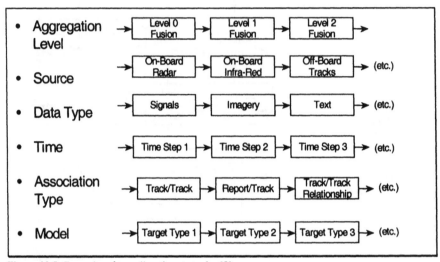

Figure 10.5 Examples of tree allocation strategies [8].

The methodology described by Bowman and Steinberg proceeds from a top-level identification of data fusion nodes (and the data and control flow among nodes), followed by allocation of functions among nodes, based on factors such as processing capability of each node, communications capacity among nodes, availability of data at each node, and other factors. This is performed at a high level of abstraction. Subsequently, the architecture is refined as the design proceeds from requirements analysis through preliminary design, detailed design, and prototyping and implementation. Details on this process are provided by [8].

10.5 ARCHITECTURE TRADE-OFFS

A fundamental issue within (and across) data fusion nodes is the choice of processing architecture. The issue revolves about the question of where (in the processing flow) to combine or fuse the data. The design choice affects the quality of the fused product, the nature of the algorithms or techniques that may be used, the complexity of the processing logic, and the bandwidth of the communications required between the sensors (or data sources) and the fusion processing. The choice of architecture is not arbitrary; instead it depends upon the nature of the sensors involved as well as the nature of the inferences sought.

Three basic architecture approaches are summarized in Table 10.2 and illustrated in Figure 10.6. These include a centralized architecture, an autonomous architecture, and a hybrid architecture. For each architecture approach, Table 10.2 summarizes the level of information fused, the applicable fusion techniques, and a summary of comments concerning each architecture approach. Figure 10.6 provides a conceptual view of the processing flow (note that two variations are illustrated for the centralized architecture). Each of these architecture approaches is described below. Additional discussion of fusion architecture concepts is provided by [9, 28-31].

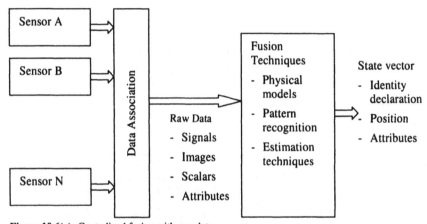

Figure 10.6(a) Centralized fusion with raw data.

A centralized architecture fuses either raw data or derived data from multiple sensors to achieve an estimate of an entity's position, velocity, attributes, or identity. If commensurate sensors are available, raw data may be fused. Commensurate sensors observe the same physical manifestation of an entity, activity, or event. Examples of commensurate sensors include identical sensors or sensors whose observational data may be closely compared or merged (e.g., an infrared image and a visual image). Centralized fusion of raw data is illustrated conceptually in Figure 10.6(a).

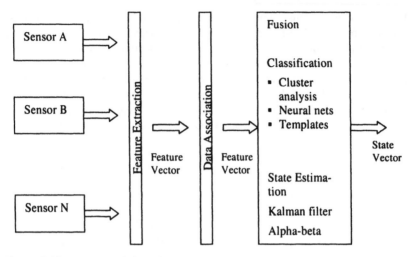

Figure 10.6(b) Centralized fusion with feature vector data.

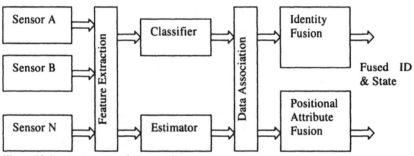

Figure 10.6(c) Autonomous fusion architecture.

Table 10.2

Summary of Alternative Fusion Architectures

Architecture Type	Description	Level of Information Fused	Applicable Techniques	Comments
Centralized	Fusion of raw data from sensors	Raw data—direct fusion of sensor data (e.g., sampled signal data, imagery)	Physical models Pattern recognition Estimation techniques	Minimum information loss Requires commensurate sensors Difficult association Extensive communications between sensors and fusion processor

Table 10.2   (continued)

Summary of Alternative Fusion Architectures

Architecture Type	Description	Level of Information Fused	Applicable Techniques	Comments
	Fusion of derived data (features) from sensors	Feature vectors— attribute features, location or kinematic parameters	Pattern recognition Estimation techniques Cluster algorithms Neural nets Parametric templates	Information loss due to feature extraction Simplified association Allows non-commensurate sensors
Autonomous	Fusion of state vectors or identity declaration	Decision-level fusion—state vectors, identity declarations	Estimation techniques Bayesian inference Dempster-Shafer's method Voting Logical templates	Information loss due to feature extraction Local optimization may prohibit global optimization Allows noncommensurate sensors Must account for statistical interdependence
Hybrid	Combination of centralized and autonomous architectures	Combination of raw data, feature vectors, and decision-level information	All of the above techniques	Combines features of centralized and autonomous architectures Complex control logic Increased communications requirements

Raw data from each sensor is associated via a data association/correlation process and subsequently fused using physical models, pattern recognition techniques, or estimation techniques. The output data consists of a joint declaration of identity, an estimate of an entity's position or velocity, or an estimate of entity attribute. Fusion of raw data requires commensurate (or near commensurate) sensor data. One advantage of such a fusion approach is a minimum information loss (i.e., the sensor data is fused directly without approximation via features, state vectors, or declarations of identity). Data association and correlation may be difficult, however, because we seek to determine if raw data (e.g., image segments or signal data) from one sensor refer to the same entity or target being observed by another sensor. Even association of identical sensors may be difficult if sensors are displaced geographically. The centralized fusion approach requires extensive data transmission between the sensors and the fusion process since all of the raw data is being transmitted from the sensors to the fusion process. The computational requirements for centralized fusion may also be high, since specific techniques may utilize complex physical models, simulations, or cross-correlation calculations.

An alternative approach to centralized fusion is illustrated in Figure 10.6(b). In this architecture, preprocessing is applied by each sensor to extract a feature vector. These feature vectors are transmitted to a central fusion process that performs data association and subsequent fusion. Identity data is fused to declare object identity using techniques such as cluster algorithms, neural networks, or parametric templates. In this approach the features from multiple sensors are concatenated into a single (potentially many-component) feature vector, which is input to a classification algorithm. Similarly, parametric data related to position or velocity is input to estimation techniques such as a Kalman filter, alpha-beta filter, or batch estimator. This approach results in an information loss from sensors, since the sensor data is represented by a feature vector. Statistical information such as a feature vector covariance may also be used to represent noise characteristics in the sensor data. The use of feature vectors, however, simplifies the data association process and reduces communications requirements between the sensors and the fusion process. Moreover, the use of feature or attribute vectors allows data from noncommensurate sensors to be fused.

A second type of architecture for fusing data is autonomous fusion, illustrated in Figure 10.6(c). In this approach, feature or attribute vectors represent sensor data. For each sensor, processing is performed to determine a state vector or declare the identity of an entity. Thus, the output from each sensor is a decision (i.e., a declaration of identity or estimate of a state vector). These decisions are input to a fusion process that performs association and either identity fusion or positional fusion (as appropriate). Autonomous fusion provides a significant information loss compared to raw data fusion, since data are represented via identity declarations or state vectors. The information processing for each sensor may result in a local optimization rather than a global optimized solution.

Autonomous fusion allows data from noncommensurate sensors to be combined. Applicable fusion techniques include Bayesian inference, the Dempster-Shafer method, Thomopoulos's generalized evidence processing (GEP) theory [32], logical templating, and voting techniques. Association for autonomous fusion is relative simple, since the process compares state vectors or identity declarations. One difficulty of autonomous fusion, however, is the requirement to fuse data such as state vectors, which are not conditionally independent. Thus, it is necessary in the fusion algorithms to account explicitly for such dependence (see for example the discussion by [33]).

The final fusion architecture summarized in Table 10.2 is a hybrid architecture that combines the centralized and autonomous architectures. Raw data, feature level data, and decision data is all input to a fusion process. The fusion process combines this data to result in fused declarations of identity or estimates of attributes, position, or kinematics. Complex logic may be required to determine the circumstances under which to combine raw data with processed data. The communications requirements for this architecture may also be high, since the link must be bidirectional in order to control sensors to send raw data when required. Similarly the computational requirements for this architecture may be larger than for any other architecture.

Selection of an appropriate fusion architecture from among these alternatives depends first on the available sensor suite. Raw data cannot be directly fused unless the sensors are commensurate. The sensor suite for a particular application may include smart sensors that output state vectors, feature vectors, or identity declarations. This data can be fused via an autonomous architecture or a hybrid architecture. A diverse sensor suite involving smart sensors and raw data sensors may require a hybrid architecture.

A basic rule of thumb for architecture selection is that fusion should be performed as close to the raw data as possible. Because of the information loss involved in feature extraction or estimation, higher inference accuracies are theoretically possible using raw data versus feature vectors or decision-level data. Hence, the closer to the sensing and detection process that data are fused, the higher the accuracy of the fused process. However, this assumes that the raw data can be properly aligned and correlated prior to the fusion (a challenging problem for image data). A cost of this fusion close to the source is an increased computational workload. Furthermore, there is an increased complexity of logic for data association and control of the processing flow. Finally, the communications bandwidth between sensors and the fusion process is increased when raw or hybrid data is utilized.

Table 10.3 provides a summary of these trade-offs for fusion architectures, listing four architectural variations and provide an assessment of the relative accuracy, the required communications bandwidth, processing complexity, and inference accuracy. Clearly, Table 10.3 is qualitative. Quantitative comparisons of these architectural alternatives depend on the specific application, sensor suite, sensor characteristics, observational environment, inferences sought, algorithms used, and other factors. Quantitative comparison may rely on techniques such as queuing analysis, Monte Carlo simulations, covariance error analysis, or numerical experiments.

Table 10.3

Summary of Trade-offs for Fusion Architectures

Architecture Approach	Relative Computational Requirements	Required Communications Bandwidth	Estimation Accuracy	Processing Complexity
Centralized— raw data	High	High	High	High
Centralized— derived data	Medium	Medium	Medium	Medium
Autonomous	Low	Low	Medium	High

Table 10.3 (continued)

Summary of Trade-offs for Fusion Architectures

Architecture Approach	Relative Computational Requirements	Required Communications Bandwidth	Estimation Accuracy	Processing Complexity
Hybrid	Very High	Very High	High	HIgh

10.6 ALGORITHM SELECTION

The process of selecting algorithms for fusion processing is complicated by the fact that the data analysis seeks to combine incomplete and missing data in a complex (and potentially deceptive) environment in near real time. Moreover, of the algorithms described in this text (and algorithms not described here), there is a many-to-many mapping between useable algorithms and processing functions within data fusion. Hence, there is not a simple, "for this process, use algorithm Y" set of rules available for choosing among techniques. In addition there is no well-disciplined "textbook" approach to fusion analysis. Instead, there are wide ranges of techniques that have potential applicability. Their success depends on the specific implementation, available a priori knowledge of entities and sensor performance, available communications and computing resources. A number of experimental and fielded systems have used, with varying success, combinations of techniques ranging from expert system technology to classical inference methods. An example of data fusion systems and associated algorithms is described by [11] for military applications.

Several authors have provided methods to aid in the selection of data fusion algorithms. Hall and Linn [4] developed a general structured approach to algorithm selection and evaluation. The method is summarized below. Bowman and Steinberg [8] describe some general systems engineering concepts for algorithm selection based on an analogy of data fusion and resource management. Finally, a study of data association and correlation algorithms was performed by Llinas, Hall, and Bowman [7], who developed decision trees to guide the selection of algorithms for data correlation.

The selection process developed by Hall and Linn is summarized below. The selection process takes an overall systems perspective that simultaneously considers the viewpoints of four major participants:

- System users: whose concerns include system requirements, user constraints, and operational preferences;
- Numerical/statistical specialists: who are knowledgeable about numerical techniques, statistical methods, and algorithm design;

- Operational analysts: concerned with the human-computer interface, transaction analysis, and operational concepts definition;
- Systems engineers: concerned with system performance, interoperability with other subsystems, and system integrity.

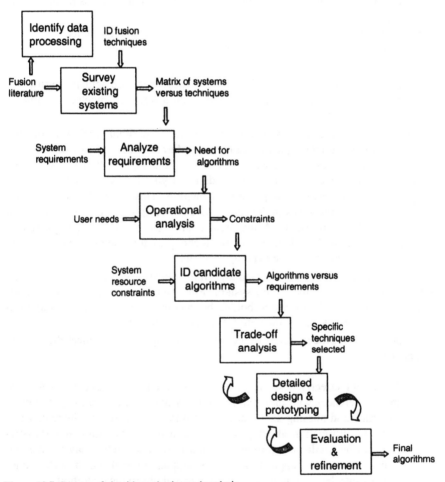

Figure 10.7 Process of algorithm selection and analysis,

Figure 10.7 illustrates a structured process for algorithm selection. The process involves eight basic steps:

- Identifying categories of data processing techniques or algorithms;
- Surveying existing prototype and fielded data fusion systems (and commercial tools);

- Analyzing system requirements;
- Analyzing and defining operational concepts for manual and automatic processes;
- Identifying preliminary (candidate) algorithms;
- Performing trade-off studies of algorithm effectiveness versus required system resources;
- Preparing detailed designs and prototypes of selected algorithms;
- Refining and tuning the algorithms.

Although shown as a sequential process, these steps are actually performed iteratively and in parallel. The process ends with a well-documented and implemented set of algorithms.

Evaluation of the performance of the selected algorithms includes both the degree to which the techniques make correct inferences as well as evaluation of required computer resources such as CPU time and input-output requirements. The latter performance evaluation proceeds at varying levels of detail from preliminary software design through implementation. This evaluation begins with initial estimates of resources in the early phases of design, and uses more detailed estimates as design proceeds. Actual benchmarking is used during implementation. Hall and Finkel [34] give specific techniques for computer performance evaluation based on a time-line queuing approach.

The effectiveness of the selected algorithms must be evaluated as part of the integration and test of the fusion system. Final tuning of parametric thresholds (e.g., gate sizes and decision thresholds), bounds for data retrieval, and other details are performed as part of the software test and integration.

The selection process seeks to identify fusion algorithms to meet the following goals:

1. Maximum effectiveness: Algorithms are sought that have the ability to make inferences with maximum specificity and accuracy in the presence of uncertain and missing data. These algorithms must deal with the reality that a priori information about probability density distributions and statistics may not be available. Moreover, even a priori knowledge may be of minimal value in actual applications. For example, nominal sensor performance bounds may be extremely inaccurate in actual observing conditions.
2. Operational constraints: Algorithm selection should consider operational constraints and perspectives for both automatic data processing and tools to assist the fusion analyst. Clearly the end product of the selected algorithms is to provide assistance to the analyst in meeting his or her mission objectives. In addition, analyst time constraints are considered. For example, in an operational environment, extensive time may not be available for the creation, selection, and modification of algorithms.

3. Resource efficiency: Selected algorithms need to minimize the use of computer resources such as CPU time, communications bandwidth, memory, and input/output.
4. Operational flexibility: In evaluating potential fusion algorithms, one concern is the need for flexibility to account for operational needs. In designing methods for higher order inferences, it should be possible to change a priori data (such as templates or prior estimates of hypothesis probability), based on field experience.
5. Functional growth: Design of data flow and choice of algorithms must also seek to allow growth in functionality as the fusion system evolves.
6. Scalability: Finally, algorithm evaluation should consider how well the algorithm adapts to large growth in data rates and amount of input data. Algorithms that perform well with a small amount of input may be totally overwhelmed when the data rate or amount significantly increases.

While this list of goals is not exhaustive, it is intended to provide a perspective on considerations to be used in selecting algorithms for fusion systems.

Because of the wide variety of fusion applications and applicable algorithms, it is not possible to define which techniques are applicable for specific fusion processing functions. Nevertheless, we can provide a tentative map to illustrate the utility of various techniques. Figure 10.8 provides such a map to level 1 fusion processing, while Figure 10.9 shows a similar mapping for knowledge-based techniques applied to level 2 and level 3 fusion.

Algorithms and Techniques	Spatial Adjustment	Temperature Adjustment	Unit Adjustment	Screening	Correlation	Assignment	Observation Predict	State Update	Uncertainty Management
	Data Alignment			Data/Object Correlation			Object postional/ Kinematic/Attribute Estimation		
Coordinate Transforms	X	X	X	X			X		
Sensor Models							X		
Physical Models							X	X	
Association Measures				X	X	X			
Assignment Logic						X			
Equations of Motion								X	

Figure 10.8 Applicability of techniques for level 1 fusion. (continued on following page)

Algorithms and Techniques	Spatial Adjustment	Temperature Adjustment	Unit Adjustment	Screening	Correlation	Assignment	Observation Predict	State Update	Uncertainty Management
	Data Alignment			Data/Object Correlation			Object postional/ Kinematic/Attribute Estimation		
Optimization Methods								X	
Kalman Filters							X	X	
Covariance Error									X
Bayesian Inference						X			X
Dempster-Shafer						X			X
Voting						X			
Pattern Recognition					X	X			
Templating				X	X	X			
Expert Systems									X
Fuzzy Sets									X

Figure 10.8 Applicability of techniques for level 1 fusion. (continued from previous page))

SITUATION ASSESSMENT

	Infer DOB	Place DOB'S in templates	Analyze environment	Analyze socio-political	I/D status of force	Estimate of likely events	Estimate enemy behaviors	Determine unknowns
Rule	X	X	X	X	X	X	X	X
Frame	X	X	X	X	X		X	X
Hierarchical Classification		X	X			X		X
Semantic Net	X	X	X	X	X	X	X	
Neural Net		X	X			X	X	
Nodal Graph	X	X	X	X	X	X	X	X
Options, Goals, Criteria	X	X			X	X	X	X
Script	X			X		X	X	X
Time Map	X			X		X	X	X
Spatial Relationships	X	X	X	X	X	X	X	X
Analytical Model			X			X		
Deduction	X	X	X	X	X	X	X	X
Induction	X	X	X	X	X	X	X	X
Abduction	X			X		X	X	
Analogy	X		X	X	X	X	X	
Plausible/Default/Inference	X	X	X	X	X	X	X	X
Circumscription	X	X	X	X	X	X	X	
Search		X	X	X		X	X	
Reason Maintenance	X	X	X	X	X	X	X	
Assumption-based TM	X		X				X	
Hierarchical Decomposition	X	X	X	X	X	X	X	X
Control Theory			X	X		X	X	
Opportunistic Reasoning	X	X	X	X	X	X	X	

*Adapted from D. Buede.

Figure 10.9 Applicability of knowledge-based algorithms to level 2 processing.

10.7 DATABASE DEFINITION

Database management support (DBMS) functions make up a major portion of a fusion system. The survey conducted by Hall and Linn [11] indicated that DBMS functions may constitute as much as 25% of an implemented data fusion system. Types of data used or developed in a fusion process include model parameters, sensor data, external database information, human inputs, environmental data, situation databases, threat data, performance data, and a priori data required of the knowledge-base fusion. These types of data are summarized in Table 10.4.

Table 10.4

Summary of Data Involved in Fusion Processing

Data	Data Description	Interpretation
Model parameters	• Sensor characteristics • Sensor locations • Physical constants • Time delays, biases, model coefficients	Characterizes model information for observation prediction, state vector propagation, sensor data processing

Sensor data	• Observations
	  o State vectors
	  o Attribute and feature vectors
	  o Raw signal and images
		Observed data from sensors, external information sources

External databases	• Previous observations • Data from external sensors • Data from other fusion nodes • Internet data	Related data from other fusion system or external databases
Human input	• Control information • Inferences and hypotheses • Request for data • Annotations • Analyst preferences to adapt the system to individual needs	Data input by humans for system control, data updates, or interpretations
Environmental data	• Geography, topography, hydrology • Weather • Environmental conditions • Man-made objects (cities, roads, networks)	Information about the static or dynamic environment in which a situation is observed
Situation database	• Location, identity of entities • Relationships among entities • Interpretation of order-of-battle	Results of level 2 processing to achieve a dynamic interpretation of entity relationships, activities, or events
Threat database	• Location, identity of enemy weapons, sensors	Results of level 3

Table 10.4   (continued)

Summary of Data Involved in Fusion Processing

Data	Data Description	Interpretation
	• Characteristics of weapons (threat envelops) and sensors • Probable courses of action • Interpretation of enemy intent, lethality, and opportunity	processing aimed at determining an enemy threat to friendly or neutral forces
Performance data	• Assessment of accuracies of fusion products o Measures of performance o Measures of effectiveness o Objective function o Constraints	Data which characterizes and allows control of the dynamic sensor observation and fusion process
A priori data	• Knowledge bases (rules, frames, scripts) or target and event behaviors, enemy tactics and doctrine • Technical data (e.g., enemy sensor performance, weapon characteristics) • Mission data (e.g., objectives and goals, strategies, tactics)	Data developed a priori analysis to provide a basis for knowledge-based fusion processing

Database functions required for fusion systems include data input, storage, routing, retrieval, archiving, and synchronization for distributed databases. These functions support automatic processing of input sensor data, updates and retrievals for automated fusion processes (e.g., target tracking, identification, situation assessment), updates and retrievals to support human inference processes, and processing for analysis such as establishing template boundaries, supervised learning of neural networks, or cluster analysis algorithms. Database management for fusion processing is difficult for a number of reasons including the following.

1. Database size and variety: A typical data fusion system requires a large and varied database involving numerous data records and widely varied data records [e.g., scalars, vectors, images, signals, matrices, map data, complex data records, and symbolic data (e.g., rules)].
2. Rapid data updates and retrievals: The DBMS must simultaneously support rapid data updates (to keep up with incoming sensor data) as well as supporting general Boolean queries by an analyst or algorithm.
3. Flexible access by human users: Data access must be flexible and user-friendly to support human queries with features such as Boolean queries, report generation, and relational retrievals.
4. Data integrity: The requirement to maintain data integrity "on the fly" as data is rapidly received and processed by automated algorithms and human-in-the-loop update. This is complicated by updates based on sensor-

corrupted data, false alarms, out-of-sequence reports, communications delays, and multiple broadcasts of the same data by multiple fusion nodes.

5. Distributed synchronization: The DBMS must maintain synchronization of local evolving databases with data being updated at other nodes or components of the fusion system.

6. Multirecord types: The DBMS must support fixed format records (e.g., from sensor data inputs) as well as free-form textual data from human input and Internet sites.

The database requirements demand the type of support provided by large-scale commercial products such as ORACLE (www.oracle.com), OMNIBASE, ADABAS, or geographical information systems such as ARCINFO. An index of open-source software systems is provided by the Association of Computing Machinery (ACM) Special Interest Group on Management of Data (SIGMOID) at the Web site http://www.acm.org/sigmoid/databaseSoftware/. However, both commercial and open-source packages are generally designed for a single type of data (e.g, text only, numerical only or images only). In addition, these commercial packages are generally designed for flexibility without regard to the utilization of computer resources. It is often necessary to adapt a commercial DBMS for use in a fusion system. In particular, one must consider the requirement to rapidly ingest data and provide rapid, general retrievals of data are contradictory from a database design viewpoint; in other words, it is not possible to optimize database updates and retrievals simultaneously. Hence ancillary software and database designs must be implemented to satisfy all of the above requirements.

Table 10.5 provides a summary of general database characteristics and the associated requirements to support fusion processing. General information on database design and implementation is provided by [15, 16]. A discussion of data required for situation and threat assessment is provided by [9]. Finally, Richard Antony provides a discussion of data representation, database design, and data characteristics for tactical systems [13].

Design of a fusion database and software to provide database services may require a majority of the effort in the development of a fusion system. It is beyond the scope of this book to provide a prescription for database implementation. Nevertheless, the information summarized here may indicate the nature of the implementation challenges. Design and implementation of a database management system for a fusion system follow the techniques required for any large-scale database design effort. Particular attention is required to satisfy the requirements summarized in Table 10.5.

Table 10.5

Database Management Requirements for Fusion Systems

DBMS Attribute	Description	Requirement
CPU/OS interface	Compatibility with specified computer configuration	User-specified
Data item	Capability to store data in a variety of formats	Character strings, decimal, integers, floating point, double precision
Data structure	The physical structure implemented by the DBMS	Network, relational, object-oriented may be required
Record structure	The types of records that may be inserted into a given file	Fixed length, variable length, repeating groups, handling
Access methods	Methods by which a retrieval from the database can be implemented	Sequential, direct randomized, index (multi-key), calculated via set
Special storage techniques	The optimization of storage by compressing fields that are not used and compressing trailing blanks; also special storage mechanisms	Data compression for text, signals, images, possible quad-tree or hierarchical storage
Database creation	The capability for easily building the database	User programs and system utilities
Database revision	Flexibility in changing the database	Automatic incremental reorganization or prevent residue buildup; new data items are to be added without the need for reorganization
Validation capability	The validation of new entries into the database	Size, type, range, duplications; may need to store and validate data pedigree
Types of backup and recovery	The capability of recovering the database	Need roll-back and roll-forward capability
Automatic backup and recovery	The capability of the DBMS to detect failures and recover	Automatic with operator recovery in the event of fatal hardware failures; automatic rollback on software failure (user); protection against data disruption and information warfare
Level of security and privacy	The level at which a lock can be placed in the database	File and possibly record level; special requirements may be induced by multi-level security and different access level by different users
Security/privacy priviledges	The capability of the DBMS to grant and revoke read and read-write access	Read; read-write
Automatic logging and journaling	The capacity to log	Journalize elements updated; support user-defined transactions – either all transaction is committed to the database or none; support selective file journalizing

Table 10.5 (continued)

Database Management Requirements for Fusion Systems

DBMS Attribute	Description	Requirement
Data dictionary and directory	Contains the definition and description of all elements in the database	An active dictionary for user and algorithm use
Select level for data manipulation	The level at which changes can be made to an element	Record and element levels
Operators, comparators, logical complexity	The criteria by which changes can be made to an element	Range search (wild card), Boolean (<, =, >), (and, or, not), thesaurus operations, possible ontological and fuzzy operations
Inquiry/retrieval capabilities	Utilities that interface with high-level languages	Higher order language interface and self-contained utilities
Mode of execution	The functional environments that a DBMS can operate in	On-line and batch
Mode of interaction	The internal functioning capability of the DGMS	Re-entrant code allowing concurrent updates and read, multi-read, deadlock resolutions
Performance estimates	The capability of determining potential bottlenecks	CPU utilization and input/output estimates
High-level language	FORTRAN, Ada, C, Java	User-specified

10.8 HCI DESIGN

Chapter 9 of this text is devoted to level 5 fusion, which includes human-computer interaction (HCI) and support to a human in the loop via computer aids, presentation tools, and user-specific adaptation. Unless a fusion system involves highly automatic applications such as robotics or monitoring of industrial processes, the HCI strongly affects the effectiveness of a fusion system. HCI functions provide the mechanisms by which the results of fusion processing are conveyed to a human operator, and also provide the means by which an operator controls and guides the fusion inference process. This section provides some general comments on HCI implementation and design. There is a very large number of texts written on HCI and interactive systems. Texts include [18-20, 35]. A brief history of HCI technology is provided by Myers [36]. A general-purpose toolkit (called Geo_Vista) for development of geographic displays is available as open-source software from the Pennsylvania State University (see www.geovista.psu.edu). An example of the Geo_Vista interface is shown in Figure 10.10. Finally, a survey of 160 user interface software tools was developed by Meyers (see [37]). A list of those tools with contact information is provided at the Web site (www.2.cs.cmu.edu).

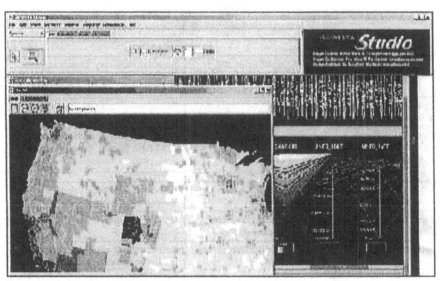

Figure 10.10 Example of Geo-Vista Development Tool (www.geovista.psu.edu)

HCI for a fusion system involving a human in the processing loop may be particularly challenging compared to ordinary computer program applications. For ordinary applications, the information flow tends to follow one of two perspectives, either user-driven applications or data driven applications.

User-driven applications such as word processing, database retrievals, or computational applications allow a human user to control the sequence, order, and rate of processing. For example, a DBMS may allow a user to formulate a query (e.g., via menu selection or query language), command the DBMS to perform a query, and with for the DBMS program to retrieve the requested data and display the results. Thus, in a user-driven (or control-driven) mode, the user chooses the sequence and timing of the operations. The human user may interrupt the process at any time and reinitiate the process at a subsequent time. Ideally, the processing flow is controlled by, and follows, the cognitive processes of the user.

Conversely, for data-driven applications, the computer program reacts to incoming data and displays processing results. An example of a data-driven process is a bulletin board in which external sources send information to be posted. A user may access the results as information is received. However, in this mode, the processing flow is directed by the sequence and timing of the data reception or interrupts.

Data fusion applications involve both modes of operations simultaneously. On one hand, data is received from sensors and processed by automated fusion processes (e.g., data association, and estimation). This information must be presented to a user in a timely fashion without overwhelming the user by continuous interruptions. Indeed, it is relatively easy for such systems to significantly reduce the effectiveness of the human decision process (an effect that M. McNeese terms fragmentosa cogminutia [38]). A fundamental design question that must be addressed is the following: What do users need to know, and when do they need to know it? For geographical displays, a constant updating of recent observations (e.g., flickering icons and moving situation display) may be distracting to a user's systematic inference process. On the other hand, a user seeks to command a fusion system to retrieve data, perform computations, and control system resources; hence both modes of information processing must be supported by a data fusion HCI. A summary of HCI design issues is shown in Table 10.6.

A second complication of data fusion HCI is simply the magnitude and variety of data to be displayed. For geographical applications, a hierarchy of data may be displayed including the following:

- Geodetic references;
- Elevation data;
- Topography features;
- Hydrology features;
- Vegetation;
- Industrial features;
- Population centers;
- Agricultural information;
- Transportation networks (and "trafficability");
- Military institutions and units;
- Dynamic data such as the following:
  - Target locations and tracks;
  - Sensors and sensor detection envelopes;
  - Weapon threat envelopes;
  - Active logistic support.
- Environmental information;
  - Seasonal data;
  - Weather information.

These data constitute a situation display for a military application. Analogous displays may be envisioned for nonmilitary applications (e.g., crisis management and environmental monitoring). Simultaneously a user may need to access command menus, retrieve reference data, and overlay information on the situation display such as functional relationships, labels, icons, and hypothesized courses of action. Additional issues include the need to present uncertainty, predictions, and the integration of human and automated analyses. An example of a three-dimensional environmental display is shown in Figure 10.11. The international society for optical engineering has conducted a series of international conferences related to visual data exploration and analysis (see for example [39]).

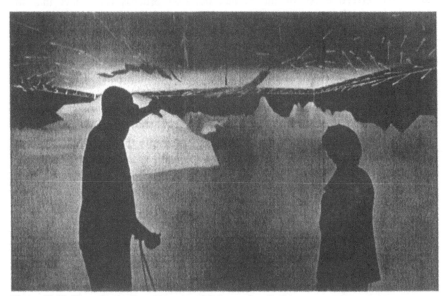

Figure 10.11 Example of a three dimension, full-immersion display.

Rapid advances in computer display technology and haptic interaction devices allow creative HCI designs to be developed. Many computer displays support high-quality graphics displays, the use of color to highlight text, graphics, icons, video, and symbols. A number of tools are available for rapid prototyping of dynamic graphic displays. At this time it has become routine for developers to use rapid prototyping tools to experiment with HCI. Examples are the tools and experiments provided in Chapter 9.

Table 10.6

HCI Design Issues

Design Issue	Description	Considerations	References
Information versus control flow	Balance of control-directed processing versus input data driven displays	Need to balance access/display of new data for a system user without undue interruption of a thoughtful analysis process; must consider the analyst's dynamic focus of attention	[40]
Decision-making style	Decision making employs different styles in information access and processing	Provide varied mechanisms to access data to support systematic versus spontaneous data	[41, 42]

Table 10.6 (continued)

HCI Design Issues

Design Issue	Description	Considerations	References
		access and external versus internal information processing	
Environmental stress	Tactical systems involve humans in a highly stressful environment	Provide mechanisms to guide humans to a successful stress-coping mechanism; balance active open-minded search for new data with deliberate inference process	[43]
Information access and processing modes	Humans use and prefer different modes of accessing and processing information	Use full range of HCI mechanisms to provide visual, kinesthetic, and auditory interfaces; user-selectable	[44]
Ergonomics	Limitations of human ability to access and process data, short-term memory limitations, perception limitations	Need to design a HCI to recognize and account for limitations in perception, physical limits, memory, etc.	[45, 46]

Finally, we note that a special NATO group (the NATO Information Systems Technology IST-021/RTG-007) conducts workshops and meetings related to visualization of large data sets, especially related to military applications. Various reports are available on tools, military applications, and technology assessments. Of special note is a model for developing HCI designs that has been developed by Martin Taylor. A description of the reports and model is available via the Web site (http://www.vistg.net/).

10.9 SOFTWARE IMPLEMENTATION

The implementation of a data fusion system typically involves the specification, design, implementation, test and evaluation of a large-scale software system. The practices and procedures for such implementation vary depending upon the sponsor, developing organization, size of the implemented system, purpose for development, and accepted practices within a particular domain. Sawyer [47] describes software development practices as being grouped into three main types: (1) formal software practices, (2) product-oriented practices, and (3) other emerging or experimental practices. These practices are summarized very briefly in this section. An overview of process models for software development is provided by MacCormack et al. [48]. Historically, data fusion systems were implemented primarily for government or military use. As such, the implementation methods tended to be specified by the sponsor and relatively formal. However, in recent years more commercial software and experimental software is being developed using product-oriented practices or experimental practices.

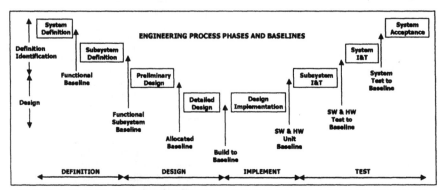

*Waltz, E. and Hall, D.L., "Requirements Derivation for Data Fusion Systems", *Handbook of Multisensor Data Fusion*, eds. David L. Hall and James Llinas, CRC Press, Boca Raton FL, 2001, p. 15-4.

Figure 10.12 Example of a waterfall model of system development [25].

Formal software development methods are typically used for government-contracted software, or internal practices used by large companies (e.g., for development and maintenance of management information systems). The formal practices tend to follow either a waterfall model or variations such as spiral development models [49]. The waterfall model of software development involves a multistep structured procedure as illustrated in Figure 10.12. Each phase of the development, from system definition through subsystem definition, preliminary design, and ultimately to system acceptance, is governed by well-defined practices and processes. The completion of each phase involves a formal review, delivery of specified documentation or products, formal sign-off or acceptance of the products by sponsor representatives, and mechanisms for review and revision. A number of variations of the waterfall model have been developed including: rapid prototyping [50], the spiral development model [51], an incremental or staged delivery model [52], adaptive development models [53], and so-called extreme programming (see http://extremeprogramming.org). The Software Engineering Institute at Carnegie Mellon University (see www.sei.cmu.edu/) provides training for organizations to improve software development, evaluation of organizations' abilities and productivity, and information on standards, procedures, and design and development tools.

In contrast to formal or process-oriented software development, product-oriented software development is more constrained by cost, speed, and profitability. In general, specific sponsors for such software do not exist, and the developer creates the requirements. Commercial software is built in anticipation of needs or wants of buyers. Hence, such software development must be rapid, flexible, and adaptable to perceived needs of buyers. The orientation of commercial software developers is not perfection but rather to create software that is "good enough" to satisfy customer needs.

Finally, there are numerous experimental or ad hoc software development communities that create software for open-source use, rapid prototypes for internal use, or special experiments about software development practices. Experiments in software development practices have been conducted for over 20 years at the NASA Goddard Software Engineering Laboratory (http://sel.gsfc.nasa.gov). Widespread discussions (and arguments) about emerging software practices may be found on the Web site (www.firstmonday.org).

10.10 TEST AND EVALUATION

In order to validate, test, and evaluate an implemented fusion system, we must first define a measure of merit to quantify the extent to which a system works or performs. Second, tools and techniques are required to perform tests against these measures. In this section we briefly address these related concepts.

In previous sections, we introduced the concept of measures of performance (MOP) and measures of effectiveness (MOE) to evaluate the performance of a fusion system (e.g., as part of the level 4 dynamic evaluation of system performance). Various measures of performance were introduced including factors such as intercept and detection probabilities, inference accuracy, response time, and utilization of system resources such as sensors and computers. Such measures of performance, clearly required in the sensor management process, are only one way of evaluating a data fusion system. Waltz and Llinas [9] discuss a hierarchy of measures including dimensional parameters, measures of performance, measures of effectiveness, and, for tactical systems, measures of force effectiveness. Such concepts can be applied to nonmilitary applications such as condition-based monitoring of complex systems. Measures of force effectiveness equate to factors such as availability of machinery to accomplish tasks, overall cost of maintenance performed, and reduced number of pieces of equipment required to accomplish a task (via reduced redundancy), etc. These four categories of evaluation measures are discussed below.

Dimensional parameters are the first and most basic measure or quantification of a fusion system. At the lowest level of evaluation, a data fusion system would be characterized by the extent to which an implemented system meets specified dimensional parameters. Dimensional parameters seek to quantify the properties or characteristics inherent in the physical components that determine system behavior. Examples of dimensional parameters include signal-to-noise ratio, bandwidth, frequency, operations per second, aperture dimensions, bit error rate, cost, and other factors. These parameters directly describe the behavior or structure of the data fusion system and provide the basic specifications for the system.

The second measure of a fusion system involves MOP. MOP are the measures that describe the behavioral aspects of the fusion system and characterize how well a fusion system performs. Examples of MOP include false alarm rates, location estimate accuracy, identification probabilities, sensor spatial coverage, target classification accuracy, and other attribute measures. In one sense, MOP characterize how well an implemented system works with respect to transforming an external physical situation into an output representation via state vectors or identity declarations. MOP characterize how well the sensor-fusion-sensor management functions perform to result in (primarily) a level 1 database. MOP involve not only sensor and system parameters, but also an interplay of those dimensional parameters with implemented techniques to perform sensor data fusion.

The third level of measure described by Llinas and Waltz are measures of effectiveness (MOE). MOE represent a hierarchical level above MOP, because MOE attempt to measure how well a sensor fusion system satisfies an intended mission. For tactical military systems, MOE examples might include target nomination rate, warning time, target leakage, immunity to countermeasures, and survivability. For an industrial monitoring system, an MOE might involve the extent to which the fusion system ensures product quality, the leakage rate of manufacturing defects, the lead time provided for repairing production equipment, robustness of the monitoring system, and related factors. MOE are more defined than MOP because they involve quantifying how well a fusion system supports an intended mission (as judged by a human user/designer).

Finally, the fourth and highest level of measure described by Waltz and Llinas for tactical systems are measures of force effectiveness (MOFE). MOFE measure how a command, control, and communication system, and the associated force (sensors, weapons, personnel) perform military missions. MOFE are derived from battle outcomes: survivability, attrition rates, exchange ratios, and measures of weapon success.

To evaluate a fusion system some hierarchy of dimensional parameters (MOE, MOP, MOFE) needs to be defined. Evaluation of system performance may then be performed while a system is being developed (and ultimately operational). Llinas [6] has described generic tools for evaluation. These are summarized in Table 10.7. While these tools are aimed primarily at evaluating military data fusion systems, the summary is instructive for nonmilitary systems. For example, constructing a data fusion tool for monitoring a mechanical system could involve simulated data (e.g., via finite element models of machinery), laboratory collection of data using seeded fault data and systems driven to failure conditions, special data collections on test stands, road or flight tests using specially instrumented systems, and collection of data with real data on fleets of equipment.

As the levels of inference for a fusion system becomes more abstract (e.g., for level 2 and level 3 processing), test and evaluation becomes progressively more difficult. Thus, while it is conceptually easy to establish tests to determine the effectiveness of increased sensitivity for a single sensor, it becomes more difficult to evaluate techniques for establishing general relationships among diverse entities. Overall it is difficult to evaluate knowledge-based systems [24] and to answer the fundamental question: to what extent does an implemented system support the mission of a user? Until tools, test beds, and measures of effectiveness become readily available, such questions will be difficult to address.

Table 10.7

Generic Spectrum of Evaluation Tools

Tool set	Characteristics
Digital simulations • Level 1: engineering models • Level 2: 1 versus N • Level 3: M against N • Level 4: organizational level • Level 5: theatre level • Numerous data fusion process models	• Relatively high fidelity; explore the physics and the one-on-one engagement problem; • Engagement models explore engagement effects (such as actions of intelligent opponents); the fidelity of these models necessarily decreases with the engagement complexity and level of engagement • Individual ad hoc simulations for tracking, identification, detection • Statistical qualification models may be used
Hybrid simulations	• Important effects of real humans and equipment; more costly than digital simulations; statistical qualification is often unaffordable
Specialized field data collection and calibration	• Real-world physics, phenomenology; relatively costly; often used to verify/validate digital simulations; good for phenomenological modeling but not for behavior modeling; statistically controlled in most cases
Test range data collection	• Real-world physics, phenomenology; relatively costly; can do limited engagement effects studies; some behavioral-affected models; statistically uncontrolled
Military exercises	Real-world physics, humans, equipment, and tactics/doctrine; costly; data difficult to collect and analyze; extended engagement affects studies at least feasible; extended behavioral effects modeled; statistically uncontrolled
Combat operations	For example Desert Storm, Desert Shield, Kosovo conflicts; actual combat with real adversaries; data difficult to collect and analyze; high fidelity enemy behavioral data; statistically uncontrolled

It is not infrequent for fusion system designers to develop a test bed to facilitate the test and evaluation of systems. Typically, these systems provide a mechanism to generate scenarios of tactical operations (namely for a user to specify targets, target locations and motion, deployment of sensors, sequences of actions), models of target characteristics and dynamics, models to predict sensor observations and performance, and tools to evaluate the fusion system (or algorithm) performance. Finally, the test beds provide tools for statistical analysis. An example of such a test bed, developed by the Mitre Corporation, is shown in Figure 10.13.

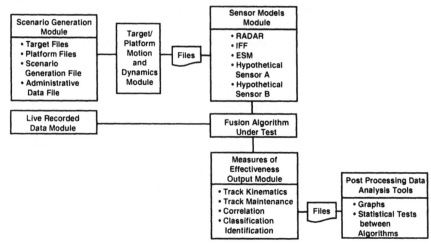

Figure 10.13 Example of a data fusion test bed (MITRE Corporation).

REFERENCES

[1]   Bowman, C. L., "The Data Fusion Tree Paradigm and It's Dual," *Proceedings of 7th National Fusion Symposium, invited paper*, Sandia Labs, NM, March, 1994.

[2]   Buede, D. M., *The Engineering Design of Systems*, New York, NY: John Wiley and Sons, 2000.

[3]   Hall, D., and R. Linn, "A Taxonomy of Multi-sensor Data Fusion Techniques," *1990 Joint Service Data Fusion Symposium*, May, 1990, pp. 593-610.

[4]   Hall, D., and J. Linn, "Algorithm Selection for Data Fusion Systems," *1987 Tri-Service Data Fusion Symposium*, Johns Hopkins Applied Research Laboratory, Laurel, MD, June, 1987, pp. 100-110.

[5]   Hall, S. A., "A Survey of COTS Software for Multisensor Data Fusion: What's New since Hall and Linn," *2002 MSS National Symposium on Sensor and Data Fusion*, San Diego, CA, June, 2002.

[6]   Llinas, J., "Assessing the Performance of Multisensor Data Fusion Systems," in *Handbook of Multisensor Data Fusion*, ed. by J. Llinas, Boca Raton, FL: CRC Press, 2001.

[7]   Llinas, J., *Engineering Guidelines for Data Correlation Algorithm Characterization*, Buffalo, NY: State University of New York at Buffalo, 1996.

[8]   Bowman, C., and A. Steinberg, "A Systems Engineering Approach for Implementing Data Fusion Systems," in *Handbook of Multisensor Data Fusion*, ed. by J. Llinas, New York, NY: CRC Press, 2001.

[9]   Waltz, E., and J. Llinas, *Multi-sensor Data Fusion*, Norwood, MA: Artech House, Inc., 1990.

[10] Steinberg, A. N., "Threat Management System for Combat Aircraft," *1987 Tri-Service Data Fusion Symposium*, Johns Hopkins University, Baltimore, MD, June, 1987, pp. 532-554.

[11] Hall, D., R. J. Linn, and J. Llinas, "Survey of Data Fusion Systems," *SPIE Conference on Data Structure and Target Classification*, Orlnado, FL, 1991, pp. 13-36.

[12] Hall, D., J. J. Gibbons, and M. Knell, "Quantifying Productivity: A Software Metrics Data Base to Support Realistic Cost Estimation," *Journal of Parametrics*, vol. VI, pp. 61-75.

[13] Antony, R., *Principles of Data Fusion Automation*, Norwood, MA: Artech House, Inc., 1995.

[14] Antony, R., "Data Management Support toTactical Data Systems," in *Handbook of Multisensor Data Fusion*, ed. by J. Llinas, Boca Raton, FL: CRC Press, 2001, pp. 18-1, 18-25.

[15] Connolly, T. M., et al., *Database Systems: A Practical Approach to Design, Implementation and Management*, 2nd ed., New York, NY: Addison Wesley Publishers, 2001.

[16] Rob, P., and C. Coronel, *Database Systems: Design, Implementation, and Management*, 5th ed., Course Technology, 2001.

[17] Raskin, J., *The Humane Interface: New Directions for Designing Interactive Systems*, New York, NY: Addison Wesley Publishing Company, 2000.

[18] Banks, W. W., and J. Weimer, *Effective Computer Display Design*, New York, NY: Prentice Hall, 1992.

[19] Newman, W. M., M. C. Lamming, and M. Lamming, *Interactive System Design*, New York, NY: Addison Wesley, 1995.

[20] Card, S. K., T. P. Moran, and A. Newell, *The Psychology of Human-Computer Interaction*, Lawrence Erlbaum & Associates, 1983.

[21] Nichols, M., "A Survey of Multisensor Data Fusion Systems," in *Handbook of Multisensor Data Fusion*, ed. by J. Llinas, Boca Raton, FL: CRC Press, 2001.

[22] Hall, D., and G. Kasmala, "A Visual Programming Toolkit for Multisensor Data Fusion," *SPIE Aerospace 1996 Symposium*, Orlando, FL, April, 1996, pp. 181-187.

[23] Keithly, H., "An Evaluation Methodology for Fusion Processes Based on Information Needs," in *Handbook of Multisensor Data Fusion*, ed. by J. Llinas, Boca Raton, FL: CRC Press, 2001.

[24] Hall, D., D. Heinze, and J. Llinas, "Test and Evaluation of Expert Systems," in *AI in Manufacturing: Theory and Practice*, ed. by A.L. Soyster: Institute of Industrial Engineering, 1989, pp. 121-150.

[25] Waltz, E., and D. Hall, "Requirements Derivation for Data Fusion Systems," in *Handbook of Multisensor Data Fusion*, ed. by J. Llinas, Boca Raton, FL: CRC Press, 2001.

[26] Kay, S. M., *Fundamentals of Statistical Signal Processing*, vol. 2, New York, NY: Prentice-Hall, 1998.

[27] Van Trees, H., *Detection, Estimation, and Modulation Theory*, New York, NY: John Wiley and Sons, 1971.

[28] Robinson, G. S., and A. O. Aboutalib, "Tradeoff Analysis of Multi-sensor Fusion Levels," *2nd National SPIE Symposium on Sensor Fusion*, Orlando, FL, March, 1989.

[29] Fox, G., and S. Arkin, "From Laboratory to the Field: Practical Considerations in Large Scale Data Fusion System Development," *1988 Tri-Service Data Fusion Symposium*, Johns Hopkins University Applied Physics Laboratory, May, 1988.

[30] White, F. E., "Data Fusion Requirements and Developments," *1989 Tri-Service Data Fusion Symposium*, Johns Hopkins University Applied Physics Laboratory, Baltimore, MD, 1989, pp. 287-320.

[31] Duren, B., "Comparison of Multi-sensor Fusion Architectures Based on Situation Assessment Principles," *1987 Tri-Service Data Fusion Symposium*, Johns Hopkins University Applied Physics Laboratory, Baltimore, MD, 1987, pp. 192-198.

[32] Thomopoulos, S. C. A., "Sensor Integration and Data Fusion," *Journal of Robotics Systems*, vol. 7, pp. 337-372.

[33] Julier, S., and J. K. Uhlmann, "General Decentralized Data Fusion with Covariance Intersection," in *Handbook of MultiSensor Data Fusion*: CRC Press, 2001.

[34] Hall, D., and D. Finkel, "Planning for Performance: Computer Performance Evaluation in the Early Stages of Project Development," *Modeling and Simulation*, vol. 14, 1983.

[35] Brewster, S., and R. Murray-Smith, *Haptic Human Computer Interaction*, Berlin: Springer-Verlag, 2001.

[36] Meyers, B. A., "A Brief History of Human Computer Interaction technology," *ACM Interactions*, vol. 5, pp. 44-54.

[37] Meyers, B. A., "User Interface Software Tools," *ACM Transactions on Computer-Human Interaction*, vol. 2, pp. 64-103.

[38] McNeese, M. D., and M. Vidulich, eds. *Cognitive Systems Engineering in Military Aviation Environments: Avoiding Cogminutia Fragmentosa*, Dayton, OH: Wright Patterson Air Force Base, CSERIAC Press, 2002.

[39] Taladoire, G., and D. Lille, "Conception and Realization of a "Living Map," *SPIE Visual Data Exploration and Analysis V*, San Jose, CA, January 26-27, 1998, pp. 214-223.

[40] Wohl, J., et al., "Human Cognitive Performance in ASW Operators," *1987 Tri-Service Data Fusion Symposium*, Johns Hopkins University Applied Physics Laboratory, Laurel, MD, June, 1987, pp. 465-479.

[41] Johnson, R., "Individual Styles of Decision Making: A Theoretical Model for Counseling," *Personnel and Guidance*, pp. 530-536.

[42] Llinas, J., J. Neal, and G. Kuperaman, "Systematic and Practical Views on Intelligent Interfacing for Data Fusion Applications," *Computing in Aerospace Conference VII*, Baltimore, MD, October, 1991.

[43] Jannis, I. I., and L. Mann, *Decision-Making*, New York, NY: Free Press, 1977.

[44] Dilts, R., et al., *Neuro-Linguistic Programming: The Study of the Structure of Subjective Experience,* vol. 1, Cupertino, CA: Meta Publication, 1980.

[45] Grandjean, E., and E. Vigiani, eds. *Ergonomic Aspects of Visual Display Terminals*, Proceedings of the International Workshop, London: Taylor and Francis, 1980.

[46] Reilly, S., and J. W. Roach, "Designing Human/computer Interfaces: A Comparison of Human Factors and Graphic Arts Principles," *Educational Technology*, pp. 36-39.

[47] Sawyer, S., *Software Development Practices*, D. Hall, Editor, State College, PA.

[48] MacCormack, A., et al., "Flexible Models of Software Development: Must We Trade Off Efficiency and Quality?," *IEEE Software*, 2002.

[49] Royce, W. W., "Managing the Development of Large Software Systems: Concepts and Techniques," *Western Electric Show and Convention (WESTCON)*, Los Angeles, CA, 1970, 1989.

[50] Connell, J. L., and L. Shafer, *Structured Rapid Prototyping: An Evolutionary Approach to Software Development*, Englewood Cliffs, NJ: Yourden Press, 1989.

[51] Boehm, B., "A Spiral Model of Software Development and Enhancement," *IEEE Computer*, vol. 21, 1988, pp. 61-72.

[52] Wang, C., "A Sucessful Software Development," *IEEE Transactions on Software Engineering*, vol. SE-10, 1984, pp. 714-727.

[53] Highsmith, J. A., *Adaptive Software Development*, New York, NY: Sorset House, 2000.

Chapter 11

Emerging Applications of Multisensor Data Fusion

This chapter provides an introduction to current and emerging applications for multisensor data fusion techniques. Surveys of Department of Defense and nondefense related applications are summarized. In addition, a cross-section of available commercial off-the-shelf (COTS) software for data fusion applications is provided.

11.1 INTRODUCTION

The multisensor data fusion discipline has advanced considerably since the first edition of this book. The basic algorithms and techniques for data fusion have evolved [1] and engineering standards are beginning to emerge for system design (Bowman and Steinberg [2]) and requirement derivation and analysis. Advances in signal processing, image processing, statistical estimation, and prototyping of expert systems have been made, and there is increased availability of COTS software to perform these functions [3]. Historically, the development of data fusion systems was hampered by a lack of infrastructure and commercial resources. This made the implementation of multisensor data fusion systems difficult for programs and initiatives other than those that were government-funded, particularly related to defense. However, newly available tools have enabled the design and implementation of modular data fusion systems based on COTS software instead of systems built from scratch. This has made multisensor data fusion techniques more available, accessible, and attractive for implementation for non-DoD applications such as law enforcement, medicine, condition-based maintenance (CBM), and others discussed in this chapter.

11.2 SURVEY OF MILITARY APPLICATIONS

Multisensor data fusion techniques were originally devised to solve complex military problems such as target tracking, automatic target recognition, and situation assessment. Three surveys of DoD data fusion systems have been published [4-6]. The purpose of these surveys was to summarize the application, development environment, system status, and key techniques utilized for identified systems. This section of the chapter provides a summary of these surveys.

The survey conducted by [4] identified 54 data fusion systems related to defense applications. The authors used several selection criteria in their survey. First, they reported on systems that could be described in an unclassified manner for open-source release. This criterion constituted a selection effect when describing the character and application of surveyed systems. Second, they selected systems for which there was published reference material—there had to be a readily accessible reference in order for a system to be included in the survey. The final criterion was a judgmental one. With the exception of two systems, they reported only on systems for which there were actual hardware or software implementations (albeit an implementation which may be only experimental or prototypical). In some cases, it was difficult from a published report to determine if a system had actually been implemented or whether it was merely conceptual. A summary of the systems identified by [4] is provided in Table 11.1. The full report describes the identified systems and summarizes their use of algorithms, software environment, and level of maturity, and provides contact information about the developers. While the survey by [4] is now dated, it provides a snapshot of the maturity of data fusion systems by the beginning of the 1990s.

Table 11.1

Summary of Systems Surveyed by [4]

System	Application	Data Types	Fusion Levels	Reference(s)
F/AATD	Navy avionic; air-to-air target tracking	Radar, passive IR, ESM	1	[7]
VIDS	Threat assessment for protection of a mobile tank	Optical, laser, passive missile detector, nuclear detector, chem.-bio sensor, millimeter wave radar	3	[8]
OPELINT	Operational electronic intelligence hull identification	ELINT reports	1	[9]
IADT	Ship defense and fire control	Multiple radars	1	[10]
NCCS(A)	Naval tactical flag command and control	Order of battle reports	1-3	[11]
PICES	Ocean surveillance tracker/correlator	Multiple infrared	1	[12]
ACDS	Navy track management and	Reports	1,2	[13]

Table 11.1 (continued)

Summary of Systems Surveyed by [4]

System	Application	Data Types	Fusion Levels	Reference(s)
	decision support			
MMFP	Air defense surveillance and tracking	Multiple radar	1	[14]
STEFIRD	Cruise missile detection	Radar, IR	1	[15]
IFFN	Tactical air-to-air IFFN	IFF identity declarations	1	[16]
LADMIES	Land-based tactical order of battle	ELINT; related reports	1	[17]
EDFA	Airborne surveillance	Pulse doppler radar, IR sensors	1	[18]
PATRIOT	SAM systems	CW surveillance radar, pulse doppler radar, CS illuminator	1,2	[19]
OSIS	Ocean surveillance	ELINT data	1	[20, 21]
P-3 UPDATE IV	Airborne ocean surveillance	Radar, ESM, acoustic, IRDS, visual, off-board	1	[22]
ERAEDF	Ship-based surveillance and self-defense	Radar and ESM	1	[23]
ESMT	Ship-based surveillance and self-defense	ESM	1	[24]
FMD-30	Automatic target recognition	Visual, thermal, radar, other	2	[25]
LFBSF	Automatic target recognition	LADAR, FLIR	1	[26]
TEDTE	Ballistic missile defense	Long wave infrared; other	1	[22]
TATR	Tactical air targeting	Reports	3	[27]
C³CM	Countermeasures assessment	Not applicable	3	[28]
PENDRAGON	Battlefield assessment	Reports	2	[29]
MASS	Ocean surveillance	Radar, ESM, IR, acoustic, IFF	1,2	[30-32]
SES	Underwater surveillance	Acoustic	1,2	[33]
AIFSARA	Tracking and situation assessment for tactical aircraft	Radar, IFF, IRST, NCTR	2	[34]
AFS	Space object status monitoring	L-band and X-band radar	1	[35]
ACHILLES	Anti-submarine warfare	Sonar, radar, MAD	2	[36]
GTSPECS	Object identification	Visible, IR imagery and radar	1	[37]
TAS	Target acquisition and tracking	Radar, IR, ESM, IFF	1	[38]
ACQUIRE	Battlefield management	Reports	2	[39]

Table 11.1 (continued)

Summary of Systems Surveyed by [4]

System	Application	Data Types	Fusion Levels	Reference(s)
KBSS	Analyst tool for algorithm development	Numerous	1,2	[40]
ICON	Identification of C^3 nodes	Order of battle reports	2	[41]
ESAU	Air order of battle situation assessment	Reports	2	[42]
KWB	AI-based prototyping tool	N/A	2,3	[43]
IPS	Indications and warnings	Reports	2,3	[44, 45]
NECTAR	Tracker correlator for ocean surveillance	Multisensor reports	1	[46]
TICKER	Antisubmarine warfare	reports	2	[47]
PATTI	Situation recognition and assessment	Images, reports	2	[48]
TMA	Ground-based order of battle assessment	Reports and human input	2	[49]
DACORR II	Combat order of battle analysis	Reports	1,2	[50]
ASSET	Various	Position and identity information	1	[51]
BETA	Tactical air-land battle management	ELINT, reports	1,2	[52-54]
LOCE	Tactical air-land battle management	ELINT, reports	1,2	[50]
LENSCE	Tactical air-land battle management	ELINT, reports	1,2	[50]
EXPRS	Same as above	Various	1,2	[55]
ANALYST I & II	Same as above	ELINT, images, reports	1,2	[56]
TCAC(D)	Same as above	ELINT, reports	1,2	[57]
INCA	Ocean surveillance	ELINT	1	[57]
AOBAA	Air order of battle analysis	Reports	1,2	[17]
TDP	Tactical navel threat analysis	Various	2	[58]
PTAPS	Ocean surveillance	Various	1	[59]
OSIF	Ocean surveillance	Various	1,2	[60]

A more recent survey of data fusion systems was conducted by Nichols [6]. She identified 79 operational systems associated with U.S. Department of Defense agencies. A list of those systems is provided in Table 11.2. Nichols analyzed the levels of fusion performed by these systems, assessed the system capabilities, and discussed the migration and legacy of theses systems. In her analysis, Nichols divided the systems into four categories: (1) migration systems that are converging to a common baseline to facilitate interoperability with other systems, (2) legacy systems that will be subsumed by a migration system, (3) government sponsored research and development prototypes, and (4) highly specialized capabilities (e. g., for a special application) that are not duplicated by any other system. Nichols concluded that there is significant coordination among military data fusion activities, which is counter to the common belief that the realm of data fusion systems is marked by multiple, duplicative activities.

Table 11.2

List of Data Fusion Systems Surveyed by [6]

Activity Acronym	Data Fusion System/Activity	Primary Service
ABI	AWACS Broadcast Intelligence	USAF
ADSI	Air Defense Systems Integrator	Joint
AEGIS	AEGIS Weapon System	USN
AEPDS	Advanced Electronic Processing & Dissemination System	USA
AMSTE	Affordable Moving Surface Target Engagement	DARPA
ANSFP	Artificial Neural System Fusion Prototype	USAF
ARTDF	Automated Real-Time Data Fusion	USMC
ASAS	All Source Analysis System	USA
ATW	Advanced Tactical Workstation	USN
ATWC	Automated Tomahawk Weapons Control System	USN
CAMDMUU	Connectionist Approach to Multi-Attribute Decision Making under Uncertainty	USAF
CEC	Cooperative Engagement Capability	USN
CEE	Conditional Events and Entropy	USN
CV	Constant Vision	USAF
DADFA	Dynamic Adaptation of Data Fusion Algorithms	USAF
DADS	Deployable Autonomous Distributed Fusion	USN
DDB	Dynamic Database	DARPA
E2C MCU	E2C Mission Computer Upgrade	USN
EASF	Enhanced All Source Fusion	USAF
EAT	Enhanced Analytical Tools	USAF
ECS	Shield Engagement Coordination System	USAF
ENT	Enhancements to NEAT Templates	USAF
ESAI	Expanded Situation Assessment & Insertion	USAF
FAST	Forward Area Support Terminal	USA

Table 11.2   (continued)

List of Data Fusion Systems Surveyed by [6]

Activity Acronym	Data Fusion System/Activity	Primary Service
GALE-Lite	Generic Area Limitation Environment Lite	Joint
GCCS	Global Command and Control System	Joint
GCCS A	Global Command and Control System Army	USA
GCCS I^3	Global Command and Control System Integrated Imagery and Intelligence	Joint
GCCS M	Global Command and Control System Maritime	USN
GDFS	Graphical Display Fusion System	USN
GISRC	Global Intelligence, Surveillance, and Reconnaissance Capability	USN
Hercules		Joint
IAS	Intelligence Analysis System	USMC
IDAS	Interactive Defense Avionics System	USAF
IFAMP	Intelligent Fusion and Asset Management Processor	USAF
ISA	Intelligence Situation Analyst	USAF
ISAT	Integrated Situation Awareness and Targeting	USA
IT	Information Trustworthiness	USA
ITAP	Intelligent Threat Assessment Processor	USAF
JIVA	Joint Intelligence Virtual Architecture	Joint
ISTARS CGS	Joint Surveillance Target Attack Radar Subsystem Common Ground Station	Joint
JTAGS	Joint Tactical Ground Station	Joint
KBS4TCT	Knowledge-based Support for Time Critical Targets	USAF
LOCE	Linked Operational Intelligence Centers Europe	Coalition
LSS	Littoral Surveillance System	USN
MDBI&U	Multiple Data Base Integration & Update	USAF
MITT	Mobile Integrated Tactical Terminal	USA
Moonlight	Moonlight	Coalition
MSCS	Multiple Source Correlation System	USAF
MSFE	Multi-source Fusion Engine	USAF
MSI	[E2C] Multi-source Integration	USN
MSTS	Multi-source Tactical System	USAF
NCIF	Network Centric Information Fusion	USAF
NEAT	Nodal Exploitation and Analysis Tool	USAF

Table 11.2 (continued)

List of Data Fusion Systems Surveyed by [6]

Activity Acronym	Data Fusion System/Activity	Primary Service
OBATS	Off-Board Augmented Theatre Surveillance	USAF
OED	Ocean Surveillance Information System Evolutionary Development	USN
Patriot	Patriot Weapon System	Joint
PICES	Processor Independent Correlation Exploitation System	USN
QIFS	Quarterback Information Fusion System	USAF
SAFETI	Situation Awareness From Enhanced Threat Information	USAF
SCTT	SAR Contextual Target Tracking	USA
SMF	SIGINT/MTI Fusion	USA
Story Teller	EP3 Story Teller	USN
TADMUS	Tactical Decision Making under Stress	USN
TAS	Timeline Analysis System	USAF
TBMCS	Theatre Battle Management Core System	USAF
TCAC	Technical Control and Analysis Center	USMC
TCR	Terrain Contextual Reasoning	USA
TDPS	Tactical Data Processing Suites	USAF
TEAMS	Tactical EA-6B Mission Support	USN
TERPES	Tactical Electronic Reconnaissance Processing Evaluation System	USMC
TES	Tactical Exploitation System	USA
TIPOFF	TIBS Integrated Processor and Online Fusion Function	Joint
IMBR	Truth Maintenance Belief Revision	USAF
TRAIT	Tactical Registration of Airborne Imagery for Targeting	USAF
TSA	Theatre Situation Awareness	USAF
UGBADFT	Unified Generalized Bayesian Adaptive Data Fusion Techniques	USAF
VF	Visual Fusion	USA
WECM	Warfighter Electronic Collection and Mapping	USA

Figure 11.1 shows a comparison of the levels of fusion performed by the surveyed systems and a comparison of the Hall, Linn, and Llinas and Nichols survey. It is not clear whether the relative distribution of systems performing levels 1, 2, and 3, fusion are representative of actual implementation or a result, for example, of a classification selection effect (namely are threat assessment systems

inherently classified compared to level 1 systems?). It is suspected that this is truly representative of actual applications, since level 1 is usually the first step that must be performed in a hierarchy of inferences involved in data fusion and battle management. From the chart it is clear that much progress has been made toward the implementation of level 2 fusion. However, there is much to be accomplished to advance the higher levels of fusion.

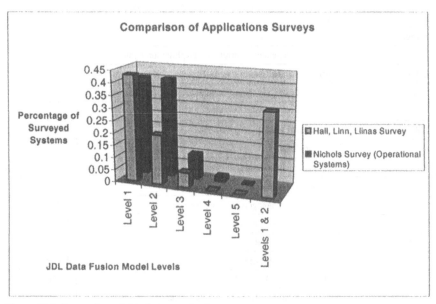

Figure 11.1 Comparison of data fusion applications surveys.

11.3 EMERGING NONMILITARY APPLICATIONS

As previously noted advances in the enabling technologies for multisensor data and the availability of commercial-off-the shelf software have encouraged the emergence of data fusion application to nonmilitary or nontraditional problems. Unfortunately, much of the work in this area has not been published due to the proprietary nature of the research and development. Research in this area has been partially funded through government grants and small business initiative research (SBIR) contracts. The applications that are described in this section represent some ongoing research. Table 11.3 shows a sample of these applications. Additional information is provided below.

Table 11.3

Nontraditional Multisensor Data Fusion Applications

Application Area	Description
Condition-based monitoring	• Monitoring the health of complex systems • Fatigue and wear of critical structures and systems (e.g., aircraft wings, and bridge supports) • Inspections of equipment (sewer systems, various containers and tanks, for example)
Medicine	• Patient monitoring • Medical condition/illness diagnosis • Remote patient treatment • Medical Information Management
Manufacturing/quality control	• Non-destructive testing (NDT) • Quality control monitoring • Assembly line monitoring/control • Product design and prototyping
Transportation	• Airfield/air traffic monitoring • Railroad/rail yard monitoring/control • Road monitoring and advanced traffic control • Shipyard monitoring
Robotics	• Autonomous control of robot movement • Control of robots in hostile environments not suitable for humans • Control of robots performing critical assembly or repair tasks • Search and rescue in confined or hazardous areas
Law enforcement	• "Roving eye" to provide situational awareness in hostile environments requiring covert sensors (e.g., hostage situations and terrorist activity) • Virtual training environments integrating the actions of multiple trainees • Identification and tracking of suspects
Environmental monitoring	• Monitoring of crop health and weather damage • Evaluation of long-term effects such as erosion • Analysis of effects due to natural disasters (e.g., hurricanes and volcanic eruptions)

11.3.1 Intelligent Monitoring of Complex Systems

Intelligent monitoring systems seek to monitor the health and status of complex systems to detect the onset of failure conditions and to predict the time to failure or remaining useful life [61-63]. In particular, intelligent monitoring systems enable the implementation of a condition-based maintenance (CBM) philosophy.

CBM is a philosophy of performing maintenance on a mechanical system only when there is objective evidence of the need for such maintenance. By contrast, time-based maintenance performs maintenance based on the utilization history or after a specified duration of time. Time-based maintenance is unduly expensive for two reasons. First, maintenance is performed on perfectly good machinery before the need for such maintenance. Second, the very act of maintenance, even when performed correctly, often induces problems in previously functioning machinery (so-called *iatropic* maintenance problems). Similarly, failure-based maintenance (i.e., the philosophy of performing maintenance after a failure has occurred) can be dangerous and lead to lack of operational readiness for critical systems.

A conceptual model for an intelligent monitoring system is shown in Figure 11.2. A mechanical system or platform such as a rotorcraft (shown at the top left-hand side) is to be monitored for status and mechanical health. Failure mechanisms for such a system may include corrosion, wear, lubricant contamination or degradation, and thermomechanical fatigue. Active and passive sensors monitor the components and subsystems of the rotorcraft. Examples of appropriate sensors include acoustics, infrared for detecting heat, laser monitoring of lubricants, accelerometers and embedded sensors. In addition to these sensors, human inputs may be available to report anomalous operating conditions. In a sense, the problem of determining the health and potential fault conditions of a mechanical system is very analogous to a multisensor data fusion system monitoring a battlefield, seeking to determine the current situation and to identify potential threats. The breadth of sensors and utilization of human inputs provide an additional analog to defense-related data fusion problems.

The importance of accurate health assessment and prediction for electrical and mechanical systems has been cited in numerous national defense plans and studies. For example, the 1996 United States Defense Technology Plan objectives included an 80% reduction in aircraft mechanical mishaps and a 35% reduction in spare parts inventories. A 1995 OASN study on safety and survivability of aircraft recommended a need to achieve a 30% reduction in life-cycle maintenance costs. Multiple safety reports have identified the impact of undetected maintenance problems that resulted in the loss of life for military personnel. A 1996 U.S. Air Force Integrated Product Team report identified 58 preventable mishaps resulting in the loss of 16 lives and an equipment cost of $365 million.

There are many challenges for detecting precursors to failure conditions. First, failures and failure mechanisms occur primarily at the materials level, but their effects are observed at a system or subsystem level. For example, a worn gear tooth in a transmission may cause undue vibration and heat. Thus, the fundamental causes of failure often cannot be observed directly. Instead, one must observe system effects such as heat, vibration, noise, or debris in a lubricant. A second problem is the many-to-many relationship between failure mechanisms and macro-observable quantities (e.g., many failure conditions cause excess vibration). Third, the observing environment can be very challenging. For example, a worn gear tooth in a helicopter transmission may result in a ¼ g change in vibration on the transmission housing. However, the in-flight vibration level for the transmission may approach 1,000 g$_s$. This is a 4,000 to 1 noise to signal ratio. Finally, the sensors used to monitor a complex system are often far more fragile than the system for which they are intended to monitor. Frequently, when a monitoring system "detects" a mechanical failure, the only thing that has failed is the sensing system.

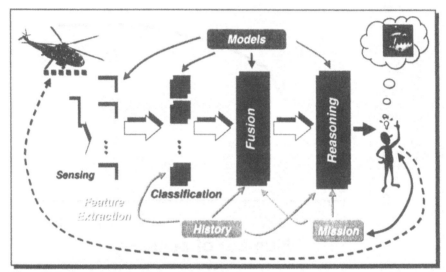

Figure 11.2 Concept of an intelligent monitoring system.

Data fusion has the potential to improve the ability to detect, characterize, and identify fault conditions. However, despite extensive research on condition-based maintenance related to sensors, mechanical models, and signal processing of individual sensor data, pattern recognition, and predictive models, relatively little research has been conducted on fusion of sensor data for CBM. McClintic [64] used a Kalman filter approach to combine vibration data for monitoring a mechanical transmission. Byington et al. [65] describe the use of automated reasoning to fuse vibration and oil debris data associated with monitoring a gearbox. Finally, Erdley [66] compared the use of voting techniques and Bayesian probability methods for improving fault identification using multiple accelerometers on a helicopter transmission. An example of the improved identification as a function of number of sensors (namely fusing data from N sensors) is shown in Figure 11.3. Note two effects. First, while increasing the number of sensors improves the fault identification, the improvement asymptotically approaches a limit beyond about eight sensors (although it should be stated that the sensors are commensurate sensors). Second, using two sensors produces worse results that using one or three sensors.

This is due to the fact that if two sensors disagree, then the results can be worse (on the average) than using only one sensor or three sensors (in which tie votes can be broken by the third sensor). It is expected that fusion of noncommensurate sensor data could produce even better results than shown in Figure 11.3.

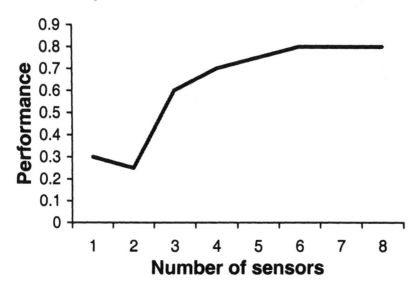

Figure 11.3 Example of Improved Fault Identification using Multiple Sensor Data [66].

11.3.2 Medical Applications

One of the most promising applications for multisensor data fusion is for use by the medical community. To care for patients, medical professionals rely on data from their own senses and sensor data such as x-rays, magnetic resonance imaging (MRI), blood tests, and tissue cultures. Patients that are hospitalized are often monitored at regular intervals or constantly by sensors, depending upon the seriousness of their condition. Health care providers use this information to make ongoing assessments for the treatment of patients. In a number of cases, multisensor data fusion can aid the health care providers by providing fused sensor reports and aiding in the diagnosis process. Figure 11.4 shows a conceptual medical information management process for remote monitoring of patients.

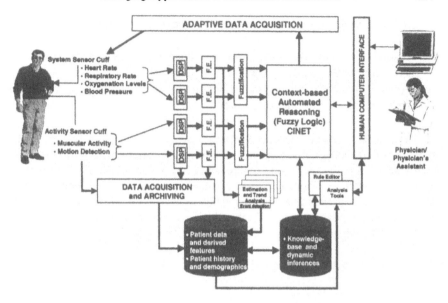

Figure 11.4 Concept of a multisensor system for remote monitoring of patients.

In Calgary, Canada, for example, a number of scientists are exploring the use of multisensor data fusion techniques implemented to aid in the detection of breast cancer [67]. The use of data from multiple sources may serve to improve the detection rate of the disease as well as to reduce the number of unnecessary biopsies. Patient self-examination data and family health history are combined with mammogram imaging, radiologist expertise, and pathology data. Currently, a number of malignant tumors go undiagnosed and numerous unnecessary biopsies are performed. The Canadian research team seeks to implement computer-aided diagnosis (CAD) systems to improve detection rates. For this application, the most difficult aspect of the implementation is to model the experience, expertise, and decision-making processes of the radiologist. However, the potential of systems of this type may greatly enhance health care processes.

11.3.3 Law Enforcement

In many respects, law enforcement requires similar situational awareness data to that required by ground military forces or small unit operations (SUO) [68]. Sensors introduced into to hostile areas prior to personnel entering the area serve to provide visualization data as to the location of hostile individuals and weapons. In the case of hostage situations, it is critical to determine the locations of the captures and the hostages to lessen the danger to innocent people should force be used. These sensors can be used to build complex visual images (in some cases in three dimensions) to support law enforcement rescue operations.

Ortolf and Saunders offer several potential capabilities for implementation:

- Sensory extension: Beyond visual range situational awareness imagery;
- Precision sensing: Readily available video sensors;
- Adaptive re-configurable systems: Sensors capable of adapting to the environment and the priorities of the user.

These sensors can be introduced into the environment using techniques such as self-attaching sensors that can be gun-launched or hand-emplaced in hard-to-reach places [68]. Another mode of operation is a sense-on-the-fly system that provides data while in flight. This system would potentially be launched in a ballistic orbit and may employ a small parachute. Other applications of data fusion for law enforcement are described in the SPIE conference on Sensors, C3I, Information, and Training Technologies for Law Enforcement [69].

11.3.4 Nondestructive Testing (NDT)

A broad application area for multisensor data fusion is nondestructive testing (NDT) and nondestructive evaluation (NDE). Two recent texts, edited by Gros, describe the application of data fusion to this area [70, 71]. NDT/NDE involves methods of determining the condition of complex systems ranging from mechanical systems to humans. The application is analogous to condition-based maintenance. Specific types of NDE include monitoring the quality of manufactured items; medical applications; determining the safety of buildings, bridges, and roads; and process control in manufacturing. Sensors for NDT span a wide range from visual inspection by human observers, to the use of X-rays, ultrasound, eddy currents, magnetic particle inspection, gamma rays, acoustic emission, infrared thermography, supercomputing quantum interference devices (SQUID), magnetometers, and laser interferometry. These are highly noncommensurate sensors. Hence, the pattern recognition and decision-level algorithms described in this text are appropriate.

11.3.5 Robotics

An application that has provided significant innovations for data fusion is the area of robotics [72]. Examples of robotic applications include: robots to explore the surface of a planet [73], robots to explore buildings [74], servant robots [75], industrial applications such as machine tending and material handling [76], and use of robots to explore areas affected by disasters such as explosions, chemical spills, or industrial accidents (see www.aaai.org/AITopics/html/rescure.html for links to papers on use of robots for rescue missions). A discussion of the use of multisensor data fusion for robot applications is provided by [77, 78]. The challenges for robotic applications include computer vision (e.g., recognition of objects and surroundings), self-navigation (movement of the robot and robot components such as arms and hands), automated situation assessment, consequence prediction, and overall resource control (including direction and utilization of the sensors and robot resources). In addition, the processing of sensor data for robots needs to be on a real-time basis in order for the robot to react to its surrounding environment in a timely basis. This is a rich area of investigation for students of data fusion.

11.4 COMMERCIAL OFF THE SHELF (COTS) TOOLS

In 1993, Hall and Linn [79] conducted a survey of COTS software to support development of data fusion systems for applications such as automatic target recognition, identification-friend-foe-neutral processing, and battlefield surveillance. In that survey, they described emerging packages containing basic algorithms for signal processing, image processing, statistical estimation, and prototyping of expert systems. Since the publication of that paper, extensive progress has been made in data fusion for both Department of Defense (DoD) applications as well as non-DoD applications. In addition, non-DoD applications such as data mining, pattern recognition and knowledge discovery have encouraged the development of commercial software tools [80] and general packages such as MATLAB (see http://www.mathworks.com) and Mathematica (http://www.wolfram.com/products/mathematica/). This chapter provides an update of the original survey conducted by Hall and Linn and provides a "survey of surveys" related to data fusion systems. This section is based on the paper by Hall and Sherry [3].

11.4.1 Survey of COTS Software

Two significant advances have shaped the evolution of COTS data fusion applications software since the publication of Hall and Linn's survey in 1993. First is the rapid development of specialized software algorithm packages used within technical computation environments such as MATLAB, Mathematica, and Mathcad (http://www.mathcad.com/). These environments provide unified platforms for mathematical computation, analysis, visualization, algorithm development, and application. The second development is the evolution of general-purpose data fusion software packages with adaptive capabilities to a diverse set of data fusion applications.

11.4.2 Special Purpose COTS Software

Table 11.4 shows a summary of specialized software packages. These are all compatible with at least one of the three mathematical environments previously described. There are literally hundreds of software packages available. Hence,

Table 11.4 is by no means a complete listing of the available commercial products. However, it provides a basis for a more in depth search for such products. Use of the below packages in conjunction with MATLAB, Mathematica or Mathcad algorithms, visualization features, and user interface capabilities can significantly reduce development time for prototype application simulations. Projects initially prototyped and implemented in MATLAB or Mathematica environment can be ported to C and C++ using programs such as RTExpress [81] to generate parallel C code from high level MATLAB programs, or MathCode C++ [82] for generation of C++ code from Mathmatica code.

Table 11.4

Examples of COTS Software for Use in Mathematical Software Environments

Category of Algorithms/ Techniques	JDL Model Level	Program Name	Source
Image processing and mapping	0	DIPimage	Delft University of Technology http://www.ph.tn.tudelft.nl/DIPlib
		Mapping Toolbox	HalloGram Publishing http://hallogram.com/mathworks/maptoolbox
Pattern recognition and classification	1,2,3	PRTPattern	Innovative Dynamics, Inc. http://www.mathworks.com/products/connections/product_main.shtml?prod_id=88
		Statistic Pattern Recognition Toolbox	Czech Technical University http://cmp.felk.cvut.cz/~xfrancv/stprtool/
		Pattern Recognition Toolbox	Ahlea Systems http://www.ahlea.com/products.html
		PRTools	Delft University of Technology http://www.ph.tn.tudelft.nl/~bob/PRTOOLS.html
Neural networks	1,2,3	Netlab	Aston University http://www.ncrg.aston.ac.uk/netlab/
		Netpack	University of Nijmegen, The Netherlands http://www.mbfys.kun.nl/snn/nijmegen/index.php3?page=36
		NNSYSID	Technical University of Denmark http://www.iau.dtu.dk/research/control/nnsysid.html
		NNCNTL	Technical University of Denmark http://www.iau.dtu.dk/research/control/nnctrl.html
		Back Propagated Neural Network	Steve Hunka http://www.mathsource.com/Content/Applications/ComputerScience/0209-078
		Simulating Neural Networks	James A. Freeman http://www.mathsource.com/Content/Publications/Book

Table 11.4 (continued)

Examples of COTS Software for Use in Mathematical Software Environments

Category of Algorithms/ Techniques	JDL Model Level	Program Name	Source
			Supplements/Freeman-993/0205-906?notables
Fuzzy logic/ control	2,3,4	FISMAT	http://www.mathtools.net/files/net/fismat.zip
		Fuzzy Model-ing	Delft University of Technology http://lcewww.et.tudelft.nl/~crweb/software/
		FMBPCTool	Delft University of Technology http://lcewww.et.tudelft.nl/~crweb/software/
Bayesian networks	1,2,3	Bayes Net	University of California at Berkeley http://www.cs.berkeley.edu/~murphyk/Bayes/request.html
		Hidden Markov Model	University of California at Berkeley http://www.cs.berkeley.edu/~murphyk/Bayes/request.html
Digital signal processing and filters	0,1	DSP-KIT	Meodat Gmbh http://www.meodat.de/PCSys/dsp-kit.htm
		DirectDSP/ DSPower Block Dia-gram	Signalogic http://www.signalogic.com/matlab.shtml
		DSP Builder	Altera http://www.altera.com/products/software/system/products/dsp/dsp-builder.html
		FIR and IIR Filter Design	Rice University http://www.dsp.rice.edu/software/rufilter.shtml
		Mango EDS	Mango DSP Ltd. http://www.mathworks.com/products/connections/product_main.shtml?prod_id=110
		DSPdeveloper	SDL http://www.sdltd.com/
		Kalman Filter	University of California at Berkeley http://www.cs.berkeley.edu/~murphyk/Bayes/request.html
		KALMTOOL	Technical University of Denmark http://www.iau.dtu.dk/research/control/kalmtool.html
		FIR Filter	University of Maryland, Baltimore County http://www.csee.umbc.edu/~dschol2/opt.html
		WarpTB	Helsinki University of Technology http://www.acoustics.hut.fi/software/warp/
		ReBEL	Oregon Health & Science University http://varsha.ece.ogi.edu/rebel/index.html

Table 11.4 (continued)

Examples of COTS Software for Use in Mathematical Software Environments

Category of Algorithms/ Techniques	JDL Model Level	Program Name	Source
		Signals Processing Packages	Brian Evans http://www.ece.utexas.edu/~bevans/projects/symbolic/spp.html
		Diehl's DSP Extensions	Adept Scientific plc http://mathcad.adeptscience.co.uk/dsp/
Tracking and positioning	1	AuTrakMatlab	AuSIM http://www.ausim3d.com/products/autrak.html
		DynaEst	Artech House http://www.mathworks.com/support/books/book1410.jsp?category=3&language=-1
		Target Tracking	Georgia Tech University http://seal.gatech.edu/onr_workshop/2001Slides/West.pdf
Wavelet analysis	0,1,2	Wavelab	Stanford University http://www-stat.stanford.edu/~wavelab/
		Rice Wavelet Toolbox	Rice University http://www.dsp.rice.edu/software/rwt.shtml
Self-organizing maps	1,2	SOM Toolbox	Helsinki University of Technology http://www.cis.hut.fi/projects/sotoolbox/
Graph theory and search	1	Graph Theory	University of California at Berkeley http://www.cs.berkeley.edu/~murphyk/Bayes/request.html
		Anneal	Jeff Stern http://www.mathsource.com/Content/Applications/ComputerScience/0205-030
		Heuristic Search Techniques	Raza Ahmed, Brian L. Evans http://www.mathsource.com/Content/Applications/ComputerScience/0208-044
Speech processing	5	VOICEBOX	Imperial College of Science, Technology and Medicine http://www.ee.ic.ac.uk/hp/staff/dmb/voicebox/voicebox.html
Database	N/A	DBTool	http://go7.163.com/energy/matlab/dbtool.htm

11.4.3   General Purpose Data Fusion Software

The second class of COTS software involves packages specifically designed to perform data fusion. A summary of these is provided in Table 11.5 and a brief summary of the packages is provided below.

The Lockheed Martin data fusion workstation [82] is directed at integrated multisensor multitarget classification based on fusing 14 types of sensor inputs including; passive-acoustic, ESM/ELINT, magnetic field, electric field, inverse synthetic aperture radar and geolocation. The data fusion workstation consists of three principal components; 1) the embedded data fusion system, 2) a graphical user interface, and 3) a data fusion input simulator. The fusion engine is fuzzy logic based with Dempster-Shafer reasoning. The software is object oriented C++ code that runs on Silicon Graphics and Sub SPARCstation platforms. The Lockheed Martin data fusion system software is utilized as an embedded function in the Rotorcraft Pilot's Associate.

Hall and Kasmala [83] at the Pennsylvania State University Applied Research Laboratory developed a visual programming tool kit for level 1 data fusion. The tool kit is modeled after the tool *Khoros* (www.khoral.com) developed for collaborative development of image processing algorithms. The top-level interface, shown in Figure 11.5, allows a user to define a level 1 processing flow by a "point and click" manipulation of graphic symbols representing types of fusion algorithms. The user can define multiple processing streams to analyze single sensor data or multiple processing streams to fuse multisensor data. A library of routines is available and can be selected for processes such as signal conditioning, feature extraction, and pattern recognition or decision-level fusion. The tool kit was originally developed for applications involving condition monitoring of mechanical systems. However, the algorithms are general purpose and can be used on any signal or vector data.

Brooks and Iyengar [84] provide a suite of C language routines as companion software to their book on multisensor fusion. These routines collectively have been labeled the sensor fusion toolkit and contain implementations of machine learning, learned metaheuristics, neural networks, simulated annealing, genetic algorithms, tabu search, a Kalman filter, and their own distributed dynamic sensor algorithm.

The Department of Energy has funded research directed at developing models for groundwater flow and contaminant transport. This work has resulted in development of the Hydrogeologic Data Fusion computer model [85]. This model can be used to combine various types of geophysical, geologic, and hydrologic data from diverse sensor types to estimate geologic and hydrogeologic properties. A commercial version of this software is available under the name HydroFACT from Fusion and Control Technology, Inc.

KnowledgeBoard [86] is a software framework for information collection, data fusion and information visualization. KnowledgeBoard is directed at the fusion of distributed high-level multisource data including XML based documents, video, Web-based content, relational and object databases, and flat files. Heterogeneous data sources are displayed within a contextual framework to support user level fusion of the information and to support improved knowledge generation.

PCI Geomatics has developed the ImageLock Data Fusion software [87], which is designed to perform semi-automatic geometric correction, fuse the corrected multisensor imagery and perform noise and artifact removal. The ImageLock Data Fusion software is a component in PCI's EASI/PACE remote sensing image processing software package.

eCognition [88] by Definiens Imaging is designed to extract features from high-resolution satellite and aerial imagery. eCognition provides for multisource fusion from a large variety of formats, sensors, and platforms at any spatial and spectral resolution.

MUSE [89] by Cambridge Consultants is a real-time data capture software environment with capabilities for fault diagnosis, condition monitoring, alarm handling, supervisory control, situation assessment, data fusion, object recognition, and knowledge-based simulation. MUSE has been used to develop applications in the following areas: robotics, financial trading, sensor data fusion, and process control.

Boeing has developed a data fusion workstation for fusing data from nondestructive examination (NDE) testing and from process monitoring [90].

In the category of defense research and development in Canada, Valcartier has developed a test bed for evaluating multisensor data fusion systems against system requirements [91]. This test bed is called Concept Analysis and Simulation Environment for Automatic Target Tracking and Identification (CASE ATTI). One of the main objectives of the development of CASE ATTI was to allow the testing and comparison of different sensor fusion algorithms and techniques without extensive recoding. CASE ATTI allows use of either externally generated sensor data or data generated by a high-fidelity simulator that emulates the behavior of targets, sensor systems, and the environment. Case ATTI runs on UNIX and Windows platforms and can be used with multiple computers across a local area network.

Table 11.5

General Purpose Data Fusion Software

Software/Product Name	Source/Developer	Description
Lockheed Martin Data Fusion Workstation [82]	Lockheed Martin	• Integrated multisensor, multitarget tool • Fuses data from 14 sensor inputs including: passive-acoustic, ESM/ELINT, magnetic field, electric field, inverse synthetic aperture radar and geo-location • Components: graphical user interface, embedded data fusion system, data fusion input simulator • Fusion engine based on Dempster-Shafer reasoning • Platform: Silicon Graphics, Sub SPARCstation
PSU Applied Research Labora-	Penn State Applied Research	• Provides a visual interface for utilizing and developing level 1 fusion algorithms

Table 11.5 (continued)

General Purpose Data Fusion Software

Software/Product Name	Source/Developer	Description
tory Data Fusion Toolkit [83]	Laboratory	• Platform: PC with Windows
Brooks and Iyengar [84]	R. Brooks & S. Iyengar	• C language routines as companion software to their data fusion book • General techniques including neural networks and Kalman filter
HydroFACT [85]	Fusion and Control Technology, Inc.	• Department of Energy-funded tool to combine data associated with geophysical, geologic, and hydro-logic data
KnowledgeBoard [86]	SAIC	• Information visualization and fusion for multisource data including video, XML-based documents, and images
ImageLock Data Fusion [87]	PCI Geomatics	• Semi-automatic geometric correction and fusion of image data
eCognition [88]	Definiens Imaging	• Fusion and preprocessing of image data
MUSE [89]	Cambridge Consultants	• Real-time S/W environment for fault diagnosis, condition monitoring and alarm handling
Boeing NDE S/W [90]	Boeing	• Data fusion workstation for nondestructive evaluation (NDE) and process monitoring
Integrated battle-space intelligence segments (IBIS)	Veridian	• IBIS is a framework for developing and integrating intelligence analysis applications; includes a tactical terrain analyzer, location refinement, movement projection, moving target tracker, and situation templating
Pharos	Sonardyne	• Acoustic navigation software for subsea acoustic geopositioning
CASE ATTI [91]	Defense R&D Canada	• Test bed for evaluating multisensor data fusion systems against requirements

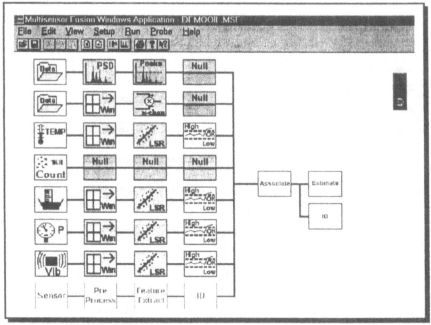

Figure 11.5 Example of HCI for data fusion tool kit [83].

11.4.4 A Survey of Surveys

In order to assist in identifying additional tools, Table 11.6 identifies sources (papers, reports, and Web sites) that provide additional surveys of relevant software. In addition to these resources, annual software surveys are available via the ACM Computing Surveys. Also, the NECI search service, Inquiris2 (see http://inquirus.nj.nec.com/i2/) provides an excellent technical search capability.

Table 11.6

Software Surveys

Developer/ Source	Focus	Reference
D. L. Hall and R. J. Linn [79]	Survey of COTS S/W to support data fusion	"A survey of commercial software for multisensor data fusion," *Proceedings of the SPIE Conference on Sensor Fusion & Aerospace Applications*, Orlando, FL, 1993
D. L. Hall and R. J. Linn and J. Llinas[4]	Survey of DoD data fusion systems	"A survey of data fusion systems," *Proceedings of the SPIE Conference on Data Structures & Target Classification*, Orlando, FL. 1991 pp 13-36

Table 11.6   (continued)

Software Surveys

Developer/ Source	Focus	Reference
M. L. Nichols [6]	Survey of DoD data fusion systems	"A survey of multisensor data fusion systems," chapter 22 in *Handbook of Multisensor Data Fusion*, ed by D. Hall & J. Llinas, CRC Press Inc. Boca Raton, FL, 2001
Aerospace Corp.	Survey and analysis of 34 tools for multisource data fusion at the NAIC	"An investigation of tools/systems for multi-source data fusion at the national air intelligence center" by C. N. Mutchler; technical report prepared for the NAIC; Aerospace Report TOR-2001 (1758)-0396e; 2001
M. Goebel and L. Gruenwald [80]	Survey of data mining and knowledge discovery S/W	"A survey of data mining & knowledge discovery software tools," *SIGKDD Explorations*, Vol. 1, issue 1, pp. 20-33, 1999
B. Duin	Web-based links to tools associated with pattern recognition, neural networks, machine learning, image processing and vision, and OCR and handwriting recognition	http://www.ph.tn.tudelft.nl/PRInfo/software.html
EvoWeb	List of software tools for pattern recognition and classification maintained by the European Commission's IST Program	http://evonet.dcs.napier.ac.uk/evoweb/resources/software/keywordklist6.02.html
Pattern recognition on the Web	Large set of links to training materials, tutorials, and software tools for pattern recognition, statistics, computer vision, classifiers, filters and other S/W	http://cgm.cs.mcgill.ca/~godfried/teaching/pr-web.html
Carnegie Mellon University (CMU)	StatLib—data, software and information for the statistics community maintained by researchers at CMU; extensive list of software and resources	http://lib.stat.cmu.edu/
Data and Analytic Center for Software (DACS)	Extensive list and comments on software for data mining, data warehousing, and knowledge discovery tools	http://www.dacs.dtic.mil/databases/url/key.hts?keycode=222:225&islowerlevel=1
Georgia Tech	Listing of commercial data mining software	http://www.cc.gatech.edu/~kingd/datamine/commercial.html
M. Wexler	Open directory of software, databases,	http://dmoz.org/Computers/Software

Table 11.6 (continued)

Software Surveys

Developer/ Source	Focus	Reference
	and reference materials on data mining; list of 87 tool vendors	/Databases/Data_Mining/
Open Channel Foundation	Open Channel Foundation publishes software from academic and research institutions; software available for development of expert systems and pattern recognition	http://www.openchannelfoundation.o rg/discipline/Artificial_Intelli- gence_and_Expert_Systems/
NRC CNRC Institute for Information Technology	Directory of software for expert sys- tems and knowledge-based systems	http://www.iit.nrc.ca/subjects/Expert .html
Compinfo – The Computer Information Center	Link to sources of information about computer software, and information technology	http://www.compinfo-center.com/
OR/MS Today	Survey of 125 vendors of statistical software	http://www.lionhrtpub.com/orms/sur veys/sa/sa-surveymain.html

11.5 PERSPECTIVES AND COMMENTS

There is a clear acceleration of the development and application of data fusion software methods across a growing number of application domains.

1. There is an explosion of commercial software tools for a wide variety of component techniques such as signal processing, image processing, pattern recognition, data mining, and automated reasoning.
2. New mathematical environments such as MATLAB, Mathcad, and Mathe- matica provide extensive tool kits (including methods for signal processing, neural nets, and image processing) that are useful for rapid prototyping and evaluation of data fusion methods.
3. A limited number of more general-purpose tool kits are available. These are quasi-COTS tools (they are available for use or purchase but do not have the general utilization or support of software such as MATLAB or Mathematica).
4. The software development environment (including new languages such as Visual Basic, Visual C^{++}, JAVA Script, and XML) provides an improved basis for development of object-oriented, interoperable software.

In general, these developments provide an increasingly efficient environment to develop and use data fusion algorithms. However, we still await the widespread availability of a standard package for multisensor data fusion.

REFERENCES

[1] Hall, D., and J. Llinas, eds. *Handbook of Multisensor Data Fusion*, New York, NY: CRC Press, 2001.

[2] Bowman, C., and A. Steinberg, "A Systems Engineering Approach for Implementing Data Fusion Systems," in *Handbook of Multisensor Data Fusion*, ed. by J. Llinas, New York, NY: CRC Press, 2001.

[3] Hall, S. A., "A Survey of COTS Software for Multisensor Data Fusion: What's New since Hall and Linn," *2002 MSS National Symposium on Sensor and Data Fusion*, San Diego, CA, June, 2002.

[4] Hall, D., R. J. Linn, and J. Llinas, "Survey of Data Fusion Systems," *SPIE Conference on Data Structure and Target Classification*, Orlando, FL, 1991, pp. 13-36.

[5] Llinas, J., and R. Antony, "Blackboard Concepts for Data Fusion Applications," *International Journal of Pattern Recognition and Artificial Intelligence*, vol. 7, 1993, pp. 285-308.

[6] Nichols, M., "A survey of Multisensor Data Fusion Systems," in *Handbook of Multisensor Data Fusion*, ed. by J. Llinas, Boca Raton, FL: CRC Press, 2001.

[7] Ditzler, W. R., et al., "A Demonstration of Multisensor Tracking," *1987 Tri-Service Data Fusion Symposium*, Johns Hopkins University, Baltimore, MD, June, 1987, pp. 303-311.

[8] Caito, S. A., and S. R. L., "Development of the Vehicle Integrated Defense System Feasibility Demonstration," *1987 Tri-Service Data Fusion Symposium*, Johns Hopkins University, Baltimore, MD, June, 1987, pp. 398-414.

[9] Kramer, F., "A Survey of New Techniques for Processing Electronic Intelligence," *1987 Tri-Service Data Fusion Conference*, Johns Hopkins University, Baltimore, MD, June, 1987, pp. 5-16.

[10] Citrin, W. I., "Integrated Automatic Detection and Tracking in Distributed Processors," *1987 Tri-Service Data Fusion Symposium*, Johns Hopkins University, Baltimore, MD, 1987, pp. 100-107.

[11] Tompson, B. B., C. T. Sutherlin, and A. Hafner, "The Composite Warfare Commander's Tactical Decision Support System," *1988 Tri-Service Data Fusion Symposium*, Johns Hopkins University, Baltimore, MD, May, 1988, pp. 79-90.

[12] Fitchek, J. J., J. P. Lee, and D. Herring, "A Multi-frequency, Multi-sensor Data Fusion System," *1988 Tri-Service Data Fusion Symposium*, Johns Hopkins University, Baltimore, MD, May, 1988, pp. 209-217.

[13] Feldman, S. J., "Real-time Automated Track Management," *1988 Tri-Service Data Fusion Symposium*, Johns Hopkins University, Baltimore, MD, May, 1988, pp. 297-305.

[14] Liggins, M. E. I., "Multi-spectral, Multi-sensor Fusion Development for Enhanced Target Detection and Tracking," *1988 Tri-Service Data Fusion Symposium*, Johns Hopkins University, Baltimore, MD, May, 1988, pp. 9-21.

[15] Yool, S. R., et al., "Small Target Enhancement with Fused Infrared and Radar Data," *1988 Tri-Service Data Fusion Symposium*, Johns Hopkins University, Baltimore, MD, May, 1988, pp. 54-57.

[16] Waltz, E., and J. Llinas, *Multi-sensor Data Fusion*, Norwood, MA: Artech House, Inc., 1990.

[17] Hall, D., and J. Linn, "Algorithm Selection for Data Fusion Systems," *1987 Tri-Service Data Fusion Symposium*, Johns Hopkins Applied Research Laboratory, Laurel, MD, June, 1987, pp. 100-110.

[18] Acquista, C., "Effectiveness of Detection Fusion in Analyzing DARPA Data Sets," *1989 Tri-Service Data Fusion Symposium*, Johns Hopkins University, Baltimore, MD, May, 1989, pp. 203-213.

[19] Regan, C. M., and J. P. Baird, "PATRIOT/HAWK Fusion Techniques," *1989 Tri-Service Data Fusion Symposium*, Johns Hopkins University, Baltimore, MD, May, 1989, pp. 373-384.

[20] Fox, G., and S. Arkin, "From Laboratory to the Field: Practical Considerations in Large Scale Data Fusion System Development," *1988 Tri-Service Data Fusion Symposium*, Johns Hopkins University Applied Physics Laboratory, May, 1988.

[21] Perry, F. A., G. Fox, and E. E. Moore, "Operating Characteristics of Probabilistic Decision Process Employed in the OSIS Baseline Upgrade System," *1989 Tri-Service Data Fusion Symposium*, Johns Hopkins University, Baltimore, MD, May, 1989, pp. 393-403.

[22] Fritz, R., "P-3 Upgrade IV Multi-sensor Correlation," *1989 Tri-Service Data Fusion Symposium*, Johns Hopkins University, Baltimore, MD, May, 1989, pp. 433-439.

[23] Yu, M., M. Meyer, and R. Carson, "Enhanced Ranging and Angle Estimation by Data Fusion of Non-stabilized Shipboard Sensors," *1989 Tri-Service Data Fusion Symposium*, Johns Hopkins University, Baltimore, MD, May, 1989, pp. 47-60.

[24] Miller, J. T., and D. I. Furst, "ESM Tracker," *Tri-Service Data Fusion Symposium*, Johns Hopkins University, Baltimore, MD, May, 1989, pp. 68-78.

[25] Chen, S., "Fusion of Multi-sensor Data into 3-D Spatial Information," *SPIE Sensor Fusion: Spatial Reasoning and Scene Interpretation*, Orlando, FL, 1988, pp. 80-85.

[26] Selzer, F. J., and D. Gutfinger, "LADAR and FLIR Based Sensor Fusion for Automatic Target Classification," *SPIE Sensor Fusion: Spatial Reasoning and Scene Interpretation*, Orlando, FL, 1988, pp. 236-246.

[27] Gorlin, D., and L. Jamison, *Toward an Expert System for Tactical Air Targeting*, Washington, DC, 1981.

[28] Hall, D., and M. Crone, "An Expert System for Tactical C3CM," *1984 Conference on Intelligent Systems and Machines*, April 24, 1984.

[29] Coursand, G., "PENDRAGON: A Highly Automated Fusion System," *1987 Tri-Service Data Fusion Symposium*, Johns Hopkins University, Baltimore, MD, June, 1987, pp. 54-63.

[30] Davis, E., "The Mission Avionics Sensor Synergism (MASS) Program," *1987 Tre-Service Data Fusion Symposium*, Johns Hopkins University, Baltimore, MD, June, 1987, pp. 366-372.

[31] Hammond, R. H., and P. S. Pentheroudakis, "Data Fusion for C3 ASW Applications," *1987 Tri-Service Data Fusion Symposium*, Johns Hopkins University, Baltimore, MD, June, 1987, pp. 517-526.

[32] Princehouse, D. W., D. G. Shankland, and M. R. Smith, "MASS Classifier," *1987 Tri-Service Data Fusion Symposium*, Johns Hopkins University, Baltimore, MD, June, 1987, pp. 566-574.

[33] Gibbons, G. D., et al., "Data Fusion in the SES: A Sonar Expert System," *1987 Tri-Service Data Fusion Symposium*, Johns Hopkins University, Baltimore, MD, June, 1987, pp. 355-364.

[34] Buede, D. M., and J. W. Martin, "Comparison of Bayesian and Dempster-Shafer Fusion," *1989 Tri-Service Data Fusion Symposium*, Johns Hopkins University, Baltimore, MD, May, 1989, pp. 81-101.

[35] Eggers, M., and T. Khrion, "Neural Network Data Fusion for Decision-making," *1989 Tri-Service Data Fusion Symposium*, Johns Hopkins University, Baltimore, MD, May, 1989, pp. 103-117.

[36] Zachary, W., and M. C. Yubritzky, "A Cognitive Model of Real-time Data Fusion in Naval Air ASW," *1989 Tri-Service Data Fusion Symposium*, Johns Hopkins University, Baltimore, MD, May, 1989, pp. 119-128.

[37] Faust, N. L., et al., "GTSPECS - Multi-spectral Sensor Simulation," *1989 Tri-Service Data Fusion Symposium*, Johns Hopkins University, Baltimore, MD, May, 1989, pp. 215-222.

[38] Fagarasan, J. T., and R. J. Jakaub, "Automatic Radar/ESM Track Association with Application to Ship Self-defense," *1989 Tri-Service Data Fusion Symposium*, Johns Hopkins University, Baltimore, MD, May, 1989, pp. 278-286.

[39] Bein, E., C. L. Tucci, and M. H. Bender, "ACQUIRE: A Data Fusion Threat-assessment System," *1989 Tri-Service Data Fusion Symposium*, Johns Hopkins University, Baltimore, MD, May, 1989, pp. 355-363.

[40] Shirey, M., and L. Converse, "A Knowledge-based Software System for Data Fusion Algorithm Selection," *1991 Joint Service Data Fusion Symposium*, September, 1991.

[41] Barth, S. W., S. A. Barrett, and K. H. Gates, "A Blackboard Architecture for Identification of Command and Control Operations Nodes," *1989 Tri-Service Data Fusion Symposium*, Johns Hopkins University, Baltimore, MD, May, 1989, pp. 385-392.

[42] Gibbons, G., et al., "Data Fusion in ESAU: Expert System Air Order of Battle Update," *1988 Tri-Service Data Fusion Symposium*, Johns Hopkins University, Baltimore, MD, May, 1988, pp. 269-275.

[43] Hill, R. W., Jr., "The Knowledge Workbench: An AI Rapid Prototyping Testbed," *1988 Tri-Service Data Fusion Symposium*, Johns Hopkins University, Baltimore, MD, May, 1988, pp. 437-444.

[44] Benoit, J. W., et al., "Integrating Plans and Scripts: An Expert System for Indications and Warnings," *1988 Tri-Service Data Fusion Symposium*, Johns Hopkins University, Baltimore, MD, May, 1988, pp. 446-452.

[45] Benoit, J. W., and S. J. Laskowski, "An Expert System for Indications and Warnings," *1990 Joint Service Data Fusion Symposium*, May, 1990, pp. 289-301.

[46] Schweiter, G. A., and W. R. Stromquist, "The Effect of Sensor Quality on Tracker/Correlator Performance," *1990 Joint Service Data Fusion Symposium*, May, 1990, pp. 197-222.

[47] Tindel, L. D., and K. J. Overton, "Event Prediction in Naval Tactical Situation Assessment Using Plan Recognition and Automated Knowledge Acquistion," *1990 Joint Service Data Fusion Symposium*, May, 1990, pp. 223-246.

[48] Gaucas, D., et al., "A Probabilistic Method for Generating Situation Descriptions," *1990 Joint Service Data Fusion Symposium*, May, 1990, pp. 351-361.

[49] Kreitzberg, T., T. Barragy, and N. Bryant, "Tactical Movement Analyzer: A Battlefield Mobility Tool," *1990 Joint Service Data Fusion Symposium*, May, 1990, pp. 363-383.

[50] Miller, J. R., and J. H. Swaffield, "An Operational User Hierarchical Knowledge-based Tactical Situation Support System," *1990 Joint Service Data Fusion Symposium*, May, 1990, pp. 511-518.

[51] Brown, D. E., C. L. Pittard, and A. R. Spillane, "A Simulation-based Test Bed for Data Association Algorithms," *SPIE: Sensor Fusion III*, Orlando, FL, April, 1990, pp. 58-68.

[52] Writer, S., "BETA - An Idea Whose Time Had Come," *Signal*, 1981, pp. 11-13.

[53] Writer, S., "Fusion Centers," *Signal*, 1984, pp. 131-133.

[54] Wright, F. L., "The Fusion of Multisensor Data," *Signal*, 1980, pp. 39-43.

[55] Writer, S., "EXPERS - A Prototype Expert System Using PROLOG for Data Fusion," *AI Magazine*, 1984, pp. 37-41.

[56] Bonasen, R. P., Jr., R. J. Antonisse, and S. J. Laskowski, *ANALYST: An Expert System for Processing Sensor Returns*, Washington, DC: MITRE, 1984.

[57] Hall, D., and J. Llinas, *Data Fusion and Multi-sensor Correlation Lecture Notes*, Technology Training Corporation, 3420 Kashiwa Street, Torrence, CA 90505, 1989.

[58] Bobrow, R., and J. Vittal, *Using AI Techniques for Threat Display and Projections including Tactical Deception Indication*, Washington, DC: NTIS, 1981.

[59] Dillard, R. A., *Higher Order Logic for Platform Identification in a Production System*, San Diego, CA: NOSC, 1987.

[60] Gillmore, J. F., "Applications of Artificial Intelligence," *SPIE*, Arlington, VA, 1984, pp. 20.

[61] Hall, D., "Intelligent Monitoring of Complex Systems," 2002.

[62] Williams, J. H. I., P. Drake, and A. Davies, eds. *Condition-based Maintenance and Machine Diagnostics*, New York, NY: Kluwer Academic Publishers, 1994.

[63] Rao, B. K. N., ed. *Handbook of Condition Monitoring*, New York, NY: Elsevier Advanced Technology, 1996.

[64] McClintic, K. T., *Feature Prediction and Tracking for Monitoring the Condition of Complex Mechanical Systems*, in *Acoustics*, University Park, PA: The Pennsylvania State University, 1998.

[65] Byington, C. S., J. D. Kozlowski, and T. A. Merdes, "Fusion Techniques for Vibration and Oil Debris/Quality in Gearbox Failure Testing," *International Conference on Condition Monitoring*, Swansea, April 12-15, 1999.

[66] Erdley, J. D., *Improved Fault Detection Using Multisensor Data Fusion*, in *Electrical Engineering*, University Park, PA: The Pennsylvania State University, 1997.

[67] Alto, H., et al., "Image Processing, Radiological, and Clinical Information Fusion in Breast Cancer Detection," *Sensor Fusion: Architectures, Algorithms, and Applications IV*, Orlando, FL, April 3-5, 2002, pp. 134-181.

[68] Ortolf, J. M., and A. Sanders, "Roving Eye: Research for Imaging Surveillance," *Sensors, C3I, Information and Training Technologies for Law Enforcement*, Boston, MA, November 3-5, 1998, pp. 2-13.

[69] Law, D. B., "Sensors, C3I, Information and Training Technologies for Law Enforcement," *SPIE*, Boston, MA, 1998.

[70] Gros, X. E., ed. *Applications of NDT Data Fusion*, NewYork: Kluwer Academic Publishers, 2001.

[71] Gros, X. E., ed. *NDT Data Fusion*, London: Arnold Publishing Company, 1997.

[72] Schenker, P. S., and G. McKee, T., "Sensor Fusion and Decentralized Control in Robotic Systems," *SPIE Conference on Sensor Fusion and Decentralized Control in Robotic Systems*, Boston, MA, November 2-3, 1998.

[73] Baumgartner, E. T., P. C. Leger, and P. S. Schenker, "Sensor-fused Navigation and Manipulation from a Planetary Rover," *SPIE Conference on Sensor Fusion and Decentralized Control in Robotic Systems*, Boston, MA, November 2-3, 1998, pp. 56-66.

[74] Martens, S., G. A. Carpenter, and P. Gaudinao, "Neural Sensor Fusion for Spatial Viusalization on a Mobile Robot," *SPIE Conference on Sensor Fusion and Decentralized Control in Robotic Systems*, Boston, MA, November 2-3, 1998, pp. 100-111.

[75] Kristensen, S., V. Hansen, and K. Kondak, "Dynamic Sensor Action Selection with Bayesian Decision Analysis," *SPIE Conference on Sensor Fusion and Decentralized Control in Robotic Systems*, Boston, MA, November 2-3, 1998, pp. 181-190.

[76] Nof, S. Y., ed. *Handbook of Industrial Robots*, 2nd ed., New York, NY: John Wiley and Sons, 1999.

[77] Abidi, M. A., R. C. Gonzalez, and R. Gonzalez, eds. *Data Fusion in Robotics and Machine Intelligence*, New York, NY: Academic Press, 1997.

[78] Mutambara, A. G. O., *Decentralized Estimation and Control for Multisensor Systems*, Boca Raton, FL: CRC Press, 1998.

[79] Hall, D., and J. Linn, "A Survey of Commercial Software for Multisensor Data Fusion," *SPIE Conference on Sensor Fusion and Aerospace Applications*, Orlando, FL, April, 1993.

[80] Goebel, M., and L. Gruenwald, "A Survey of Data Mining and Knowledge Discovery Tools," *SIGKDD Explorations*, vol. 1, 1999, pp. 20-33.

[81] West, P., et al., "Development and Real-time Testing of Target Tracking Algorithms with AN/SPY-1 Radar Using Matlab," *Fourth ONR/GTRI Workshop on Target Tracking and Sensor Fusion*, Monterey, CA, 2001.

[82] Pawlowski, A., and P. Gerken, "Simulator, Workstation and Data Fusion Components for Onboard/off-board Multi-targeted Multisensor Data Fusion," *IEEE/AIAA Digital Avionics Systems*, 1998.

[83] Hall, D., and G. Kasmala, "A Visual Programming Toolkit for Multisensor Data Fusion," *SPIE Aerospace 1996 Symposium*, Orlando, FL, April, 1996, pp. 181-187.

[84] Brooks, R. R., and S. S. Iyengar, *Multi-sensor Fusion: Fundamentals and Applications with Software*, Upper Saddle River, NJ: Prentice-Hall, 1998.

[85] Technology, I., *Hydrogeologic Data Fusion*, 1999.

[86] SAIC, *KnowledgeBoard*, 2003.

[87] Geomatics, P., *ImageLock Data Fusion Package*, 2002.

[88] Imaging, D., *eCognition- Object-Oriented Image Analysis*: Definiens Imaging, 2002.

[89] Cambridge Consultants, L., *MUSE: Real-time Knowledge-based Data Capture Software*: Cambridge Consultants, Ltd., 2002.

[90] Bossi, R., and J. B. Nelson, *NDE Data Fusion*: Boeing Corporation, 2002.

[91] Valcartier, *CASE ATTI: A Test Bed for Sensor Data Fusion*: Valcartier (Defence R&D Canada), 2002.

Chapter 12

Automated Information Management

This chapter provides an information management perspective that encompasses data fusion, but is characterized by a significantly broader scope. Information management spans all levels of the Joint Directors of Laboratories (JDL) fusion model as well as sensor tasking, information needs management, sensor scheduling, communications allocation, and collaborative optimization of the entire set of operational resources. Automation of the higher level information management process is required to manage this diverse enterprise. This chapter describes an automated targeting and data fusion application developed for the U. S. Air Force C^4ISR Center in Langley, VA. This chapter was contributed by R. K. Young, M. P. Mahon, and P. S. Wyckoff of the Pennsylvania State University Applied Research Laboratory.

12.1 INTRODUCTION

As discussed in Chapter 11, multisensor data fusion continues to rapidly advance and expand its scope. Enterprise-level information management is a natural extension for multisensor data fusion operations. As Figure 12.1 displays, most current data fusion research and operational systems focus on data correlation, track creation/amplification/management, and estimation of track parameters. Another standard approach is multisensor or multiphenomenology overlays; "track files" from different sensors visually overlaid in the same geographic registration process.

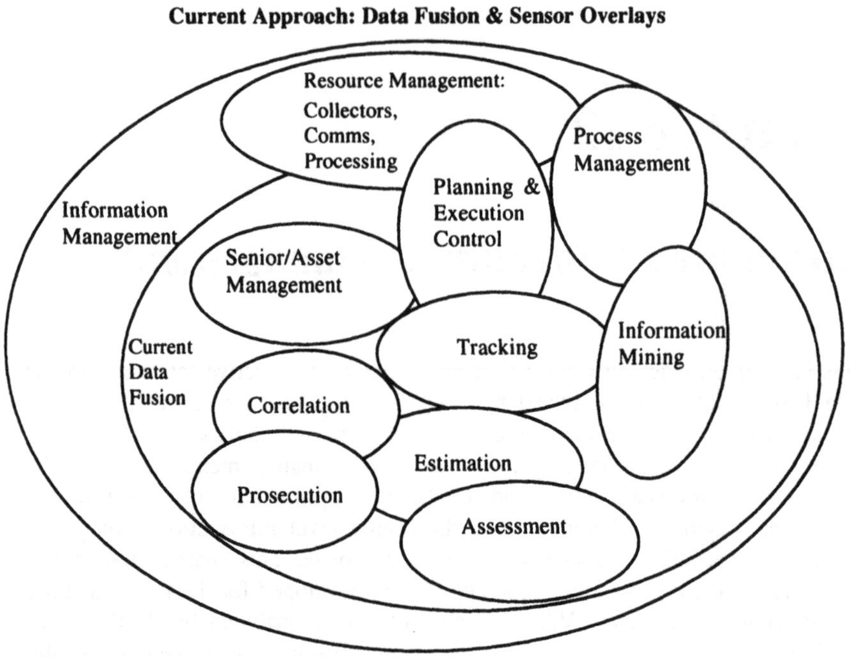

Current Approach: Data Fusion & Sensor Overlays

Figure 12.1 A context for information management and data fusion.

Automated information management addresses the entire enterprise-level operation to broaden the scope of information management as well as the allocation of resources, whether they are allocated to fuse information or reschedule a collection opportunity. Additionally, automated information management necessarily involves human decision makers. The automated information manager only supplements users; it is not the decision maker. Figure 12.2 attempts to scope the processes that are addressed by an automated information manager. Obviously, the enterprise scope changes for each set of users and for each global mission; however, this is a representative scope of functions required to manage the flow of information throughout an enterprise.

Automated Information Management for Many Information Users

Rapid information management requires boundaries to disappear and consistent use of "Automated Information Management" process that is common to most of the information managers

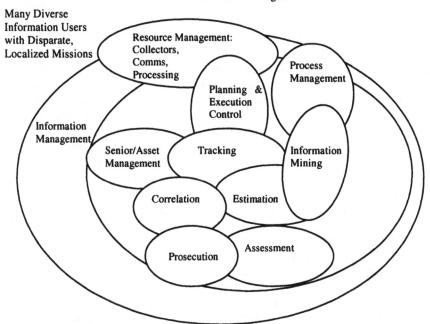

Figure 12.2 Information management spans the entire enterprise.

Although the functions identified in Figure 12.2 represent the core components of the information manager, many other functions are implied. Communications and the management of the communications process is a necessity for efficiently managing available resources and is assumed to be available and properly structured when optimizing other resources and operations; however, realizing enterprise-level communications sufficient for each and every user is a tremendous task and assumption. Thus, an automated information manager (AIM) must treat communication as one of its variables and the AIM must be scalable to operate regardless of a user's available and dynamic communications. Users in a command center will likely have significantly higher available communication data rates as well as significantly better reliability than a deployed user who is crawling in the mud. However, each of their automated information managers (that are personalized to their individual missions) must scale to their accessible resources and appropriately alert them. Again the beeping, flashing, and broadcasting in an operations center to try to get attention is not appropriate for the mud-laden special operations user. The automated information manager must be tailorable for each user's utility and mission(s), as well as being scalable for each user's available communications, processing, and alerting and acceptable resources/environment.

The personalization or tailoring of the automated information manager enables the allocation of its processing, communications and accessible resources to focus on (AIM at) each user's particular mission(s). Figure 12.3 shows the automated information manager for each user. The goal is to automatically close the information management loop for every user and for each of the user's missions.

Each user configures his or her automated information manager; this configuration must be easy, rapid, complete and scalable. Users specify their mission (e.g., collection management, targeting, or image analyst), their area-of-interest, their target/event-categories-of-interest, their temporal ingestion window for each information/intelligence source, their priorities, and their alarm structure and thresholds. It is desirable for this configuration to take less than five minutes the first time and under a minute for subsequent configuration modifications. Thus, a user does not invest much time or effort. After configuration the AIM application performs its functions as a background software agent. The application consumes intelligence and information products (such as reports and messages) and rapidly associates these products based upon the user's criteria and constraints. This association or fusion process creates sets of associated intelligence/information products; these sets are then compared to the user's "target nomination criteria" to determine if the set of products meets all of the user's target criteria. If the criteria are met, then the target is posted into the assessed situation or target nomination list. As these sets of associated intelligence/information products are compared to the user's target criteria, an associated set of "intelligence/information needs" is identified for that set of products that would transform that set of products into an automatic target nomination. These specific, prioritized, information/intelligence needs for each set of associated information/intelligence products are then ranked or scored according to the user's priorities; then the prioritized, consolidated list of information needs is posted/published for potential resolution.

This set of user-centric operations or automated information management displayed in Figure 12.3 closes the information management loop for each user. This loop can be executed in the background, without user interaction, continuously and as rapidly as processing and communications resources will allow (some users in a command center may have resources sufficient to execute this loop multiple times a second, while a forward deployed warfighter may only have resources to execute this loop once a minute, for example).

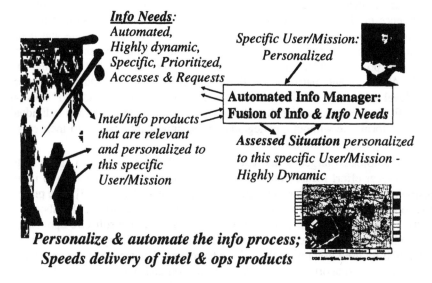

User-Centric Ops: *Exploit Networks to Automatically Close the Loop*

Info Needs:
Automated,
Highly dynamic,
Specific, Prioritized,
Accesses & Requests

Specific User/Mission:
Personalized

Automated Info Manager:
Fusion of Info & *Info Needs*

Intel/info products
that are relevant
and personalized to
this specific
User/Mission

Assessed Situation *personalized*
to this specific User/Mission -
Highly Dynamic

Personalize & automate the info process;
Speeds delivery of intel & ops products

Figure 12.3 Automated Information management fuses and manages information needs.

12.2 INITIAL AUTOMATED INFORMATION MANAGER: AUTOMATED TARGETING DATA FUSION

Although current multisensor data fusion techniques address complex military problems such as target tracking, automatic target recognition, and situation assessment [1], the next generation of information fusion and management systems must address the entire information enterprise. However, the entire information enterprise must be made to work for each individual user as well. For time-critical decision making, automated augmentation/supplementation of decision makers is a standard approach. Expert systems and decision aids routinely provide "user external" information and intelligence to aid a human decision maker. In the targeting of time-critical-targets (TCTs) this decision aid process must rapidly supply available information to diverse decision makers. It is the diversity of decision makers (users) and the compartmentalization of each user's tools that causes serious and automated corroboration efforts to currently fail; interoperability is difficult to achieve without commonality and scalability. For a diverse set of users with widely varying missions, the common/scalable application must also be transparent; otherwise it will interfere with one of the user's missions and then it might lose its wide user participation.

With very different mission criteria and perspectives, the diverse decision makers in the TCT prosecution process have difficulty rapidly and automatically consolidating their decisions as well as the basis or "pedigree" of information and intelligence products that caused them to make that particular decision. This section discusses the automated targeting data fusion (ATDF) architecture and realization for enhancing and speeding up the decision-making process for the set of users who prosecute TCTs in an air operation center (AOC). This particular example, ATDF, of automated information management is the introduction of enterprise-level automated information management, but allows the reader to focus on a specific problem and realize the difficulties of trying to enhance each individual user's performance without any degradation or significant changes to their existing and on-going operations.

The automated targeting and data fusion (ATDF) project was a single year, one man-year effort focused on creating and delivering a new approach to information and intelligence flow in an air operations center (AOC) [2, 3]. The project was conducted by the Pennsylvania State University Applied Research Laboratory in 2002. The product of this one-year effort was a prototype software application that multiple users could evaluate for its potential as a TCT targeting decision aid. The initial focused users were the AOC's targeteers. The targeteer's primary mission is to build support (bring together information and intelligence products) for potential "priority targets" in their assigned operational area, nominate targets (to their commander) that have sufficient support, and assign courses-of-action to prosecute those targets if appropriate. The experience of warfare has educated commanders and targeteers that concealment and deception are most adversaries' first tools. Thus, multisensor and multiphenomenology information and intelligence association are critical countermeasures; it is difficult to conceal and deceive against many phenomenologies simultaneously. A targeteer utilizes "all-source" analysis (and even a dedicated all-source analyst in some situations) to attempt to garner and associate any available target support; automated multisensor fusion is highly desirable for these users!

However, automated all-source data fusion primarily fuses analyst generated messages (messages or reports that are usually of sufficiently high confidence to be widely communicated/disseminated). The ATDF effort created an application that rapidly associated very diverse information sources automatically. In recent military conflict situations, enormous amounts of multisource data have been made available to warfighters (especially targeteers). These data include imagery, signals intelligence, acoustic information, input from human analysts, and other sources. The significant magnitude of collected intelligence overwhelms analysts, operators, collection managers, commanders, and targeteers. Complicating matters further is that the data is collected, transmitted, processed, exploited, and disseminated via multiple stovepipe architectures involving different types of intelligence and support personnel. The term stovepipe architectures refers to systems that process only a single type of data, without considering other types. It is difficult for diverse users to maintain situational awareness over the battle-space. This is an extremely crucial issue with rapid targeting decisions. This ATDF capability offers an alternative to automatically support each individual user in the decision making process, as well as command-level decision makers who must use all of the decisions of their subordinates or colleagues. ATDF rapidly associates diverse intelligence products as they are collected and fuses them into actionable knowledge in near real time. The ATDF does *not* infer decisions nor attempt to automatically make the decisions assigned to the human users; ATDF was designed as a supplement for multiple diverse users who each have their own specific mission, but who must also consolidate their individual decisions to form a global/composite prosecution decision.

ATDF had to be "tailorable" so that users could personalize their ATDF realization for their particular mission(s). This is a crucial issue for many hierarchical decision makers. Each decision maker has his or her own mission and performance criteria as well as access to different levels of information/intelligence. Some users will require raw, signals/imagery level inputs while others just want to evaluate where ships should be sent within the next 48 hours.

Due to the criticality of targeting in current military operations, targeting decisions—especially those for TCTs—have to be made in a very timely fashion (several requirements list this targeting decision cycle in minutes). Consequently, the driving requirements behind the ATDF were to develop a solution centered around an automated, near real-time process that sorts through the accessible information sources and intelligence data as it is collected, fuses it with other related data, and reports the associated (fused) information/intelligence products as a target nomination.

Additionally and simultaneously, "specific, prioritized, information needs" can be developed for each user. Since each user specifies his or her own criteria/priorities/constraints (ATDF configuration) that are required to automatically nominate a target, the ATDF application offered the opportunity to determine what specific information/intelligence products were still required to transform the existing set of associated information/intelligence products into a set that meets all of the criteria necessary to be placed onto this user's target nomination list. These "specific, prioritized, information needs" are ambiguous in the sense that many different products can satisfy the need, so the effort to satisfy these information needs is multifaceted and broad. Existing data could be reprocessed with lower thresholds now that the "information need" is only asking for corroboration that a target may exist—and the existing set of associated information/intelligence products will have a target identity and at least a coarse geolocation. Thus, the process of rapidly associating information/intelligence products also leads to the rapid identification of "specific, prioritized information needs." These information needs are identified individually for each user, so each user's identity/priority/membership helps qualify and consolidate these information needs. Additionally, since these information needs are really based upon existing information/intelligence products, each of these information needs actually has a pedigree of information/intelligence products themselves; the set of personalized, specific, prioritized, information needs comes with a predigree. Although the ATDF identified the opportunity to identify and manage information needs, that information management effort was not within the scope of the ATDF.

These personalized information needs provide battlespace-wide collection managers with a perspective from each user as well as communities of specific users. Thus, higher level tasking of information collection assets and communication assets can be more efficiently and optimally tasked. Thus an information manager can perform multi-INT, multisensor, multiuser information fusion, as well as automated multiuser, multisensor, multi-INT tasking analysis. By simultaneously addressing the fusion of diverse information sources as well as the tasking of information collectors with automatic information managers across many users, the information assessment, tasking and collection loops can be augmented/supplemented with a set of automated information management tools— ATDF was an initial tool that started enabling this automated information management process.

ATDF supported the current intelligence cycle (tasking, collection, analysis, and dissemination) by producing TCT targeting/collection nomination lists that could lead to further required real-time ISR sensor tasking. However, ATDF is a precursor to the much preferred task, post, process and utilize (TPPU) intelligence cycle that is currently emerging as a net-centric and military transformation architecture. This data-centric approach enables appropriate users access to any available information sources as soon as it becomes available. The ATDF-enabled TPPU begins the progression toward data-focused, net-centric operations on a wide scale across many diverse users. A related capability, automated image corroboration, addresses the utility of rapidly posting unanalyzed imagery for automated image corroboration for an automated information management application. This automated image corroboration application in conjunction with ATDF-like fusion closes the intelligence cycle with a TPPU perspective. The following discussion details this automated exploitation of unreported (not necessarily unanalyzed) intelligence and information products for an automated information manager.

More general than just imagery, any external intelligence, surveillance and reconnaissance (ISR) data/collects that a particular user is not actively employing or analyzing can be utilized to create potential targets and to direct resource allocation to the resolution of high-priority information needs. Figure 12.4 identifies the functional block diagram of the process to both fuse information products as well as simultaneously identify information needs. The process starts by accessing, acquiring, and filtering information/intelligence products that the user is willing to accept for target/event nomination (e.g., filter products that are not in the area of interest, not for the target/event set of interest, not collected in the time window of interest, and not from an INT source of interest). The automated information manager then performs rapid association of the information/intelligence products to group them into potential targets/events sets (current personalized situation assessment). Next, each set of associated information/intelligence products are compared to the user's target/event criteria; if all criteria are satisfied, then the target/event is placed on the user's potential target/event nomination list. The entire set of identified potential target/event nominations is then scored relative to the user's scoring criteria and priorities. The highest scoring/ranked targets/events are reported to the user at the rate that they request. Part of the user's criteria is to specify the rate and minimum score of targets/events to be reported; for example, the user could specify: *provide me with the 6–8 highest priority targets/events every 10 minutes.* With this configuration, the user is not bothered with any other interaction and simply has the top 6–8 targets/events posted to their personal operational picture every 10 minutes; and these 6–8 targets/events are only posted if they meet all of the user's constraints and criteria (otherwise fewer or none will be posted).

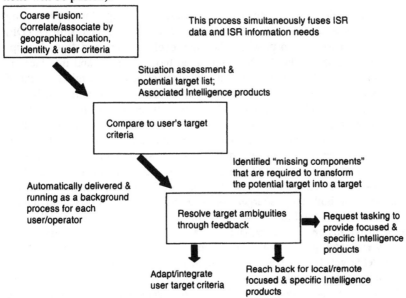

Figure 12.4 Simultaneous personalized fusion and identification of prioritized information needs.

When a set of associated information/intelligence products do *not* meet all of the user's constraints and criteria, but the target/event that is identified by the existing products is a high priority, then the automated information manager attempts to acquire corroborating information/intelligence products to transform this set of associated products into a target/event nomination. Since the AIM application knows which component information/intelligence products are already associated, AIM can identify the set of "missing information/intelligence products" that are yet required to create a high-priority alert. AIM identifies these as specific, prioritized, information needs (SPINs). These information needs can be resolved through multiple resolution paths: adapting/adjusting user criteria, processing/reprocessing local/external collects specifically for this target/event, or task collection/sensor resources. The resolution of these SPINs may be achieved by processing weak/raw information/intelligence sources with different processing thresholds/criteria, now that a specific target/event is identified, ambiguous results are acceptable, and only corroboration is required. Thus, the AIM application would significantly benefit from the posting of weak/raw information/intelligence and can justify its delivery, cataloging and storage. TPPU is the right approach to information utilization in the future.

12.3 AUTOMATED TARGETING DATA FUSION: STRUCTURE AND FLOW

The following sections detail the data structures, processing flows, processing algorithms and user interfaces for the automated targeting data fusion (ATDF) implementation. This section provides the high-level structure and flow description. A block diagram of the ATDF's functional block diagram and interfaces is provided in Figure 12.5. The information management box in the middle-right of Figure 12.3 is the ATDF processing segment, including the filter/controls segment. Outside of the ATDF, the multi-INT messages/reports are parsed and formatted to create the primary input to the ATDF, the multi-INT databank subsequently referred to as the Databank. Thus, the input to the ATDF is a database of reports/messages that are preprocessed by a parsing application. The other databases accepted as input to the ATDF are the target and intelligence source (INT) "equate associations" databases. These databases define the equivalent identifications (aliases) for both targets/target component/target observables, and intelligence sources' names (and their associated ambiguous names/aliases) to allow for a more inclusive fusion. These "equate or alias" information-bases are essentially the identity and intelligence source ontologies. For the target equate associations, the high-level target IDs, such as a tank, can be *equivalenced* to its component and subcomponent systems, such as its tracked vehicle. Databank elements with these *equivalenced* component systems as the ID are essentially equivalent to the target ID, thus enabling the association of all of those "equivalenced" databank elements.

ATDF STRUCTURE & FLOW

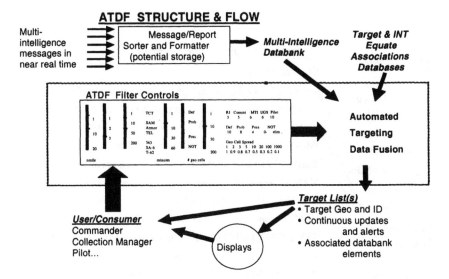

Figure 12.5 Functional block diagram of the ATDF.

With the databank and equate associations as input databases the core ATDF executes its filtration and fusion according to specifications/controls provided by the user. These ATDF filter controls specify which databank elements will be fused and how the fusion/association processes operate. The ATDF filter controls are configured through Windows-based GUI interfaces. A series of graphic user interfaces (GUIs) lead a user through the configuration of the filters/controls and offer default setups. The output of the ATDF is the prioritized target list. The ATDF output prioritized target list and the alarm GUIs are also integrated with these filter/control GUIs. All of the user interfaces originate from the primary ATDF GUI that displays the current prioritized target list and offers all of the configuration screens as pull-down menu options. Figure 12.6 displays the elements of the prioritized target list. The fused targets are listed according to their *score*. Only targets meeting all of the specified filter/control criteria will be listed. Figure 12.6 displays a list of fused targets and then underneath each fused target it lists all of the databank elements that were associated to create that fused target. Figure 12.6 only represents the functionality that will be provided and not the exact format; the list of associated databank elements are being referred to as the fused target's *pedigree*. This pedigree list may only be displayed upon a user's request to review these supporting details. The operational system that integrates the ATDF is envisioned to map this target list to a geodisplay.

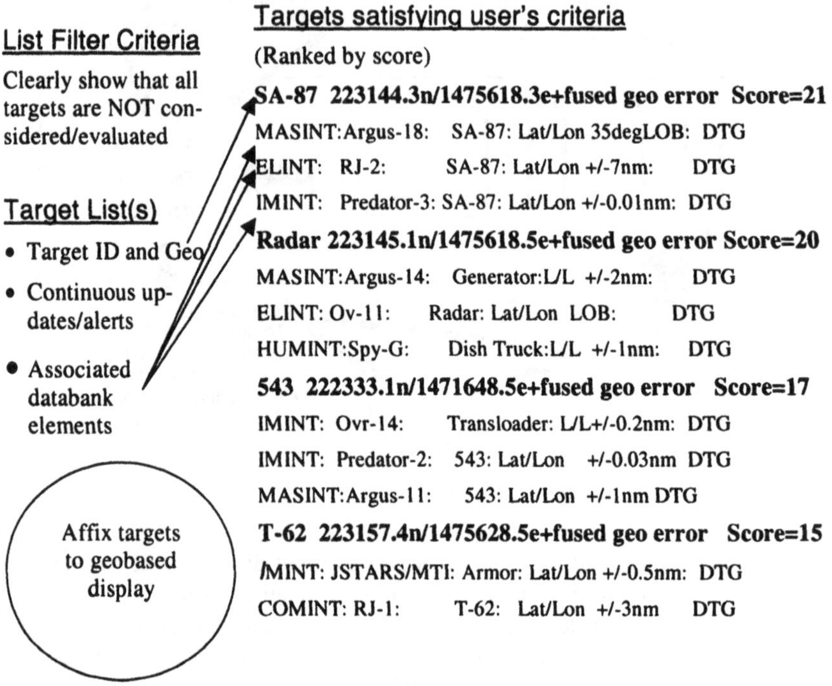

List Filter Criteria

Clearly show that all
targets are NOT con-
sidered/evaluated

Target List(s)

- Target ID and Geo

- Continuous up-
  dates/alerts

- Associated
  databank
  elements

Affix targets
to geobased
display

Targets satisfying user's criteria

(Ranked by score)

SA-87 223144.3n/1475618.3e+fused geo error Score=21

MASINT:Argus-18:   SA-87: Lat/Lon 35degLOB: DTG

ELINT: RJ-2:   SA-87: Lat/Lon +/-7nm: DTG

IMINT: Predator-3: SA-87: Lat/Lon +/-0.01nm: DTG

Radar 223145.1n/1475618.5e+fused geo error Score=20

MASINT:Argus-14:   Generator:L/L +/-2nm: DTG

ELINT: Ov-11:   Radar: Lat/Lon LOB: DTG

HUMINT:Spy-G: Dish Truck:L/L +/-1nm: DTG

543 222333.1n/1471648.5e+fused geo error   Score=17

IMINT: Ovr-14:   Transloader: L/L+/-0.2nm: DTG

IMINT: Predator-2: 543: Lat/Lon +/-0.03nm DTG

MASINT:Argus-11:   543: Lat/Lon +/-1nm DTG

T-62 223157.4n/1475628.5e+fused geo error Score=15

IMINT: JSTARS/MTI: Armor: Lat/Lon +/-0.5nm: DTG

COMINT: RJ-1:   T-62: Lat/Lon +/-3nm   DTG

Figure 12.6 ATDF output target/event nomination list.

Thus, the overall functionality is to accept a multi-INT message databank
and, according to a user-specified set of filters/controls and equated targets/INTs,
fuse and associate supporting databank elements to generate a prioritized target
list with each target in that list satisfying all of the specified criteria. Further, it is
envisioned that multiple, parallel versions of the ATDF will be executing simultaneously.

The functional block diagram description of Figure 12.5 maps to the software
implementation structure of Figure 12.7. This diagram provides the structure, interactions, and flow of the primary software components in the ATDF. The one
set of databases are the required two hierarchical equate thesauri and the input
databank of message elements. The primary processing loop is in the middle of
the diagram: It uses the control filter parameters from the left of the figure to
process the databank elements entering from the top right to produce the prioritized target list in the center of the bottom of the figure. The filtering and message
association components are detailed next.

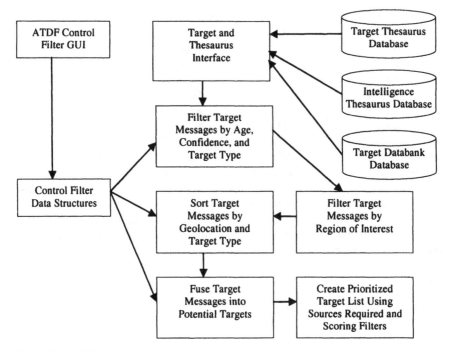

Figure 12.7 ATDF software functional block diagram.

Without discussing the graphical user interfaces (GUIs) that accept the ATDF configuration and control its operation, the basic functional processing algorithms will be detailed. As depicted in Figure 12.8, the first process is to read in the databank elements (messages). Then the target-ID thesaurus is read and all target-IDs are compared to those in the thesaurus. If a target-ID is not in the thesaurus, then that message is ignored and no longer processed. If the target-ID is in the thesaurus but it is an ID that is equated to a higher level in the target-ID thesaurus, then the target-ID in that message is replaced with the highest level target-ID in its equate chain (replace all target-IDs that are accepted—in the thesaurus—with a target-ID at the highest level in the hierarchy of the thesaurus); thus, the output of this process is a list of messages that must contain target-IDs at the highest level in the target-ID thesaurus. The original message/databank element data is not overwritten. Instead, the *copy* of the databank being processed is edited with the equated target-IDs.

Next in Figure 12.8, the filtering and control parameters are applied. The region-of-interest, minimum required message confidences, maximum age, and targets-of-interest databank element filters are applied to all databank elements to only pass through messages of interest to the subsequent association processing. The region-of-interest filter is only applied to messages with a geo-error that is an ellipse. Messages with a line of bearing (LOB) geo-error are *not* filtered out based upon region-of-interest (the latitude/longitude in these messages is the *location of the sensor* and not the estimated target location as in the elliptical-based geo-error messages). The resulting filtered messages are then sorted by target-ID, including equated target-IDs. Then for each particular equated target-ID (looping over all equated target-IDs), each message is checked for geographic association with any of the other messages with that equated target-ID. This geo-association process is depicted in Figure 12.9. However, this geo-association process only applies to messages with elliptically described geo-errors. Messages with LOB geo-errors cannot be associated with the following technique. The geo-association of LOB-based geo-error messages is deferred. The geo-association of elliptical-base geo-error messages is referred to as the *stake-in-the-ground* geo-association because it takes the geo (lat/lon) of the message being processed as the center of a square geo-cell and looks for any other messages (with the same target-ID) within that geo-cell. If any messages are within that geo-cell, then they are geo-associated (added to an associated message list). The lat/lon offset that determines the size of this geo-cell is referred to as the geo-cell-size and is accepted in nautical miles. The geo-cell-size is an offset from the center of the cell, so it acts as a radius rather than a diameter for the geo-cell.

Databank Element (Message) Flow & Association

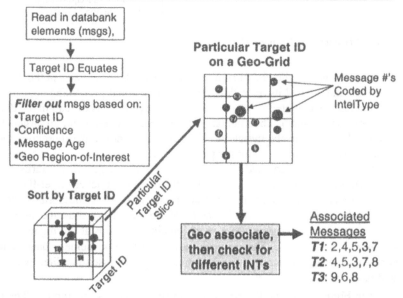

Figure 12.8 Application of filtering and control parameters.

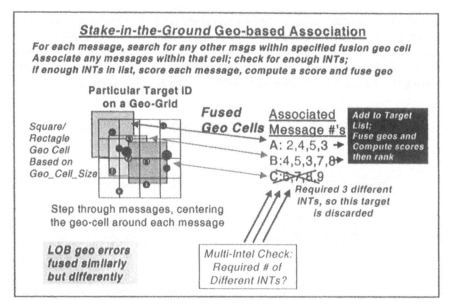

Figure 12.9 User-controlled geo-association process.

In Figure 12.9, the shaded overlays represent the geo-cells for message association. These overlays are on top of the *message locations* on a geo-grid (lat/lon).
For message 2 the top overlay geo-cell covers messages 2, 3, 4, 5 and 7; thus,
these five messages are associated based upon target-ID and geographical proximity. These five messages form a set of associated messages. As other messages
are processed for association, the *location of the stake* shifts to the center of that
particular message being evaluated; Figure 12.9 shows the geo-cell box for associating message 7. For message 7, messages 4, 5, 3, 7 and 8 are associated and
added to the associated messages list. The shade-coding of the message numbers
is to indicate the INT of a particular message. In this example ELINT could be
represented by light gray, MASINT by a darker shade and IMINT by black. Thus,
for the first set of associated messages, two MASINT messages, one IMINT and
one ELINT messages are combined. Figure 12.10 demonstrates the potential geo-
associations for three messages in the same vicinity.

The next component of the processing is shown in Figure 12.9 where the
large light gray *X* overlays the third element of the associated messages list. Here
each set of associated messages is evaluated to determine if it meets the potential
target criteria. One element of the potential target criteria is the number of different intelligence data types (INTs) that contribute to the set of associated messages;
if the criteria is a minimum of three different INTs for a potential target declaration, then as in Figure 12.9, the third element (C) only has two different INTs contributing and, thus, that set of associated messages is eliminated from consideration as a potential target. The first two sets of associated messages (A and B) each

have three INTs contributing and are passed along for further evaluation and processing and are added to the *prioritized target list*. Finally, each set of these associated messages is scored and its geo-locations and geo-errors are fused. This discussion is deferred. First the addition of messages that only possess line-of-bearing (LOB) geo-error is addressed in the message association process.

Stake-in-the-Ground Geo-Smeared Association

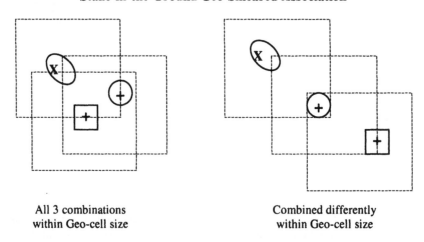

All 3 combinations   Combined differently
within Geo-cell size   within Geo-cell size

Three 3-associated message lists but   One 3-associated message lists
redundancy check creates just one   Two 2-associated messages lists

Figure 12.10 User-controlled geo-location-based message association.

Since a message with a LOB geo-error specifies the *sensor's* lat/lon location (and NOT the lat/lon estimate of the *target* as in the elliptical geo-error description), the stake-in-the-ground geo-association cannot be easily applied; the location estimate of the target is highly ambiguous—anywhere along the LOB line. So LOB-based messages are handled separately. A LOB-based message is geo-associated with a message that has an elliptically based geo-error by finding the point on the LOB that is closest to the estimated target position from the elliptical-based message; see Figure 12.11–Figure 12.13. Figure 12.11 shows the elliptical-based geo-error description; the description is relative to the target (not the sensor). The angle describes the orientation of the ellipse relative to north. The SMA and sma are the lengths (in nautical miles) of the semimajor and semiminor axes of the ellipse, respectively. The center of the ellipse is the estimated target location. Figure 12.12 shows the LOB-based geo-error. This description is highly ambiguous; it specifies that the target is somewhere along the LOB angle relative to the sensor's location.

The geo-association process for ellipse and LOB-based messages is to find the location on the LOB that is closest to the center of the ellipse and compare that closest distance to the geo-cell size. If the distance (slant range) is less than the geo-cell size, then the messages are geo-associated otherwise they are not associated. The evaluation begins by computing a distance between the sensor location (LOB-based message) and the estimated target location (ellipse-based message). Then the LOB is used to find a point at that distance along the LOB line (referred to as the lat/lon near target point). Now a slant range is computed between that point and the center of the ellipse. Then that slant range is compared to the specified geo-cell-size for message association. If the INT of the LOB-based message is ELINT and the slant range is less than the geo-cell size, then the messages are associated. If the INT of the LOB-based message is MASINT and the slant range is less than the geo-cell size and the distance is less than 2 nautical miles, then the messages are associated.

Geo-Error Specification – 95% Confidence Ellipse

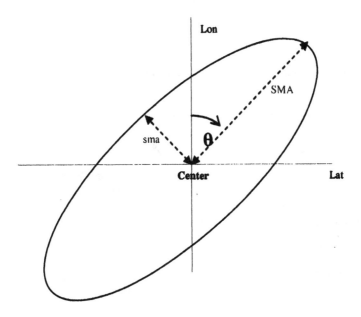

Figure 12.11 Elliptical-based geo-error description.

Geo-Error Specification – Line-of-Bearing (LOB)

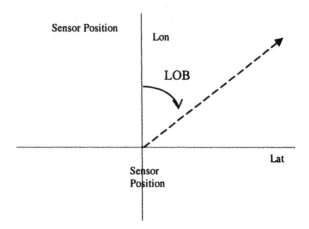

Figure 12.12 Line-of-bearing (LOB)-based geographical error.

Fusing LOB Geo-errors with Message Center Points

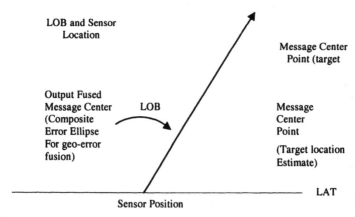

Figure 12.13 Fusing LOB geo-errors with message center points.

When both messages have LOB-based geo-errors, then association is considered only if one of the messages is a MASINT message. If two LOB ELINT messages are geo-associated, then the resulting intersection would always be *fused* (and the messages associated) as long as the intersection is in the region-of-interest *without any consideration of the geo-cell-size.* That association is inconsistent with the requirement that the estimated target location be within the geo-cell-size. However, if one of the messages is a MASINT message, then the estimated target should be within several nautical miles of the sensor's position, and, thus, should be at least close to the specified geo-cell size. The association proceeds identical to the ellipse/LOB geo-association test, but the MASINT sensor location is simply used as the ellipse center. If the slant range is less than the geo-cell size the MASINT and ELINT (or MASINT) messages are geo-associated. Thus, the geo-fusion of any set of messages is estimated. The overall ATDF application rapidly associates many diverse intelligence products and creates target nominations with fused geolocations.

12.4 AUTOMATIC INFORMATION NEEDS RESOLUTION EXAMPLE: AUTOMATED IMAGERY CORROBORATION

To effectively utilize the rapid association of reported information/intelligence products, the automated information manager must also rapidly resolve the priority information needs that are identified and prioritized. However, since existing intelligence products have some reasonably high confidence before they are widely communicated, the emphasis in "information needs resolution" functionality is to discover/extract/create/task "corroborative" information/intelligence. The distinction between reported information/intelligence products as "creative intelligence" and the automatically derived/tasked intelligence products as "corroborative" intelligence products both implies and allows a lower threshold of confidence and a higher acceptable level of ambiguity in the corroborative intelligence products. Typically, the automatically generated corroborative intelligence products will be from "weaker" or unanalyzed information/intelligence products or intelligence products that a human user could not report due to a higher confidence or ambiguity threshold. Since the automated information manager is continuously and constantly processing to derive more knowledge, it accepts significantly more uncertainty and ambiguity than most human analysts (it is doing a different job with different criteria). However, the key concept is that since only weaker, uncertain, and more ambiguous information/intelligence is acceptable for corroboration, then maybe a computer is an acceptable processor/decision-maker for corroboration of existing human-generated "creative" information/intelligence products. The automated imagery corroboration application is a realization of automated "information needs resolution" through automated weak/ambiguous corroborative intelligence product generation.

Automated imagery corroboration uses the set of associated information/intelligence for a high priority target to establish the area of interest (AOI). If available in imagery storage facilities, this AOI is used to find and establish a reference to recent existing imagery (of any or multiple types) from this AOI, a set of image "chips." Then, knowing a target/event identity from the associated information/intelligence products, a set of "size/shape/orientation" parameters are accessed for that particular target/event. These "size/shape/orientation" parameters are then mapped to image processing filter coefficients. These coefficients and the imagery chip reference(s) are tasked to either local or remote image processing; the target/event filter coefficients are used to filter the image chip(s) and detect any potential objects-of-potential-interest in this imagery. The image processing then sends back any potential detections, along with a "confidence coefficient" (its correlation score), and the geographical coordinates of the detection. Note that all of this processing is automatic and that no imagery analysts need to be involved; however, if they were available and accessible, they could also be involved and further enhance the image processing and corroboration process.

By automating this "information needs resolution" process and minimizing the transfer of data/imagery (only filter coefficients, imagery references, and potential target/event detections with locations need to be communicated), this corroboration can occur many times per second. Again, this functionality is scalable to each particular user's available resources; however, no processing or communications resources will be wasted; the automated information manager will employ all available resources to corroborate potentially high-priority targets/events that meet all of the user's constraints, criteria, and priorities.

Existing multisensor data fusion primarily fuses analyst generated reports and messages. These intelligence products represent a small portion, albeit a highly reliable segment, of the available information/intelligence products. See Figure 12.14. Figure 12.14 displays the range of available information/intelligence products, from raw sensor collects at carrier frequencies to polished/finished intelligence products that are published worldwide. However, analysts are overwhelmed. Task saturation as well as communications and computational limitations often prevent volumes of raw collection data from reaching and being processed by an analyst in a timely manner. Valuable intelligence information that at least corroborates an important time critical target may be present in unanalyzed raw data files. Additionally, the thresholds that an analyst applies for reliability and ambiguity resolution are typically very high, to minimize false alarms and wasted responses by others in the intelligence and operations processes. The automated imagery corroboration application utilizes analyst generated reports/messages to direct a precision search to corroborate the existence of a high-priority target/event in a small, highly focused spatial-temporal segment of raw data. Automatic corroboration processing simply attempts to derive further evidence for the presence of a "particular set of target/event features" in a particular space-time location.

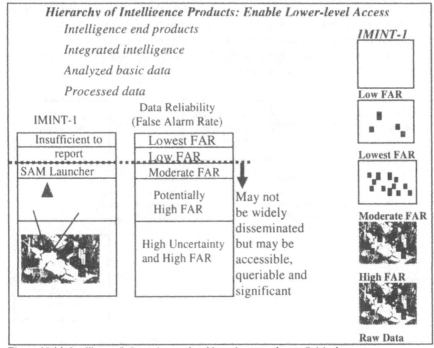

Figure 12.14 Intelligence/information product hierarchy—raw data to finished report.

Corroboration is a much simpler process than automatic target recognition (ATR), and corroboration requires significantly less processing and fidelity. The existing associated information/intelligence products coarsely "detect and identify" the potential presence of a priority target/event and enable the corroboration to act in a lower fidelity "confirmation role." This corroborative processing enables the utilization of unanalyzed, but available, raw data processing. This section concentrates on imagery as the raw data product, but other unanalyzed (but posted) INT/sensor information could be similarly utilized for corroboration. The new approach transforms previously disregarded raw data into associated corroborative information without increasing analyst tasking. Existing software fuses the corroborative information with analyst messages. An example demonstrates raw data corroboration using imagery. The approximate location, time, and target identity are determined using two associated analyst messages. Raw imagery processing confirms the target and associates an additional message. The fused target priority is amplified by the corroborative message.

Corroboration means to strengthen or support with other evidence and make more certain. Corroboration using unanalyzed/raw data attempts to strengthen the pedigree support for a high priority target/event. Multisensor data fusion that exploits raw data for corroboration is summarized in Figure 12.15. The "initial message association and fusion" is the rapid existing information/intelligence product association performed by ATDF. ATDF creates a set of associated information/intelligence products for high-priority targets/events. The "fused" identity and geolocation of these associated information/intelligence products are now used to find and task the processing of raw data which might corroborate the existing associated information/intelligence products.

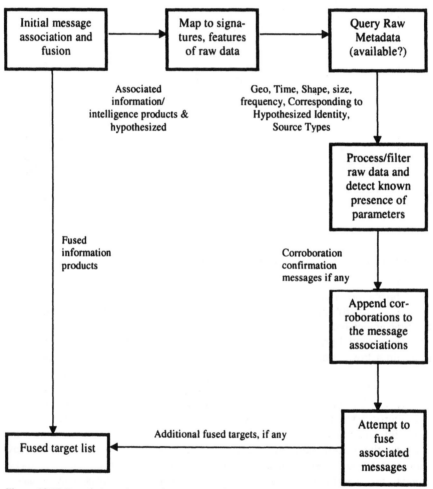

Figure 12.15 Data fusion using raw data corroboration.

12.4.1 Automated Image Corroboration Example

Automated image corroboration (AIC) is a specific example of automated "raw data" corroboration. AIC seeks corroborating imagery evidence for a set of previously associated information/intelligence products. Corroboration of imagery with other associated information products improves the automated target/event declaration process reliability *without any human interaction.* Rather than considering extensive imagery, automated image corroboration focuses the image availability, image processing, and target/event evidence accrual processes onto a prioritized and relatively small spatial-temporal segment of relevant imagery that might support a particular target/event. Prioritized focusing reduces the required processing and communications while simultaneously enhancing the automated detection reliability. Focusing is derived from the previously associated information/intelligence products using identity, location, and time of these products. Environmental parameters may guide the raw data selection process toward specific hypothesized target features that could be emphasized in one or more available imagery sources.

A software prototype was created to demonstrate automated image corroboration as an extension to ATDF. For example, a user might configure the ATDF to require that automated target/event alerts must be supported by at least three distinct INTs/sources.

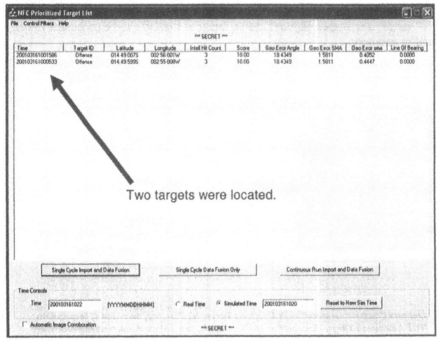

Figure 12.16 ATDF fusion without AIC only yields two targets.

Refer to Figure 12.16. The first fused hypothesized target list contains two targets. This example uses targets that are professional football teams or units (components) within these teams. Both targets in list have the identity of

"Offense" and they are each supported by messages/reports from three different INT/source disciplines since the "intelligence hit count" field is three.

The lower left corner of Figure 12.16 includes a checkbox that allows the user to enable automated image corroboration; some users may be unwilling to accept anything other than imagery analyzed by a human IMINT analyst. Once the AIC is activated by a user, sets of associated information/intelligence products for high-priority targets that require imagery as one of their additional source/INT disciplines will automatically seek to corroborate raw imagery to transform this set of information/intelligence products into an automatically declared, high-priority target/event (see Figure 12.17).

In Figure 12.18, intelligence products were associated to indicate a target named "Steelers." ("Steelers" is the name of a professional U.S. football team.) However, this target was not shown in Figure 12.16 since there are only two distinct supporting INT disciplines. For example purposes, assume that "Steelers" targets are approximately 200' x 30' rectangular shaped objects. Existing imagery databases are searched for metadata on imagery in the location of the vicinity of the two previously associated information/intelligence products, within a user specified time window. The image shown in Figure 12.17 is determined to be available. Now the image processing/image filter coefficients are known (200' x 30' rectangular shaped objects) and an image is available to be processed (locally and/or remotely).

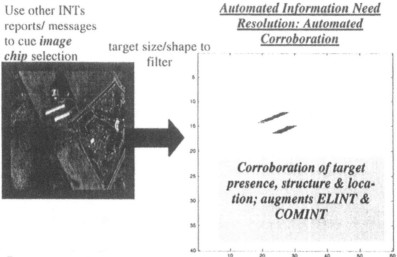

Delivers Precision Geolocation & Confirms Target ID Filtering, detection & localization; NOT Automated Target Recognition (ATR); Simple, fast, robust & automated processing!

Figure 12.17 Automated image processing to extract corroborating target support.

In this case the size/shape of components of the "Steelers" target is known (it may also be known that both rectangles will be oriented in parallel or with an approximate separation, etc.). The imagery is filtered/searched for objects of interest. In this case, the automated image processing detects two locations that might belong to a component of the "Steelers" target; or objects that might corroborate the presence of the Steelers target in that particular geolocation.

Note that the detections in Figure 12.17 include points of two rectangular shaped objects. In practice, several false alarm detections could be encountered. False alarms are acceptable since any automated detection provides corroboration, not target recognition. Additionally, the objects could be correctly detected by the image processing, but may be buses instead of transporter-erector-launchers (TELs) of the same size and shape; the identity ambiguity of the correct "detection" may still be high. This automatically extracted corroborative information may be posted for additional analyst processing if warranted and appropriate. Minimal false alarm rates can be obtained by processing very small image segments in space and time. The image processing provides useful corroborating data, but the confidence of this data is low compared to human analyst decisions. Any image processing algorithm could be used, provided it reliably detects some useful feature of the hypothesized target. Image features that can be used include size, shape, texture, heat, orientation, reflective spectrum, and color.

The resulting detection report/message (corroborating intelligence product) is now associated with the two existing intelligence products that initiated the automated image corroboration. When the associated messages are fused the target list contains the original two targets plus an additional "Steelers" target. The "Steelers" target appears in the list since there are now three supporting disciplines. However, the "intelligence hit count" remains at two since there were two confident messages supported by a weaker corroboration.

The target "Steelers" from Figure 12.18 is supported by the pedigree data shown in Figure 12.19. The two initial messages are shown with "High" confidence while the automatic image corroboration message is a "Low" confidence message.

Automated image corroboration can also effectively utilize communication bandwidth (see Figure 12.20). The only wideband communication requirement is between the image processing algorithm and the image storage. Raw data is only required for high probability spatial-temporal locations.

The end-user application sends information about the time, location, and target identity that is sought. This information transfer is a small text message. The catalog of raw data is the directory that provides pointers to the image data and search parameters for the image processing algorithm.

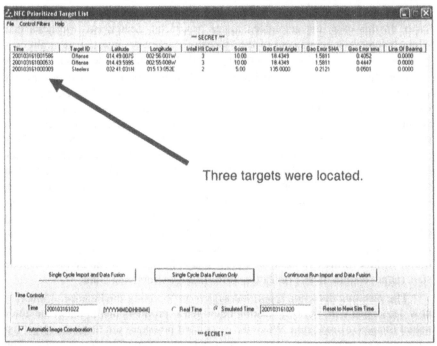

Figure 12.18 Fusion with AIC yields three targets.

Reference	Intel Type	Intel ID	Time	Target ID	Confidence	Latitude	Longitude	Score
1	COMINT	C-OVR2	200103161000132	Steelers	H	032:42:030N	015:12:054E	0.00
2	ELINT	E-OVR2	200103161000309	Steelers	H	032:42:030N	015:12:054E	0.00
1000	IMINT	AIC-OVR1		Steelers	L	032:42:030N	015:12:054E	0.00

Figure 12.19 Pedigree supporting corroborated target.

The image processing algorithm must retrieve the image from the imagery source. This is the most bandwidth-intensive data transfer. The imagery can be transferred after applying a coarse threshold operator to reduce bandwidth since searching for a particular shape may not require the highest fidelity.

Once the image processing algorithm is complete, text messages are sent back to the end-user application to enable fusion. This architecture allows the end user to benefit from raw imagery despite severe bandwidth constraints from the end users' platform.

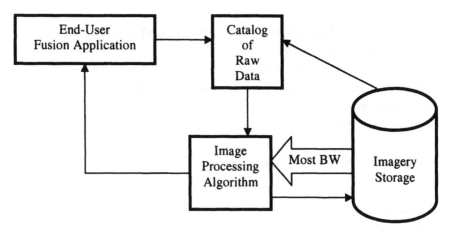

Figure 12.20 Automated image corroboration with communications bandwidth considerations.

12.5 AUTOMATED INFORMATION MANAGER: UBIQUITOUS UTILITY

The vision of automated information management is for as many users as possible to be employing (AIMing) the available resources for their personalized and individual missions. The vision for future operations is to have the assets in coherent coordination with each other and all of them complete with ad hoc awareness of both their current and projected future situation (their intent)—see Figure 12.21. Then resources can be more efficient and effectively optimized to better achieve all missions. Recall that the automated information manager only provides situational awareness, information needs, and decision *recommendations,* not the decision itself. Thus, the automated information manager cannot automatically task assets, but can better make the collection and/or asset managers aware of current information needs and their consolidated priorities and pedigrees—"the automated information manager doesn't make the decisions, it just makes the decisions better."

Tactical, Coherently Coordinated Cooperative Operations Vision
Ad Hoc Communications Network
Ad Hoc Sensor/ISR/Information Source network
Automated Personalized, Prioritized Information Fusion
Automated Personalized, Prioritized Information Needs Management
Multiple Users, Same Intelligence/Information Portal; Personalized by User

Constraints to Realizing Vision
- Must deliver significant performance enhancements
- Capable of employing all legacy systems and operators
- Relatively low-cost and low-impact implementation
- Little added work/interference for normal user operations
- Enhances the missions of all decision makers in the joint decision process
- Delivers easy access to pedigree information for each decision
- Enhances _automated, prioritized resource allocation_ of people, sensors, communications, platforms, weapons ...

Figure 12.21 Vision and constraints for realizing automated information management.

Referring to Figure 12.21, although the desired vision is lofty and preferred, practical and tactical constraints limit the pursuit of "blue-sky concepts." The "constraints to realizing vision" bullets state the bounds of any acceptable solution—no new serious investments or impacts to existing systems, but benefit to everyone. Ad hoc networks will only exist when many users believe that the network benefits them personally. The Internet is a good example of networking success; the users believe that they benefit by using this network. Other networks have failed due to their high costs or lack of sufficiently foreseeable benefits by the intended users. Since the realization of an automated information manager requires wide, enterprise-level participation to work, almost every user and/or decision-maker must perceive potential benefits if they utilize the automated information manager and its associated network(s).

Therefore, the user cannot be significantly impacted if they choose to use the automated information manager. The automated information manager cannot interfere with the user's normal operations. The automated information manager can only supplement each user's current operations. To be cost-effective and easy to integrate across an entire enterprise, the automated information manager must be a set of portable software that is easy to install, configure, and operate on a wide variety of computer systems. Finally, the software/firmware/hardware realization should scale its operations to the resources available to each particular user, regardless of that user's communications, security access level, available processing, and access/willingness to communicate to external information sources. The scalability is a requirement for widespread utility and enterprise-wide realization. This scalability also requires both a thin-client and thick-client realization, so that users with network access can exploit the network, while "disconnected" users can still exploit local and broadcast information and intelligence. These extreme constraints actually limit the potential realizations to being a simple software plug-in agent that configures itself and adapts to the available resources to execute its automated information manager mission(s). The ATDF was an initial prototype software system that can be evolved to a widely scalable automated information manager.

Automated information management is a lofty goal, especially with the constraints of Figure 12.21, but the right architecture and a focus on each user will help guide us to a user supplement that will help users accomplish their mission(s). If the AIM application is scalable, transparent, and useful, then there is hope that we can permeate enough users to create a community of AIM users that gets stronger as the membership grows. Thus, individual users are motivated to use the application, but are also motivated to have other users utilizing the AIM application as well. This will enable and create a "network of users" rather than the current "users of the network."

REFERENCES

[1]   Hall, D., and J. Llinas, eds. *Handbook of Multi-sensor Data Fusion*, New York, NY: CRC Press, 2001.

[2]   Wyckoff, P., and R. Young, "Automated Imagery Corroboration," *Proceedings of the SPIE AeroSense 2003 Symposium*, 2003.

[3]   Young, R., P. Wyckoff, and J. Wise, "Automated Target Data Fusion," *Proceedings of the SPIE AeroSense 2003 Symposium*, 2003.

About the Authors

David L. Hall

David Hall is the Associate Dean for Research, School of Information Sciences and Technology at The Pennsylvania State University. He has a M. S. and PhD in astronomy and a B. A. degree in physics and mathematics. He has more than 25 years of experience in research, research management, and systems development in both industrial and academic environments. Dr. Hall has performed research in a wide variety of areas including celestial mechanics, digital signal processing, software engineering, automated reasoning, and multisensor data fusion. During the past 15 years, his research has focused on multisensor data fusion. He is the author of over 175 technical papers, reports, book chapters, and books. Dr. Hall is a member of the Joint Directors of Laboratories (JDL) Data Fusion Working Group. He serves on the Advisory Board of the Data Fusion Center based at the State University of New York at Buffalo. In addition, he has served on the National Aeronautics and Space Administration (NASA) Aeronautics and Space Transportation Technology Advisory Committee. In 2001, Dr. Hall was awarded the Joe Mignona award to honor his contributions as a national leader in the Data Fusion Community. The Data Fusion Group instituted the award in 1994 to honor the memory of Joseph Mignona. Dr. Hall was named as an IEEE Fellow in 2003.

Sonya A. H. McMullen

Sonya A. H. McMullen is a Captain in the United States Air Force. She is a Space and Missile Operations Officer with over seven years of active duty service. She has served as a Delta II Training Officer, Minuteman III Intercontinental Ballistic Missile Combat Crew Commander, Flight Commander, Deputy Director Collection Systems Evaluation Task Force, and C2 Chief of Training. Captain McMullen has a Masters of Aeronautical Science from Embry-Riddle Aeronautical University and a Bachelor of Science in Industrial Engineering from The Pennsylvania State University. Her awards include the Defense Meritorious Service Medal, Air Force Commendation Medal, Joint Service Achievement Medal, Air Force Achievement Medal, Combat Readiness Medal, and the National Defense Service Medal with the Bronze Star. Her Research interests include sensor test and evaluation, human-computer interfaces, and learning theory. She is married to Captain J.J. "Mac" McMullen, USAF.

Index

The Artech House Information Warfare Library

For further information on these and other Artech House titles, including previously considered out-of-print books now available through our In-Print-Forever® (IPF®) program, contact:

Artech House	Artech House
685 Canton Street	46 Gillingham Street
Norwood, MA 02062	London SW1V 1AH UK
Phone: 781-769-9750	Phone: +44 (0)20-7596-8750
Fax: 781-769-6334	Fax: +44 (0)20-7630-0166
e-mail: artech@artechhouse.com	e-mail: artech-uk@artechhouse.com

Find us on the World Wide Web at: www.artechhouse.com

Lightning Source UK Ltd.
Milton Keynes UK
UKHW02n1333051217
313927UK00004B/331/P